Current Topics in Pathology

Continuation of Ergebnisse der Pathologie

64

Pulmonary Hypertension Related to Aminorex Intake
DNA Injuries, Their Repair, and Carcinogenesis
Soft Tissue Tumors in the Rat
Visceral Candidosis

Contributors

K. Küchemann, J. L. Van Lancker, D. Maas, E. L. Quiroga,
U. N. Riede, K. Salfelder, J. Schwarz, H. J. Steinhardt,
C. Thomas, K. Ueda, S. Widgren

With 107 Figures

Springer-Verlag Berlin Heidelberg New York 1977

E. Grundmann, Professor Dr., Pathologisches Institut der Universität, Westring 17, D-4400 Münster/Westf., Germany

W.H. Kirsten, Professor Dr., Department of Pathology, The University of Chicago, 950 East 59th Street, Chicago, IL 60637, USA

ISBN-13: 978-3-642-66583-7 e-ISBN-13: 978-3-642-66581-3
DOI: 10.1007/978-3-642-66581-3

Softcover reprint of the hardcover 1st edition 1977

Library of Congress Catalog Card Number 56-49162.

Contents

List of Contributors

Küchemann, Dr. K.	Pathologisches Institut des Städt. Krankenhauses, Dhünberg 60, D-5090 Leverkusen
Van Lancker, Prof. Dr. J.L.	University of California, Dept. of Pathology, School of Medicine, The Center for the Health Sciences, Los Angeles, CA 90024, USA
Maas, Dr. D.	Abt. für klinische Immunpathologie an der Medizinischen Universitätsklinik, Hugstetter Straße 55, D-7800 Freiburg
Quiroga, Dr. E.L.	Department of Pathology, University of Cincinnati, OH, USA
Riede, Dr. U.N.	Pathologisches Institut der Universität, Albertstraße 19, D-7800 Freiburg
Salfelder, Prof. Dr. K.	Department of Pathology, Universidad de Los Andes, Merida, Venezuela
Schwarz, Dr. J.	Clinical Laboratories, Jewish Hospital, Cincinnati, OH 45229, USA
Steinhardt, Dr. H.J.	Pathologisches Institut der Universität, Albertstraße 19, D-7800 Freiburg
Thomas, Prof. Dr. C.	Pathologisches Institut der Universität, Albertstraße 19 D-7800 Freiburg
Ueda, Dr. K.	Department of Pathology, University of Cincinnati, OH, USA
Widgren, Dr. S.	Institut de Pathologie de l'Université, 40, Bvd. de la Cluse, CH-1211 Genève 4

Pulmonary Hypertension Related to Aminorex Intake

Histologic, Ultrastructural, and Morphometric Studies of 37 Cases in Switzerland

S. WIDGREN

Introduction

In 1967 cardiologists in Switzerland were struck by the sudden increase in the number of cases of so-called primary pulmonary hypertensive disease (PPHD). Up until 1966, the cardiology centers in Switzerland had not seen more than one or two cases a year, but this number suddenly multiplied by ten or twenty times. At the June 1968 meeting of the Swiss Society of Cardiology, *Gurtner* et al. (1968b) presented a paper which raised the question whether the vascular types of cor pulmonale had increased. They also raised the question about the possible responsibility of environmental factors, such as toxins or drugs, in causing this increase.

Krähenbühl et al. (1968) first suggested the possible responsibility of anorexigenic drugs, but no proof was provided. Soon after this, the cardiology team in Bern (*Gurtner* et al., 1968a) published their report, which resumed and completed the above-mentioned preliminary communication; 31 cases of PPHD, among which 17 had taken an anorexigenic drug, aminorex fumarate (Menocil, Cilag), were reported. The importance of this work was soon recognized and it was followed by several meetings (in Vienna, Hannover, and Bürgenstock) and two round tables sponsored by the Swiss Society of Cardiology (in Montreux and Basel).

Following the publication of a paper by *Jornod* et al. (1970) we were asked with Professor Kapanci by the Swiss Society of Cardiology–with the support of Cilag AG–to undertake a pathologic study of PHD cases deceased in Switzerland. A preliminary report was given to the Swiss Society of Pathology in the fall of 1969 (*Widgren* and *Kapanci*, 1969), in order to inform members of this Society of the problems and aims of the work which had been undertaken. Material was provided by most of the pathology departments in Switzerland. Thus it was possible to study the lungs taken at the necropsy of 37 patients deceased from PHD related to the absorption of aminorex, and of 5 persons who had suffered from PPHD with no history of drug intake. Preceeding papers (*Kapanci* and *Widgren*, 1970; *Widgren* and *Kapanci*, 1970a, b, 1972) have given the preliminary results obtained by optical and electron microscopy, numerical evaluation of the lesions of pulmonary arteries according to the classification established by *Heath* and *Edwards* (1958), and by morphometric studies based on the principles of *Weibel* (*Weibel*, 1963, 1970; *Weibel* and *Gomez*, 1962; *Weibel* et al., 1966). The purpose of the present work is to resume and complete these various preliminary studies.

It must be emphasized that the lesions of pulmonary muscular arteries observed in this study, did not differ, in any way, from those codified by *Heath* and *Edwards* (1958) in cases of PHD secondary to cardiac malformations. Although it was not possible to provide anatomic proof of the responsibility of aminorex in the development of PHD, some of the findings allow for the suggestion of a hypothesis regarding the action of the drug. Our studies have shown, in fact, that muscularization of the intima often seen under the light microscope, was revealed by the electron microscope to be constituted by myo-intimal cells as described by *Buck* (1961) in experimental arteriosclerosis. The evolution of the lesions toward dilatation and fibrosis is characterized by a thickening of the constituents of the arterial wall, the consequence of which is a segmental obliteration of the lumen, which contributes to the fixing or even the worsening of the muscular spasm induced by the direct or indirect action of the drug.

I. Aminorex: First Pharmacologic and Clinical Tests; Acute Toxicity

1. Pharmacology

Aminorex (2-amino-5-phenyl-2-oxazoline) was synthesized by *Poos* et al. (1963). Its chemical structure has some similarities to that of amphetamine, norepinephrine, and ephedrine (Fig. 1). Like other anorexigens it is a sympathomimetic drug with which it shares certain vasomotor effects. Introduced by the McNeil Corporation, it was commercialized in 1965 by Cilag Chemie, Schaffhausen, as a fumarate (Menocil) in tablets containing 14 mg of the product. The recommended daily dose was 1-2 tablets.

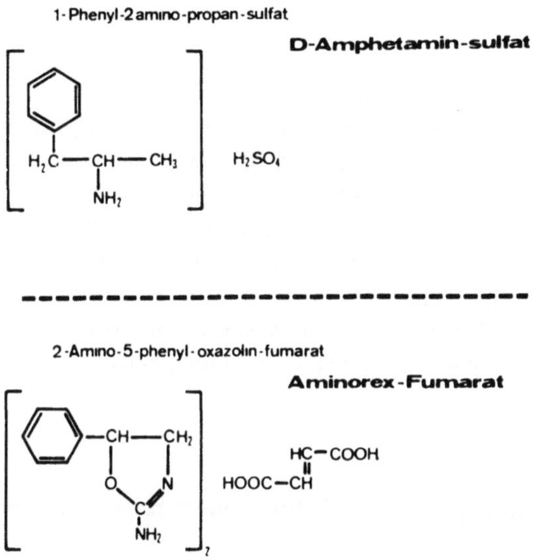

Fig. 1. Chemical structure of aminorex fumarate, compared to that of amphetamine sulfate (with the kind permission of *Stepanek* and *Zolliker*)

The first published experimental tests of aminorex were made by *Yelnosky* (1966) on dogs under phenobarbital anesthesia. The contractile strength of the right heart measured with open-chest, the systemic arterial pressure, and the heart rate were transitorily increased; the injection of a β-blocking agent abolished or strongly reduced this cardiac effect, suggesting that aminorex stimulates β-recptors; the injection of an α-blocking drug was found to suppress or reduce the effect on the systemic arterial pressure. Prior treatment with reserpine, which depletes catecholamines, was found to significantly decrease the effect of aminorex, suggesting that the sympathomimetic effect of aminorex might be due to a liberation of catecholamines. The effect on the lesser circulation was not studied in this paper.

2. Clinical Tests

Clinical tests were generally run on small groups of patients (*Hadler*, 1967; *Hotz*, 1967; *Kew*, 1970; *Krueger*, 1968; *Roos* et al., 1970; *Sandoval* et al., 1971; *Wood* and *Owen*, 1965);

they showed that aminorex, administered as a base or as fumarate, had the advantage of producing a regular weight loss. Unlike amphetamine, it was found to have no excessively stimulating effect but did have a euphorizing one (*Gadd* and *Gunne*, 1970). It has been tested on obese people without any other known health problems (*Kew*, 1970; *Roos* et al., 1970; *Sandoval* et al., 1971) as well as on obese patients suffering from other diseases (arterial hypertension, cardiac failure, arteriosclerosis, diabetes), and no severe side-effects were reported. *Sandoval* et al. (1971), however, noted that one case complained of dizziness and another of fatigue. *Hadler* (1967), after testing two groups of patients taking aminorex and comparing its effects with those of either amphetamine sulfate or phenmetrazine, observed 3 cases in one group and 2 in the other with cardiovascular side-effects. Precise details were not given.

3. Acute Toxicity

Acute intoxication with aminorex (accidental or suicidal) has been reported in three papers. Among 48 cases of intoxication due to the ingestion of various types of anorexigens (*Pasi*, 1967), 25 had taken aminorex. *Borbely* et al. (1970) studied 30 cases of acute intoxication due to aminorex. This intoxication begins with a state of psychomotor restlessness accompanied by mydriasis, redness of the skin, tachycardia, rise in blood pressure, and hyperpnea. In severe cases, convulsions and a central respiratory depression may be noted. No fatal case was reported. Late sequellae were not mentioned. The case reported by *Schuster* et al. showed the same symptoms. Repeated check-ups, and in particular cardiac catheterization could not objectivate any pulmonary vascular consequences.

II. Epidemiology; Clinical and Pathologic Features of Pulmonary Hypertension in Connection with Aminorex Intake

1. Epidemiology

As stated in the introduction, an increase in the number of cases of so-called primary pulmonary hypertensive disease (PPHD) was noted by Swiss cardiologists in 1967. Following the report by *Gurtner* et al. (1968a), a possible causal relationship with the intake of an anorexigenic drug, aminorex fumarate (Menocil, Cilag) put on the market during the fall of 1965 in Switzerland, was suggested. An increase in the incidence of this disease was observed in the Federal Republic of Germany and in Austria, where this drug was also sold. Results of cooperative studies published in these three countries (Germany: *Loogen*, 1970; Austria: *Kaindl*, 1969, 1971; Switzerland: *Wirz* and *Arbenz*, 1970) indicated the approximate number of patients affected with PHD in connection with the intake of aminorex, as well as the number of fatalities (see chart Fig. 2):

Germany: 35 deaths out of 215 cases = 16.2%
Austria: 16 deaths out of 134 cases = 11.9%
Switzerland: ca. 44 deaths out of 244 cases = 18%.

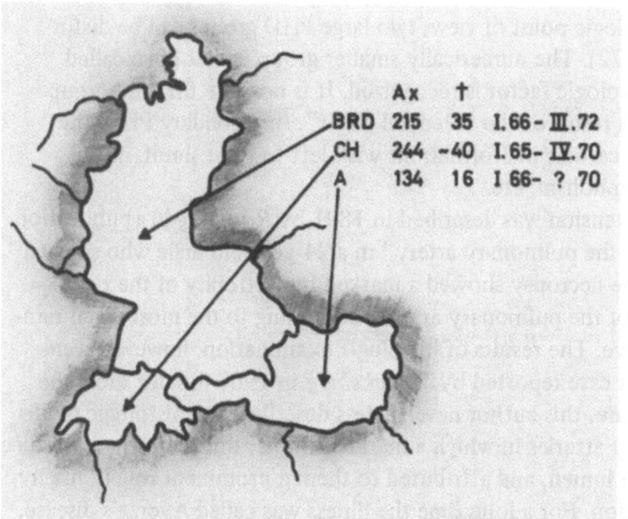

Fig. 2. Geographic distribution of cases of PHD in connection with aminorex intake. Number of patients *(Ax)* and of deceased cases (+) according to *Loogen* (BRD = German Federal Republic), *Wirz* and *Arbenz* (CH = Switzerland), and *Kaindl* (A = Austria) and the duration of the studies are indicated

The mortality rate is thus somewhere between 12 and 18% of the number of patients registered in the larger cardiology centers of these three countries, but it is quite possible that the actual figure is much higher, since, for various reasons, many cases may not have been included in the surveys. *Bass* and *Gurtner* (1973) indicate a mortality rate of 20%. On the other hand, in the countries where aminorex was sold, the incidence of PHD did not change during these years, according to the inquiry of *Ferrero* (1970). Outside these three countries, only two papers have reported PHD related to aminorex intake, one case in France (*Berthou*, 1972) and the other in Denmark (*Norderφ* and *Müller*, 1970).

The studies undertaken in Switzerland, Germany, and Austria (*Loogen*, 1972; *Rivier* et al., 1972; *Kaindl*, 1969, 1971) showed that the incidence of illness in proportion to the total number of persons who had taken the drug, was between 1 and 2‰. It is evident, therefore, that other factors, as yet unknown, must be involved (*Blankart*, 1970). No evidence, however, of risk factors such as varicose veins, phlebitis, pulmonary embolism, abdominal, gynecological or varicosis operations, was found in the course of an extensive epidemiologic study (*Blankart*, 1974). After the drug was withdrawn from the market in 1968, the frequency of PHD suddenly diminished, and since 1969 (*Loogen*, 1972) it has again become as rare as it was before 1967 (*Rivier*, personal communication).

2. Clinical Features

The clinical features of PHD related to the intake of anorexigens, especially aminorex fumarate, are not fundamentally different from those of primary or secondary PHD.

From the etiologic and nosologic point of view, two large PHD groups can be distinguished (*Harris*, 1970; *Trell*, 1972). The numerically smaller group is that of so-called idiopathic PHD, in which no etiologic factor is recognized. It is possible that this group might become even smaller, as a result of the "Menocil affair". In secondary PHD, the etiologic factors are numerous: cardiac malformations with left-to-right shunt, mitral stenosis, pulmonary thrombo-embolism, etc.

Idiopathic pulmonary hypertension was described in 1891 by *Romberg* in a publication reporting a case of "sclerosis of the pulmonary artery" in a 24-year-old male who showed all the clinical signs of PHD. The necropsy showed a marked hypertrophy of the right ventricle and a severe sclerosis of the pulmonary arteries extending to the most distal ramifications visible to the naked eye. The results of histologic examination, however, were not reported. Although the first case reported by *Mönckeberg* in 1907 did not show the typical clinical signs of the disease, this author nevertheless described the histologic modifications of the small pulmonary arteries in which a marked intimal thickening had produced a very important stenosis of the lumen, and attributed to them a prominent role in the hypertension of the lesser circulation. For a long time the illness was called Ayerza's disease, but very little precise clinical and anatomic data were available.

So-called primary PHD is a rare disease, most commonly affecting relatively young women. Its etiology is unknown and its evolution severe. It has been observed in several members of the same family (*Spencer*, 1968) and may sometimes occur very early in life. It is characterized by an increasing and often sudden exertion dyspnea, thoracic pains sometimes of anginal type, fainting, attacks of coughing, and later of signs of right cardiac failure. Among these, the most frequent are cyanosis and accentuation or the splitting of the second pulmonary tone. The electrocardiogram usually shows a pulmonary P and right axial deviation. The chest radiography reveals a moderately enlarged heart and a widening of the pulmonary hili. The positive diagnosis is of course based on cardiac catheterism which makes it possible to quantify the right ventricular systolic pressure, the pulmonary arterial pressure (systolic, diastolic, and mean), which are highly increased, the wedge pressure remaining normal. The pulmonary resistance is greatly increased, sometimes to more than 1000 dynes/sec/cm^{-5}. The evolution is severe and rapid, the time between diagnosis and death rarely exceeding two years.

Most of the cases of PHD related to aminorex intake (*Gurtner* et al., 1968a; *Kaindl*, 1969, 1971; *Loogen*, 1972; *Rivier* et al., 1972; *Follath* et al., 1971; *Gahl* et al., 1970; *Wirz* and *Arbenz*, 1970; *Chardonnens*, 1972) are usually characterized by a sudden onset of the disease. It affects mainly women between 30 and 50 years of age, moderately obese or believing themselves to be so, who have taken a variable quantity of aminorex. No relationship has been observed between the severity of PHD and either the quantity of the drug absorbed or the duration of the treatment. *Wirz* and *Arbenz* (1970) noticed further that there is no direct relationship between age, amount of excess weight at the beginning of the treatment, or the importance of weight loss. On the other hand, the commission of the Deutsche Gesellschaft für Kreislaufforschung (*Loogen*, 1972) has established that the risk of morbidity is a function of the dose. The duration of treatment as well as the quantity of ingested tablets can vary greatly as reported in section V. The first symptoms may appear during the course of the treatment or only a certain time after the treatment is ended (*Gahl*: 1 1/2-16 months; *Wirz* and *Arbenz*: 8.6 months; *Blankart*: 6 months). The clinical signs were those of PPHD. We would like to emphasize some peculiarities.

Only *Gahl* et al. (1970) mentioned an elevation of hematocrit. In the other studies, the laboratory data were normal, especially those related to the blood crasis. With regard to the results of right cardiac catheterization, all of the studies reported a noticeable increase in pulmonary arterial pressure with the mean pulmonary arterial pressure near 50 mm Hg. On the other hand, the wedge pressure is not modified, indicating the absence of any post-capillary obstruction. The right ventricular systolic pressure exceeds 80 mm Hg (*Kaindl*, 1971; *Gurtner* et al., 1968a). The cardiac index and the cardiac output are decreased to varying degrees. The pulmonary arterial resistance is extremely high, but has a very wide range. While the mean reported in the various papers is above 800 dynes/sec/cm^{-5}, the range extends from 150 to 3000. Although the pulmonary functions are generally not modified, hyperventilation at rest (*Gurtner* et al., 1968a; *Rivier* et al., 1972) or a slight decrease of the function (*Follath* et al., 1971; *Gahl* et al., 1970) are sometimes observed. The latter authors have observed a reduction of the pO_2 which they attribute to an opening of precapillary arterio-venous anastomoses. In the series of *Gurtner* et al. (1968a) selective pulmonary angiography, which was not performed in every case, showed a marked widening of the pulmonary trunk in 9 cases and an early venous phase in some cases, suggesting the existence of arterio-venous shunts. This peculiarity was also observed by *Kaindl* (1969, 1971) in 10 cases. *Kotscher* et al. (1973) reported that of 27 cases, there were 20 with a sudden narrowing of the arteries and 17 without any capillary phase; they also noted the early appearance of the opacifying material in the pulmonary vein, suggesting the existence of shunts.

The treatment proved to be disappointing. Intra-arterial injections of acetylcholine had only a temporary effect. Fibrinolytic treatment was found to be less effective than had at first been hoped, and was not without any risk to the patient (*Bouvier* et al., 1970). Some patients benefited, at least subjectively, from a treatment consisting of bed rest, digitalization, diuretics, anti-coagulation, and Persantine. Complete cures, however, like that of the taxi driver of *Gertsch* and *Stucki* (1970) are exceptional. *Kaindl* (1971) noted 30 cases in his series with a subjective improvement. *Bass* and *Gurtner* (1973) observed a lethality of 20% in a series of 23 cases followed for between 1 and 5 years; patients under 50 years of age may show various degrees of remission, but those over 50 showed either no improvement or a slight worsening. *Corrodi* and *Bühlmann* (1973), in 39 cases treated by anti-coagulation, noted 7 deaths, 12 stationary cases, 9 worsened, and 9 improved. Most patients are stabilized; the others showed a more or less rapid but inexorable evolution toward death due to right cardiac failure. We have noted the geographic incidence of the published fatal cases on the map shown in Figure 2.

3. Pathology

The necropsy data of the published cases are analyzed in the section on histologic modifications (Section VI). These data, similar to ours, have been interpreted in various ways. The lesions of the muscular pulmonary arteries are obviously similar to those codified by *Heath* and *Edwards* (1958) in cases of PHD in connection with cardiac malformations. This classification includes six grades of severity, which are discussed in Section VI.

Since the "Menocil affair" could have had medico-legal implications, clinical tests were submitted to extremely critical examination, some repeated, and new pharmacologic tests were performed. These studies are the subject of the next section.

III. New Clinical and Pharmacologic Tests

A review of the papers reporting the clinical tests of aminorex shows that secondary effects are infrequent and not very serious. At the symposium in Bürgenstock *Peters* and *Gourzis* (1970) reviewed the clinical tests run by various research groups at the request of the McNeil Corporation, the firm that had developed and tested the drug in the USA. *Peters* and *Gourzis* reported that, out of 4400 subjects, there was only one case of well-documented PHD. They noted that the side-effects were slight and reversible, but one of their tables raises some questions since the following side-effects were mentioned: tachycardia: 74; shortness of breath: 10; edema: 27; chest pain: 19; ECG changes: 3; fainting: 3.

These, in our opinion, are clinical signs of PHD and should have been investigated more thoroughly. In the light of the conflicting reports about aminorex, it is astonishing that *Steim* (1969) gave i.v. infusions of 5 mg in half an hour to subjects while checking the pulmonary arterial pressure, which was found either to remain normal or to show only a slight and transitory increase. Catamnestic information was not provided.

In the group of obese patients of *Sandoval* et al. (1971) one case of PHD developed after 6 months of treatment. *Hadler* (1970), in a second paper published after the responsibility of aminorex had been suggested, mentioned one patient who complained of weakness and fainting and another who had suffered from bouts of palpitations. It is of course easy to criticize these studies retrospectively, but the question remains as to whether the clinical tests were conducted with the necessary care and rigour.

Following the clinical reports published in Switzerland (*Gurtner* et al., 1968a) on the possible relationship between aminorex intake and the sudden increase in the number of cases of pulmonary hypertension, new and more thorough pharmacologic studies were undertaken. At the symposium in Vienna, *Kraupp* (1969) communicated his observations on the effect of aminorex on the pulmonary circulation of anesthetized open-chest dogs, maintained under artificial respiration; a single injection produced an increase in the pulmonary resistance and the pulmonary arterial pressure, accompanied by a reduction of the pulmonary arterial flow, which normalized slowly. These modifications were dose-dependent. Subsequent papers by *Kraupp* and his team (*Kraupp* et al., 1969, 1970; *Stühlinger* et al., 1967, 1970, 1971) confirmed these preliminary data; moreover, an i.v. perfusion performed under the same experimental conditions (chloralose anesthesia) caused a sustained rise of the pulmonary arterial pressure, the aortic pressure, the pulmonary arterial resistance, and the aortic resistance, which regressed but slowly after the perfusion was terminated. The administration of an α-blocking agent (phentolamine) inhibited this effect, whereas a β-blocker (propranolol) potentiated it, indicating that aminorex is an α-sympathomimetic drug. Thus, there seems to be a contradiction with *Yelnosky*'s results, probably because the experiments were performed on two different circulations and with different doses. *Brunner* and *Stepanek* (1970), *Stepanek* and *Zolliker* (1971), *Stepanek* and *Zak* (1975), have made similar observations on dogs; under chloralose anesthesia, an i.v. injection caused a dose-related rise in the pulmonary arterial pressure; in the non-anesthetized animals, injections repeated on 4 consecutive days resulted in a sustained increase in the pulmonary arterial pressure for a period of one week. Under prolonged administration, *Stepanek* (1973) noticed in the dog a raise of the mean pulmonary arterial pressure of 61 or 71%, according to the dose; likewise, the total pulmonary arterial resistance increased by 62 or 96%.

Will and *Bisgard* (1972) reproduced the experiment, but in nonanesthetized dogs. They noted a significant increase in the mean pulmonary arterial pressure, but this was transient and had no cumulative or residual effects; the other parameters checked at catheterization were not significantly modified.

Engelhardt and *Hort* (1970) reported that, in the rat, the pulmonary arterial pressure increased after repeated high i.p. doses, whereas low doses had no effect; nevertheless, no histologic modifications of the pulmonary vessels were demonstrated. Also experimenting with rats, *Mielke* et al. (1973) observed a 30% increase in the right ventricular pressure; moreover they noticed an increase in pulmonary 5-hydroxy-tryptamine, to which they attributed, without question, the rise in pressure. No histologic modifications were reported.

Peroral administration (gavage) of aminorex to rats for 3-6 months (*Engelhardt* et al., 1971) was found to cause no modification of the thickness of the media nor of the internal diameter of muscular pulmonary arteries. *Leuschner* et al. (1971) treated dogs orally for 13 weeks and observed no significant alterations of the pulmonary arterial pressure checked under closed-chest anesthesia after 6 and 13 weeks. At necropsy neither macroscopic nor microscopic alterations of the lungs and heart were found. *Heath* et al. (1971) and *Kay* et al. (1971) found no histologic modifications of the muscular and elastic pulmonary arteries in rats and dogs having received aminorex fumarate orally for prolonged periods. The opinion of these authors, based on their experimental results, is that there is no proof that aminorex might be the cause of PHD in humans.

Smith et al. (1973) administered aminorex base orally to two monkeys *(Erythrocebus patas patas)* for 347 days. At the end of the experiment no hemodynamic abnormalities (wedge pressure, pulmonary arterial pressure, right ventricular pressure, and right auricular pressure) were observed. At necropsy no macroscopic modifications were found and histology did not reveal any alterations of muscular arteries, either in measurement or morphologically. The media of one of the animals, however, was found to be slightly thicker than that of either the other animal or the control, a modification which the authors tended to minimize. *Smith* et al. (1973), like the authors of the previously mentioned paper, feel that great caution is required in the evaluation of the role of aminorex in PHD. They do admit, however, that their experimental results were obtained from a very small number of subjects, even taking in account the low rate of PHD among humans. In another experiment they submitted rats to a high fat diet and administered aminorex over a prolonged period. They observed no histologic modifications of pulmonary arteries nor right cardiac hypertrophy. In their opinion PHD has never been successfully produced in animals and the suspected relationship between aminorex and PHD remains to be established.

In calves maintained at an altitude of 1600m, *Byrne-Quinn* and *Grover* (1972) noted an increase, although it was not significant, of the mean pulmonary arterial pressure and of the pulmonary arterial resistance after daily i.v. injections for 28 days. Under hypoxia, the increase of pulmonary arterial pressure is not aggravated by aminorex nor by amphetamine used for purpose of comparison.

Peters and *Gourzis* (1970) reported the first experimental study undertaken at the McNeil Corporation: only a single injection in anesthetized animals produced an increase of the blood pressure and pulmonary arterial pressure, comparable to that due to serotonin, norepinephrine, or epinephrine. On the other hand, the increase of pulmonary perfusion pressure due to microemboli of diatomaceous earth was accentuated by the injection of aminorex or amphetamine.

Lüllmann et al. (1972) compared the effects of various anorexigens to those of *Crotalaria spectabilis.* All of these drugs caused an increase in the pulmonary arterial pressure of varying degrees: *Crotalaria spectabilis* more than 100%, chlorphentermine more than 50-70%, and aminorex more than 30%. After *Crotalaria spectabilis* they found fresh capillary thrombi. The other drugs caused no vascular modifications that could be detected by the light microscope. The only lesion was an accumulation of intraalveolar foamy cells, which was massive after chlorphentermine and slight after aminorex. This phenomenon is thought to be reactive, nonspecific, and unrelated to PHD.

Finally, *Backmann*'s experimental results with rats (1970) should be mentioned: perivascular edema and slight proliferation of adventitial cells, which are considered a pathogenic phenomenon. In two subsequent studies, also on rats, *Backmann* et al. (1972, 1974) noted an increase in the number of mast cells but did not observe any arterial lesions.

These contradictory experimental results suggest the following remarks:

1. The method of anesthesia in dogs, under experimental conditions not really physiological (e.g., *Kraupp* and his team) could cause significant hemodynamic modifications; *Will* (personal communication) points out that chloralose anesthesia, by itself, induces an increase of the pulmonary arterial pressure.
2. The number of animals used was insufficient, for if the low incidence of PHD related to aminorex intake in humans (1-2‰) is taken into account, a comparable experimental rate could only be obtained by using thousands of animals for a single experimental model.
3. It may be that aminorex has only a potentiating effect upon some other mechanism, which is still unknown or has been neglected. Did not *Peters* (1970) mention that the increase in the pulmonary perfusion pressure following injections of diatomaceous earth was accentuated by injections of aminorex?

IV. Material and Methods

Our study was made on the lungs taken during the necropsy from 37 patients who have developed PHD related to aminorex intake, from 5 patients with PPHD without any known intake of anorexigenic drugs, and from 5 patients without any pulmonary vascular disease, used as controls. The material was placed at our disposal by various pathology departments in Switzerland.

1. Material

The following cases were used for the present study (the material obtained from the Geneva Department of Pathology are in parentheses):

PHD + aminorex	37 necropsies (12)
PPHD	5 necropsies (4)
controls	5 necropsies (5)
secondary PHD	2 biopsies (2)

a) PHD Related to Aminorex Intake (PHD + Ax): 37 Cases

The pulmonary material was fixed as follows:

Formaldehyde fixation by intratracheal instillation:	19 cases
Formaldehyde fixation by immersion:	15 cases
Glutaraldehyde fixation by intratracheal instillation:	3 cases

Of these cases,

–all were submitted to light-microscope evaluation,

–29 were received within a reasonable time to permit a numerical evaluation of arterial lesions,

–14, for whom complete and accurate hemodynamic data were available and which were accurately fixed, were used for morphometric studies,

–3, fixed with glutaraldehyde, were studied with the electron microscope.

b) PPHD: 5 Cases

Formaldehyde fixation by intratracheal instillation:	4 cases
Glutaraldehyde fixation by intratracheal instillation:	1 case

Of these cases,

–all were examined with the light microscope,

–4 were used for numerical evaluation of lesions,

–3 were used for morphometric studies,

–1 case was studied with the electron microscope.

c) Secondary PHD

Two lung biopsies fixed by immersion in glutaraldehyde were studied both optically and with the electron microscope.

d) Controls: 5 Cases

Formaldehyde fixation by intratracheal instillation:	3 cases
Glutaraldehyde fixation by intratracheal instillation:	2 cases

Of these cases,

–all were examined with the light microscope,

–3 were used for morphometric studies,

–2 were studied with the electron microscope.

2. Methods

a) Light microscopy: All specimens were fixed with 10% buffered formaldehyde or glutaraldehyde, dehydrated and embedded in paraffine according to the usual techniques.

The following stains were made on 5μ thick sections: hematoxylin and eosin, van Gieson elastine or Goldner elastine or Verhoeff elastine, periodic acid Schiff (PAS), Mallory's phosphotungstic hematoxylin for fibrin, Giemsa, and Prussian blue.

b) Electron microscopy: For ultrastructural studies we gathered material from necropsies performed in our Department very soon after the death of 3 patients who had taken aminorex, 1 case of PPHD and 2 control cases without pulmonary vascular disease; we also studied lung biopsies from 2 cases of PHD secondary to a cardiac malformation. Necropsy lungs were fixed *in situ* by intratracheal instillation of 2% S-collidine-buffered glutaraldehyde adjusted to a pH of 7.4 and 340 mosm. After necropsy they were immersed for at least 5 h in the same solution. Specimens were then postfixed with 1% S-collidine-buffered osmic acid. Semithin sections, stained with toluidine blue, allowed any interesting lesions to be detected. Ultrathin sections, obtained by means of an LKB ultramicrotome (Ultrotom), were contrasted with uranyl acetate and lead nitrate, and examined with a Philips EM 300 electron microscope at an 80 KV tension.

Biopsy material was fixed by immersion in glutaraldehyde and submitted to the same treatment as necropsy material.

c) Numerical evaluation of arterial lesions: Muscular arteries, the external diameter of which ranged between 100 and 1000μ, were counted in 29 cases of aminorex intake and 4 cases of PPHD. The lesions of these arteries were classified into 6 grades according to a modification of the classification established by *Heath* and *Edwards* (1958) and used in previous studies (*Kapanci* and *Widgren*, 1970). This classification is presented in Section VI. Lung specimens, 10 from each lung, were taken at random according to a grid with points distributed as hexagons. Randomly chosen paraffin sections, stained with van Gieson elastine, were examined by means of a Wild M 20 automatic stage microscope, on a screen with test lines according to the method of *Weibel* et al. (1966). The counting was done by a Fistronic 630 electronic counter with automatic tabulator recorded to an IBM electric typewriter. A mean of 91.7 ± 6.5 arteries was obtained for a total of 29 cases of PHD + Ax and 100.5 for PPHD cases.

d) Morphometric studies: The measurements of the various parameters listed below were made according to *Weibel's* point counting method described in the preceding paragraph. Only 14 cases of PHD + Ax fixed under adequate conditions and for whom complete hemodynamic data were available could be used for this study, which also included 3 cases of PPHD and 3 controls. Van Gieson elastine stained sections were projected on a test grid with 168 points. For each of 20 tissue blocks per case, a mean of 20 microscope fields were evaluated. The following points and sections were registered for the evaluation of muscular arteries:

PL: points within the arterial lumen
PI: points on the arterial intima
PM: points on the media
PX: points outside of the arteries (adventitia included)
II: intersection with the intima
IE: intersection with the lamina elastica interna
IM: intersection with the media

The following *measurements* were calculated according to *Weibel*'s data:

mean thickness of media: $m = \frac{Z \times PM}{2 \times IE}$ (where Z = length of test line: 0.015 mm for a 40 x objective enlargement)

mean thickness of intima: $i = \frac{Z \times PI}{2\ IE}$

volume density of lumen: $= \frac{PL}{PE}$ (where PE = sum of arterial points = PL + PI + PM)

circumference of lamina elastica interna: $Ce = \frac{\pi \times Z \times 3 \times IE}{2}$

V. Principal Clinical and Anatomic Data

The details of the main data are given in Table 1 for the 37 cases and summed up in Table 2. It can be seen immediately from these data that the *predominant sex* affected by PHD is female, since our series comprises 34 females. *Mean age* was 50.5 years ± 2.3. The youngest subject was 20 years old, the oldest, 77. Three data concerning *treatment* are particularly important:

1. The *daily dose* varied from 1 to 2 tablets (14-28 mg) and the *total dose* from 30-600 tablets. The average dose was 249 ± 33 tablets.
2. The *duration* ranged from 30-720 days with an average of 271 ± 40 days.
3. Table 1 shows that there is no relationship between the treatment (dose and duration) and the severity of hemodynamic changes (mean pulmonary arterial pressure and pulmonary arterial resistance), as has been pointed out in the literature. *The mean pulmonary arterial pressure* (MPAP) was 52.1 mm Hg, the range being 35-70. The highest systolic PAP was 119, the lowest 55.

The *pulmonary arterial resistance* was between 430 and 2000 dynes/sec/cm^{-5}, with a mean of 1258.7 ± 80.4.

Since the laboratory data at our disposal were too fragmentory, they are not shown in the table. In view of the observations of *Gahl* et al. (1970) on the modifications of the coagulation, it is interesting to note that they were also found in 7 of our cases; in 6, hematocrit was ⩾ 50% and in 2 an abnormal platelet adhesivity was observed. Other modifications, e.g., those reported by *Inglesby* et al. (1973) concerning a family in which 7 out of 10 members suffering from PPHD had abnormally high antiplasmin levels, were not found in this study.

VI. Histologic Findings

1. Personal Observations

The description of the histologic modifications is based on that given by *Heath* and *Edwards* (1958), who established a classification with 6 grades corresponding to the severity of lesions

Table 1. PHD + Ax–Principal clinical and anatomic data of 37 cases with aminorex intake

Case	Sex	Age	Treatment		ECG	PAP (mm Hg) S/D/M	PAR dyn/sec/cm^{-5}	Heart		Cause of death	Other necropsy findings	Remarks
			Duration	Dose (tablets)				Weight g	Thickness RV mm			
1	♀	25	5 m.	120-160	RVH	100/44/64	1250	390	9	Recent pulmonary emboli		
2	♀	35	?	?	RVH	55/38/48	1470	450	9	Right ventricular failure		
3	♀	65	?	120-160	RVH	∅	∅	505	10	RVF		
4	♀	20	ca. 1 y.	1 t/d	RVH	∅	∅	330	8	RVF		
5	♀	44	3 y.	90	RVH ANT. ISCH.	60/25/36	430	470	5	RVF Myocarditis		
6	♀	52	14 m.	400	RVH ANT. ISCH.	80/40/52	1264	320	6	RVF		
7	♀	47	40 d.	40	RVH	55/30/40	1056	320	4	RVF		Saridonism
8	♂	45	1 1/2 y.	1 t/d	RVH	80/48/60	∅	450	7	RVF		
9	♀	57	1 1/2 y.	1 t/d	RVH	∅	∅	560	9	RVF		
10	♀	51	4 m.	120-150	RVH	82/27/47	1240	?	?	Recent pulmonary emboli		
11	♀	33	?	?	RVH	+	+	?	+	RVF		
12	♀	48	ca. 3 m.	1 t/d	RVH	82/43/59	1275	350	15	RVF		
13	♀	44	7 m.	1 t/d	RVH	70/33/46	1115	470	12	RVF		
14	♀	45	ca. 8 m.	1-2 t/d	RVH	104/44/60	1696	?	?	?		

15	♀	50	8m.	1 t/d	RA overload	83/38/53	1404	?	?	?		Polytoxi-comania
16	♀	62	7 m.	1 t/d	?	∅	∅	520	18	RVF		
17	♀	35	2-3 m.	?	RVH	73/34/45	1200	380	5	RVF		
18	♀	64	ca. 2 y.	270	RVH	∅	∅	400	8	RVF		
19	♂	50	?	240	?	∅	∅	480	?	RVF		Tentamen Barbiturates
20	♀	46	?	320	RVH	?	?	440	?	Broncho-pneumonia		
21	♂	60	?	?	?	?	?	550	8	Retroperitoneal hematoma		
22	♀	31	?	?	RVH	? /? /46	920	?	6	RVF		Amphetamine abuse
23	♀	38	7 m.	205	RVH	70/40/49	900	350	6	RVF		
24	♀	62	?	40	RVH	∅	∅	440	5.5	Bronchopneumonia, lymphangitis carcinomatosa	CA cecum META ++	
25	♀	33	ca. 1 y.	300	RAH	58/28/35	823	390	6	Cerebral edema	Hodgkin treated encephalitis	
26	♀	68	2 y.	?	RVH	85/38/56	1440	345	6	RVF		
27	♀	61	1 m.	30	RVH	82/40/55	1184	330	6	RVF		
28	♀	68	8 m.	2 t/d	LVH+R	90/42/60	2000 (T.)	320	5	RVF	Acute Glo-merulo-nephritis	
29	♀	60	20 m.	1 t/d	RVH	115/45/70	?	460	7	RVF		

Table 1 (continued)

Case	Sex	Age	Treatment		ECG	PAP (mm Hg) S/D/M	PAR dyn/sec/cm^{-5}	Heart		Cause of death	Other necropsy findings	Remarks
			Duration	Dose (tablets)				Weight g	Thickness RV mm			
30	♀	71	6 1/2 m.	?	RVH	∅	∅	535	7	RVF	CA thyroid	
31	♀	77	3 m.	1 t/d	RVH	119/38/60	1510	530	5	Recent pulmonary emboli		
32	♀	53	3 m.	80	RA	110/35/55	?	?	10	Acute pyelonephritis		
33	♀	25	Several months	3 t/d	RA Partial R. block	∅	∅	280	8	Recent thrombo-emboli	Multiple pulmonary emboli	+ Other anorexigens
34	♀	57	1 y.	1 t/d	?	?	?	410	4	RVF		
35	♀	65	1 m.	40	RVH	90-100/?	1700-2000	465	8	RVF	Thyroid atrophy	
36	♀	68	1 y.	?	RVH	100/32/?	?	570	9	RVF	Carcinoid jejunum + mesenteric meta	
37	♀	53	?	120	?	75/35/50	1146	?	+	RVF	chronic hepatitis	

? = unknown
∅ = no cardiac catheterism
\+ = increased, value unknown

d. = days
m. = months
y. = years

RVH = right ventricular hypertrophy
RVF = right ventricular failure
RA = right axial deviation

Table 2. Pulmonary hypertensive disease and aminorex– summary of main clinical and anatomic data (37 cases)

Sex	34 F; 3 M
Age	50.5 ± 2.3 (20-77)
Duration of treatment	271 ± 40 d (30-720) (27 cases)
Average dosis	249 ± 33 t (30-600) (25 cases)
E.C.G.	RVH (24 cases)
MPAP (mm Hg)	52.1 ± 1.9 (35-70)(22 cases)
PAR (dyn/sec/cm^{-5})	1258.7 ± 80.4 (430-2000) (20 cases)
Heart: weight (g)	427.0 ± 15.1 (280-570) (30 cases)
Heart: RV thickness (mm)	7.7 ± 0.6 (4-18) (30 cases)
Cause of death	RV insufficiency: 26/35 cases

Summary of Table 1. The first figure gives the mean and S.E. The first parenthesis indicates the range, the second the number of cases with available data.

observed in cases of pulmonary hypertension, primary or secondary to a cardiac defect. A modification of this grading was adopted since the numerical and morphometric data, which will be reported in the next section, revealed that grade III (fibrosis and diffuse dilatation) was too large a group. It was therefore divided into III A (intimal fibrosis) and III B (diffuse dilatation). The histologic modifications will be analyzed according the following *classification:*

I	Medial hypertrophy	V	angiomatoid lesions with hemosiderosis
II	intimal hyperplasia	VI	arterial necrosis
III A	intimal fibrosis		
III B	diffuse dilatation		
IV	plexiform lesions		

Figure 3 shows the schematic representation of the lesions which will be described here in greater detail.

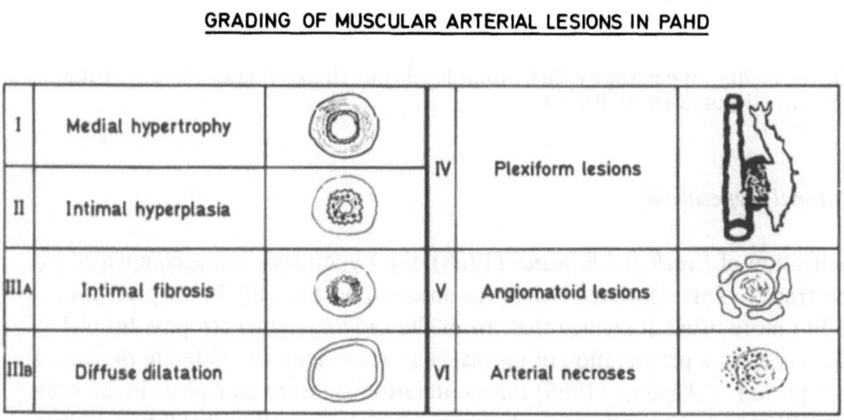

Fig. 3. Schematic representation of lesions of muscular arteries according to our classification

Grade I: Medial Hypertrophy

As will be seen in the numerical evaluation of the morphologic data, this type of lesion is seldom encountered; it might be one of the first modifications of pulmonary arteries, probably reflecting a vasospasm. It is characterized by a thickening of the media, which is usually homogeneous, much more rarely somewhat anarchical. This thickening is uniform for the entire arterial circumference (Fig. 4). The intima does not produce any reaction at this stage. Exceptionally, a moderate fibrosis of this layer is observed. The adventitia is often thickened and fibrotic and may contain more or less dilated lymphatic vessels, apparently denoting right cardiac failure. This thickening is also observed in the lesions of other grades.

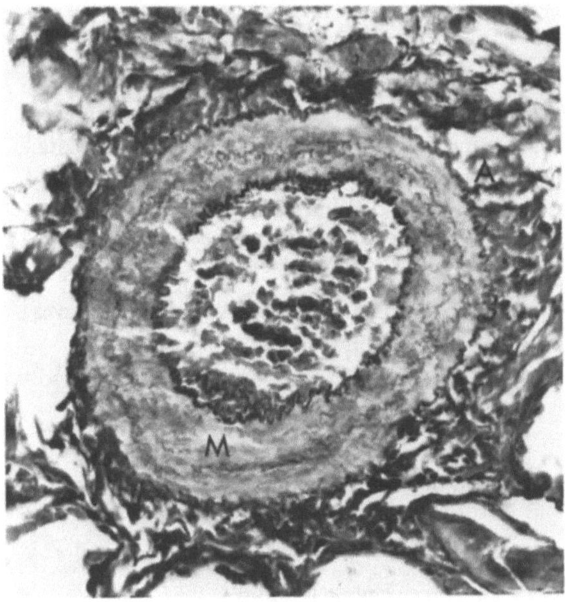

Fig. 4. Uniform medial hypertrophy *(M)*. Notable fibrous thickening of the adventitia *(A)*. (Case 8; van Gieson elastin; 205 x)

Grade II: Intimal Hyperplasia

In the classification of *Heath* and *Edwards* (1958), this hyperplasia was accompanied by medial hypertrophy. This associated lesion was observed occasionally in the present investigation, but more often, it seemed that the media had undergone atrophy. Intimal hyperplasia consists of a proliferation of intimal cells which tend to obliterate the lumen (Fig. 5). As reported by *Kapanci* (1965) this obliteration seems to take place in the vicinity of PAS-positive material, rarely stained as fibrin, which is, in all probability, a "fibrinoid" material, i.e., degraded fibrin. Upon contact with this material, intimal cells proliferate; some of them may contain such material (Fig. 6). In some places this cellular pro-

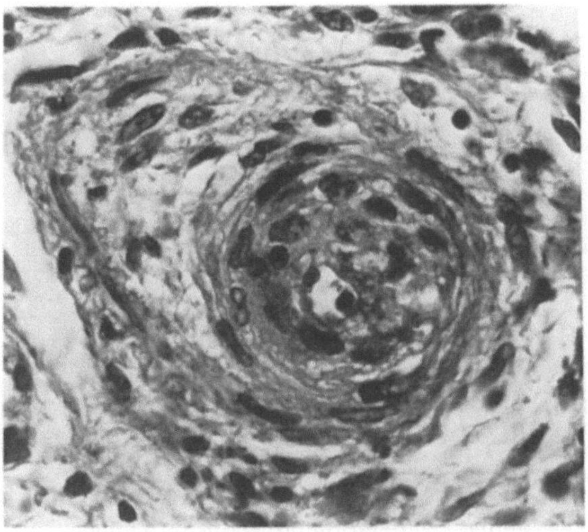

Fig. 5. Intimal hyperplasia: almost complete obliteration of the lumen by cellular proliferation. (Case 1; hematoxylin and eosin; 512 x)

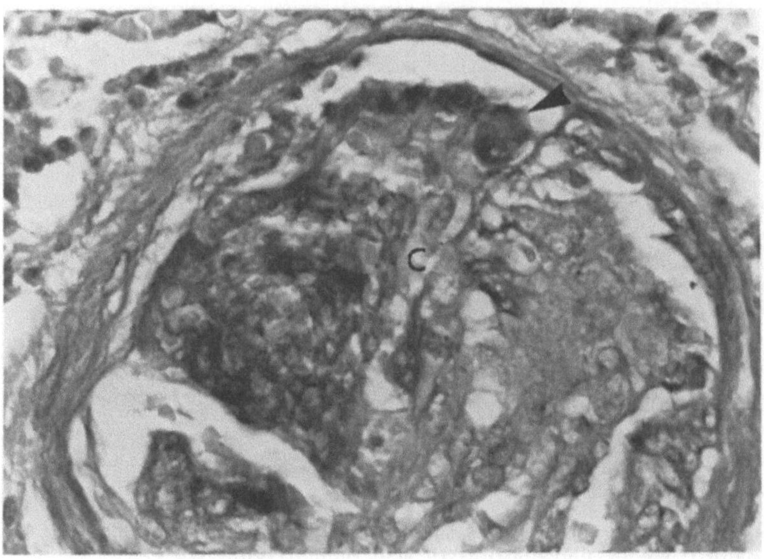

Fig. 6. Intimal hyperplasia: cellular proliferation *(C)* in the vicinity of fibrinoid material, sometimes intracellular (arrow). (Case 16; PAS; 512 x)

liferation may assume the shape of mushrooms or small clubs projecting into the lumen (Fig. 7). In others, it forms a crescent or a concentric ring tending to narrow, and later to obliterate the lumen. These cellular proliferations may be seen at arterial branchings. Later, these lesions may undergo two types of modifications:

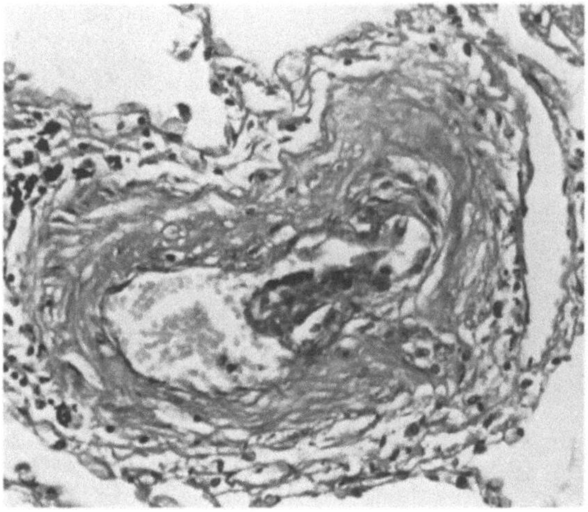

Fig. 7. Intimal hyperplasia: mushroom-like cellular proliferation projecting into lumen. (Case 7; PAS; 205 x)

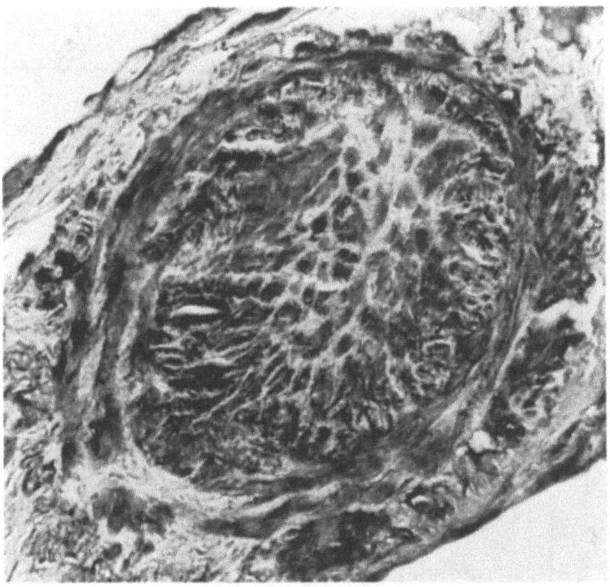

Fig. 8. Muscularized intimal hyperplasia obliterating lumen: fusiform appearance of SMC, sectioned longitudinally. (Case 28; phosphotungstic acid hematoxylin; 512 x)

1. *Muscularization* (Fig. 8) which will be studied in greater detail in the chapter on electron microscopy; in this intimal hyperplasia, smooth muscle cells (SMC) which may have a radial or concentric disposition toward the lumen are observed. A muscularization of this kind may be seen in the obliterating lesions as well as in intimal rings or crescents. It seems plausible, therefore, that these cells contribute to increase the arterial wall tone;

2. intimal hyperplasia undergoes a *progressive fibrosis* leading to grade III A.

Grade III A: Intimal Fibrosis

As will be seen in the chapter on numerical evaluation, this type of lesion was frequent. The following three substages could be distinguished:

1. A concentric, sometimes obliterating fibrosis, often associated with elastosis and preservation, even hypertrophy, of the media (Fig. 9); this lesion is not, in our opinion, the result of the organization of a thrombus;
2. A later stage in which these lesions lose their cells (Fig. 10) and assume a completely hyaline appearance, the lumen being more or less preserved, sometimes stenosed, while the media is clearly atrophic;

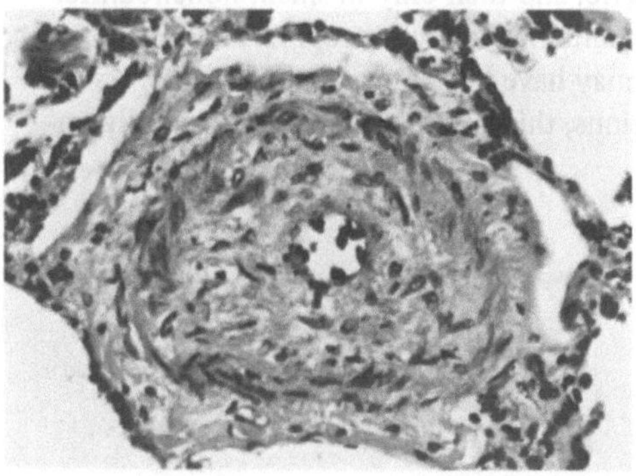

Fig. 9. Intimal fibrosis stenosing the lumen, containing a few cells. (Case 6; Goldner; 205 x)

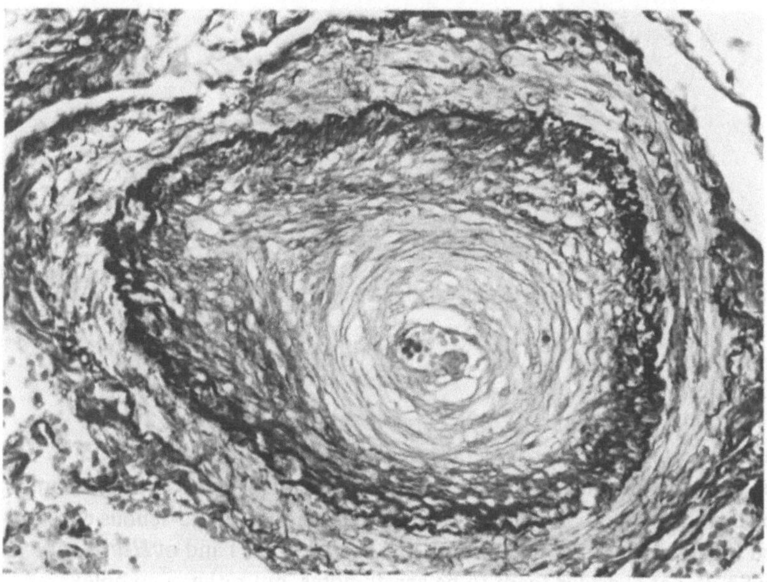

Fig. 10. Acellular intimal fibrosis; nearly complete obliteration of lumen; elastosis. (Case 16; van Gieson elastin; 205 x)

3. In the final stage after the presumed discontinuation of the drug intake, aspects of both the first two substages can be observed, comprising a more or less marked elastosis with or without obliteration of the lumen. The lamina elastica interna, often doubled or multiplied, takes on a very sinuous appearance, giving the lumen a withered aspect; sometimes, only an elastotic skeletic formation remains.

Grade III B: Diffuse Dilatation

In the opinion of *Heath* and *Edwards* (1958) these lesions were micro-aneurisms of which there are four principal categories: plexiform lesions, vein-like ramifications of hypertrophied muscular arteries, angiomatoid lesions, and cavernous lesions.

In our classification, we have reserved the designation of diffuse dilatation for a peculiar type of lesion, that of a single artery cut transversely, with a regular, homogeneous circular appearance; the intima is only rarely hyperplastic, and then only in small, noncircumferencial areas. The media has a varying thickness and may be hypertrophic. Arteries of various sizes, distributed over the whole section, may have this appearance (Fig. 11). As will be shown in the results of quantitative evaluations, this lesion is the most frequently encountered one in our study.

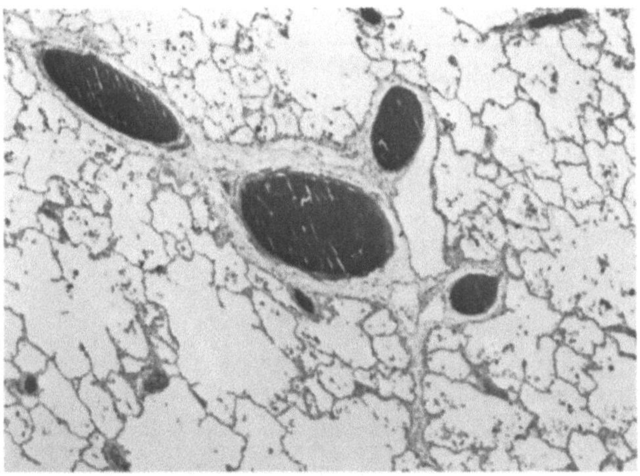

Fig. 11. Diffuse dilatation of several arteries of varying size. (Case 7; Goldner; 32 x)

Grade IV: Plexiform Lesions

Although classified by *Heath* and *Edwards* (1958) as dilatation lesions, these should, in our opinion, be separated from other varieties of dilatation lesions. Because of their complexity, they have been interpreted in various ways: glomic lesions, arterio-venous anastomoses, and organized thrombi. We share the opinion of *Kapanci* (1957) and of *Heath* and *Edwards* (1958), who do not consider them to be anastomoses, but rather a complex lesion of vascular dilatation and endothelial proliferation, with possible thrombosis or

Fig. 12. Plexiform lesion: schematic drawing modified according to *Naeye*, *A* = afferent artery; *B* = origin of arterial branch; *C* and *D* = vascular slits; *E* = poststenotic dilatation of the efferent branch

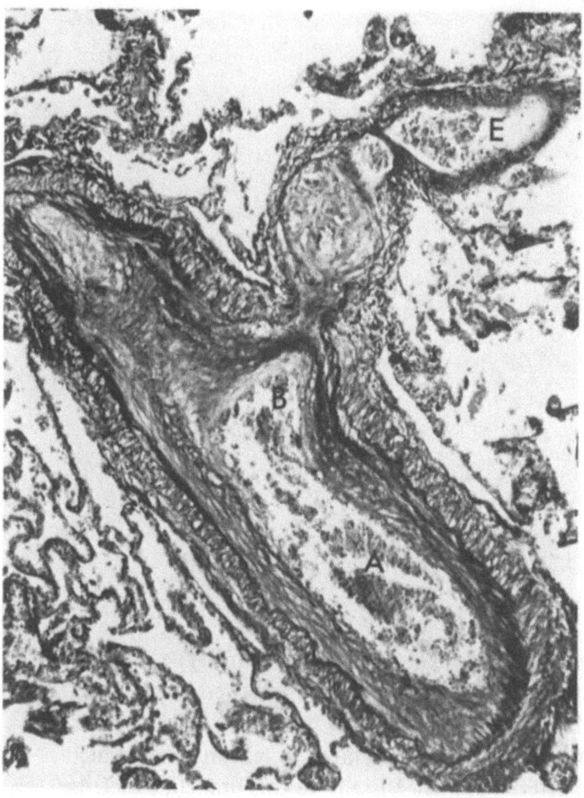

Fig. 13. Plexiform lesion: afferent artery *(A)* with medial hypertrophy and intimal fibrosis; obstruction of origin of arterial branch *(B)*; poststenotic dilatation *(E)*. (Case 16; van Gieson elastin; 80 x)

apposition of fibrin-like material. A good model has been suggested by *Naeye* (1968) which is reproduced in Figure 12. Although we did not perform serial sections or reconstructions, we agree with the hypothesis suggested by *Heath* and *Edwards* (1958), among others, but have modified it slightly: an afferent arterial branch, the lumen of which is slightly dilated, with medial hypertrophy and more or less regular intimal hyperplasia associated with irregular fibrosis, is continued by an arterial branch which is partly or completely obliterated by a proliferation of intimal cells (Fig. 13). These cells may form either solid plugs or more or less loose networks containing thin vascular slits in which red blood cells are sometimes found (Fig. 14). Within these intimal proliferations, masses or filaments of PAS-positive material (fibrinoid) comparable to that described in grade II are

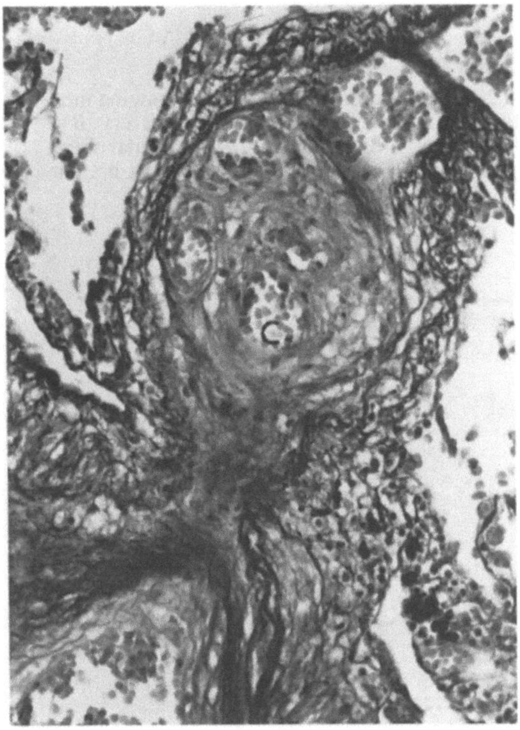

Fig. 14. Detail of Fig. 13: obstruction of vascular lumen by intimal proliferation; transverse section of vascular slits *(C)*. (Case 16; van Gieson elastin; 205 x)

occasionally observed. It is exceptional to find a real thrombosis in this type of lesion. The deposits observed do not, in our opinion, represent real thromboses, but rather are simply deposits of fibrinoid material at the site of an arterial branching which have induced, as described in grade II, intimal cell proliferation, more or less anarchic, which may finally cause obliteration of the arterial branch. Distally to this branch, a saccular arterial dilatation can be seen which represents a poststenotic dilatation of the arterial branch which is obliterated at its origin. The dilatation may induce an atrophy of the wall such as to make this branch appear to be a vein, which might explain why a lesion of this type could have been thought to be an arterio-venous anastomosis. The atrophy of the media may, in fact, be so marked that the lamina elastica interna and externa actually touch or become fused (Fig. 15). Intimal proliferations at the beginning of arterial branches, as has already been stated, may be very loose or highly cellular. Sometimes, in later stages, they seem to undergo fibrosis which may obliterate the arterial lumen and may be accompanied by atrophy of the media.

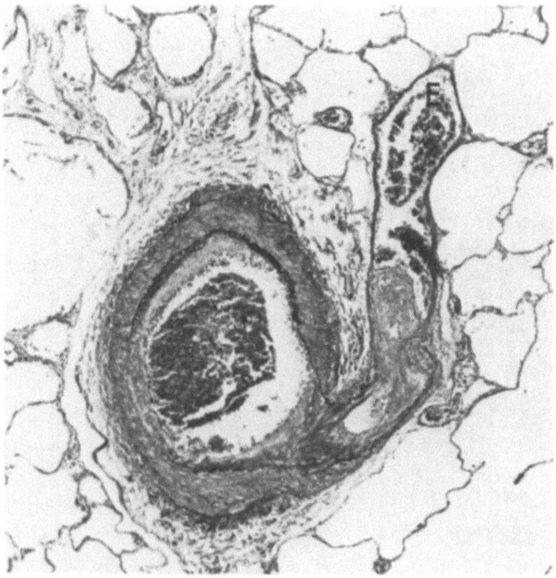

Fig. 15. Plexiform lesion simulating an arterio-venous anastomosis: marked poststenotic dilatation with atrophic wall *(E)*. (Case 16; van Gieson elastin; 32 x)

Grade V: Angiomatoid Lesions with Hemosiderosis

According to *Heath* and *Edwards* (1958), this type of lesion is a consequence of intimal fibrosis with thin vessel walls, sometimes with a lumen subdivided by fibrous tracts. Lesions of this kind, were, in fact, encountered in the present study. In other cases, and, in our opinion, more often, they seem to be a complication of a plexiform lesion. The intimal proliferation occupies not only the arterial branch, but even the artery itself, where it produces cushions or cellular mushrooms which may or may not be associated with deposits of fibrinoid material; in some places, these proliferations delimit vascular slits like those seen in partially recanalized thrombosis; this might explain why angiomatoid lesions have been sometimes thought to be organized thrombi. In our opinion, however, these lesions are almost certainly intimal proliferations in contact with fibrinoid material, which subdivide the arterial lumen (Fig. 16). As a consequence of fibrosis, these intimal proliferations are reduced to slender intraluminal tracts. Around these "main lesions", there are numerous highly dilated arterial branches, resembling efferent branches of plexiform lesions.

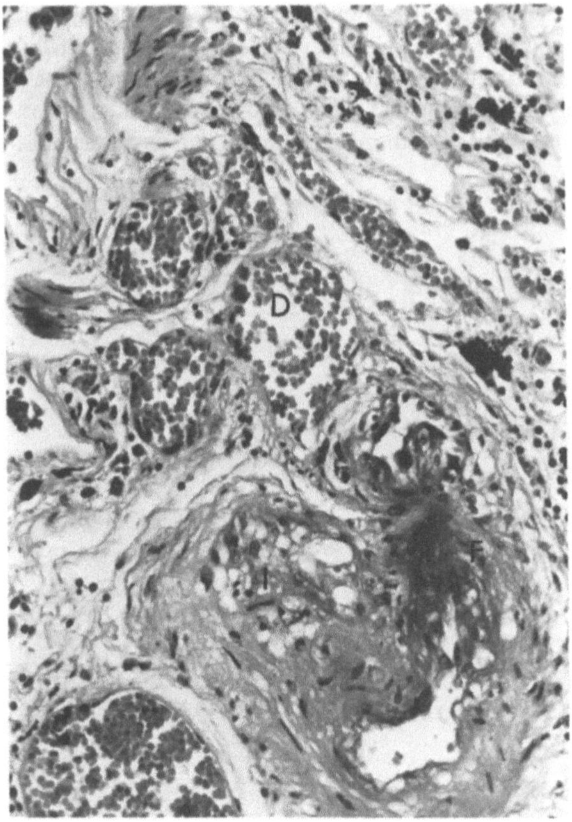

Fig. 16. Angiomatoid lesion: intimal proliferation *(I)* associated with fibrinoid deposits *(F)*; dilatation of distal branches *(D)*. (Case 16; Giemsa; 205 x)

Angiomatoid lesions, like the other lesions described above, instead of atrophying into small fibrous tracts, may undergo fibrosis of varying degrees, which could cause more or less severe stenosis of the arterial lumen. Only the distal part will finally remain, appearing as dilated vascular cavities with highly atrophied walls. Hemosiderosis may be seen in the form of grains of pigment within intra-alveolar macrophages or within macrophages situated in the septa or perivascular or peribronchial connective tissue (Fig. 17). It is generally not severe and very irregularly disseminated in the same section.

Grade VI: Arterial Necrosis

This type of lesion which was observed only rarely in our study, seems to be related to a severe evolution of pulmonary hypertension. It most often appeared in the form of more or less disseminated humps of polymorphs in the various layers of the arterial wall, mainly in the intima and the media (Fig. 18). Thus it is more of an "arteritis" than a necrosis in the strict sense of the word; it is in fact very rare that the arterial wall appears to be PAS-positive with a blurring of its structure. In other cases, the inflammatory infiltrate is com-

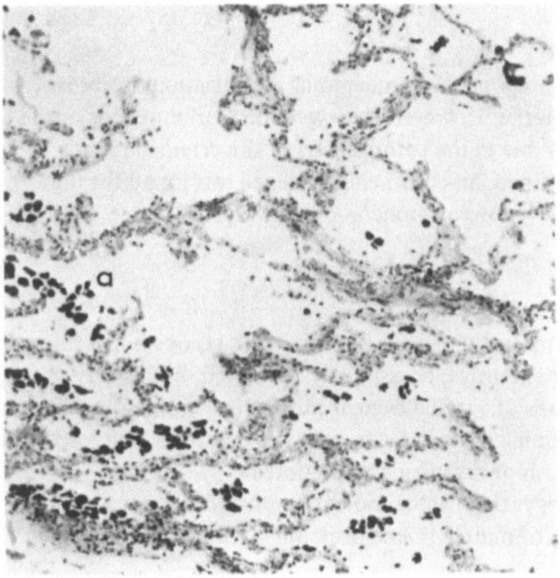

Fig. 17. Hemosiderosis: pigment mainly in alveolar macrophages *(a)*. (Case 29; Prussian blue; 80 x)

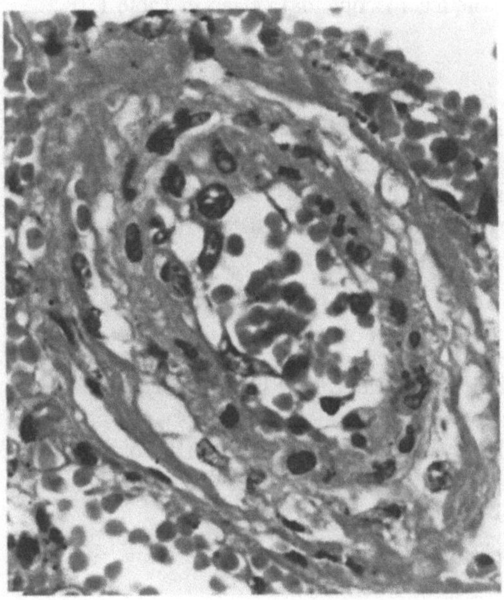

Fig. 18. Arteritis: presence of a few polymorphs in arterial wall. (Case 5; hematoxylin and eosin; 512 x)

posed of lymphocytes mainly accompanied by some plasma cells, macrophages, eosinophils, and mast cells. No rupture of vascular walls with consequent hemorrhages was observed. In only one case, an excentric parietal fibrotic lesion suggestive of a cicatricial aneurismal lesion was observed.

Other Arterial Lesions

In a few cases, thromboses or emboli, either fresh or undergoing organization or fibrosis, were observed in muscular or elastic arteries. However, they were neither numerous nor systematized enough to be attributed a role in the pathogenesis of the arterial lesions. Among the great number of slides examined, no convincing evidence was found for the existence of anastomoses, either arterio-venous or broncho-pulmonary.

Lesions of Elastic Arteries

Elastic arteries frequently reveal, even macroscopically, atheromatous lesions in the form of yellowish plaques, sometimes rather extensive, occasionally indurated, but usually not calcified. The existence of atherosclerosis of varying degrees of severity, in the form of fibrous or fibro-elastotic plaques containing nests of foam cells was confirmed microscopically. In some cases parietal thrombi, fresh or organizing, were noted.

Veins showed no significant pathology; they were moderately dilated, their walls thin, often hyalinized and practically devoid of muscle fibers; occasionally they had duplicated or chipped elastic lamina.

Other Pulmonary Parenchymatous Findings

There were no histologic findings modifying the macroscopic data given in Table 1.

Myocardic Lesions

In nine cases, it was possible to verify the histology of the heart. The macroscopic features are mentioned in the recapitulating tables of clinical and anatomic findings (Tables 1 and 2). Histology showed unmistakable perivascular fibrosis, in some places diffuse, which occasionally formed small scars. In one of these cases (case 26), a focal necrosis with polymorphs in the left ventricle and a fresh infarct in the left auricle were observed.

Lesions of Other Organs

Chronic and acute congestion of the splanchnic region, especially in the spleen and liver, were routinely observed or mentioned in the protocols. Beside the accessory necropsy findings shown in the recapitulative table, there were no other lesions worthy of mention; there were, however, several cases with gall-stones.

Primary Pulmonary Hypertension without Known Drug Intake

In these five cases, the lungs had lesions that were comparable in every way to those described above, except that no grade VI lesions (necrosis) were observed.

Secondary Pulmonary Hypertension

The findings in the two lung biopsies were also comparable with those described in the two groups just mentioned. The first (a three-month-old girl with persistent ductus arteriosus) showed muscular arteries which had retained their fetal appearance in some places; however some arteries had a marked intimal hyperplasia or a medial hypertrophy. The second (a 5-year-old girl with situs inversus) showed grade IV, i.e., plexiform lesions.

To summarize the *subjective impression* of the histologic examination of the cases, it can be stated that the lesions observed were not all of the same grade. Since PHD is commonly classified according to the most severe lesion encountered, the distribution of our 37 cases of PHD related to aminorex intake would be as follows:

grade III B	(diffuse dilatation)	4
grade IV	(plexiform lesions)	13
grade V	(angiomatoid lesions)	17
grade VI	(necrosis)	3

This classification of cases according to the severity of the lesions is thus different from the numerical evaluation of the various types of lesions that was made in the majority of our cases, as will be seen in section IX.

2. Review of Literature and Discussion

We have seen that the lesions observed in muscular pulmonary arteries in patients who had taken aminorex, did not differ from those of subjects with pulmonary hypertensive disease, either primary or secondary to a cardiac defect. The lesions described by *Bredt* (1932) are entirely comparable to those codified by *Heath* and *Edwards* (1958).

A search of the literature has revealed 31 necropsy cases and 28 biopsies of patients suffering from PHD after aminorex intake. The lesions reported seem, in general, comparable to those we have described, although they have often been interpreted in a different way.

The intima is the site of lesions that have been described and interpreted in various ways. *Baghirzade* and *Donath* (1969) have observed a nonspecific arteritis predominantly in the intima, which they consider the initial lesion. Several authors speak of a heavy cellular infiltration of the intima (*Meessen*, 1969), of obliterating angeitis (*Schultz*, 1969), of endangeitic productive processes (*Lang* et al., 1969), of endarteriopathy, of endarteritis (*Weissel*, 1969), or of obliterating thrombangeitis (*Dienstl*, 1969). All of these somewhat confusing designations refer, in our opinion, to arteritic lesions, to more or less cellular intimal proliferations, or to plexiform lesions seen on transverse sections. Moreover, the intima may show fibrosis of various degrees of intensity including obliteration (*Baghirzade* and *Donath*, 1969; *Kaindl*, 1971; *Norderø* and *Müller*, 1970; *Ostertag*, 1969; *Gahl* et al., 1970; *Backmann*, 1970; *Brandt* and *Preussler*, 1969). Obliteration would correspond to our grade III A (intimal fibrosis leading to total obliteration). *Laissue* also observed cases of grade III of *Heath* and *Edwards* (1958) as well as multiple thrombotic obliterations, as did *Baghirzade* and *Donath* (1969), *Schultz* (1969) (thrombo-embolism), *Gahl* et al. (1970) (recent or organized, or partly recanalized, thrombi) and *Nager* (1970) who reported recanalized emboli, which, judging from his illustrations, correspond to a plexiform lesion for the one (Fig. 2), to stenosing intimal fibrosis for the second (Fig. 1). This case figures in our series as case number 10. Intimal proliferations which we have quite often observed to form cushions or mushrooms have been interpreted by some authors as parietal thromboses (*Gurtner* et al., 1968a; *Brandt* and *Preussler*, 1969; *Höfer* and *Paeprer*, 1969; *Obiditsch-Mayer*, 1969). Occasionally mentioned in the literature are the presence of fibrinous deposits in the lumen, interpreted as thromboses (*Georgii*, 1969), recent thrombi (*Norderφ* and *Müller*, 1970; *Gahl* et al., 1970; *Ostertag*, 1969) or thrombotic material (*Baghirzade* and *Donath*, 1969). All of these are probably deposits of fibrinoid material, fresh or organized, as previously stated. All of the lesions described above and interpreted in various ways by the different authors, are actually, in our opinion, plexiform lesions or intimal hyperplasia. Thromboses or emboli, recent or organized, are really too infrequent in the muscular arteries to be considered as important pathogenetically as some authors have done.

Modifications of the media are quite frequently mentioned. A more or less marked hypertrophy was reported by *Norderφ* and *Müller* (1970); *Ostertag* (1969); *Obiditsch-Mayer* (1969); *Jahrmarker* et al. (1969); *Gahl* et al. (1970); *Backmann* (1970); *Brandt* and *Preussler* (1969); *Höfer* and *Paeprer* (1969). Other parietal lesions are mentioned such as fibrinoid deposits, necrosis (*Jahrmärker* et al., 1969), and imbibition (*Pendl*, 1969).

The network-like intimal proliferations described by *Obiditsch-Mayer* (1969) and the concentric, sometimes septate intimal proliferations reported by *Brandt* and *Preussler* (1969) should be considered plexiform lesions.

Plexiform or angiomatoid lesions are often interpreted as "glomangiosis" (*Baghirzade* and *Donath*, 1969; *Jahrmarker* et al., 1969). These lesions, the definition of which is highly controversial, have, in fact, been interpreted in various ways; glomangiosis–which is considered malformative in nature (*Masshoff*, 1964); glomic arterio-venous anastomoses (*Rutishauser* and *Blanc*, 1950); idiopathic obliterating endofibrosis (*Rutishauser* and *Feuardent*, 1952; *Feuardent*, 1953); organized thromboses or emboli, etc. We share the opinion of *Kapanci* (1957) and of *Heath* and *Edwards* (1958) that these lesions are the consequence of intimal proliferations or thromboses on branchings in dilated vessels. *Naeye* (1968) clearly demonstrated that these phenomena occur at the branching place of a vessel (see schematic drawing of his Figure 110 which is reproduced in Figure 12). *Obiditsch-Mayer* (1969) described arterio-venous anastomoses, which judging from her Figure 4, would undoubtedly correspond to our plexiform lesions. *Pendl* (1969) has also described them in two necropsy cases. In the abundant material we examined, we found no proof of the existence of pulmonary arterio-venous or broncho-pulmonary anastomoses.

The lesion of muscular arteries called micro-aneurismal arteries (*Baghirzade* and *Donath*, 1969) appears to correspond to the angiomatoid lesions described in the present study.

Sclerosis and thickening of the adventitia, which *Jahrmärker* et al. (1969); *Backmann* (1970); *Höfer* and *Paeprer* (1969), consider rather important, should also be mentioned.

Most of the authors reported lesions of *elastic arteries*: marked arteriosclerosis, more or less marked intimal fibrosis, and dilatation.

The *heart*, when mentioned, is characterized by lesions of cor pulmonale (*Baghirzade* and *Donath*, 1969; *Gahl* et al., 1970), signs of right ventricular failure (*Dienstl*, 1969), or right ventricular hypertrophy (*Nager*, 1970).

It can be stated in conclusion that our histologic studies have revealed all the classical features of PHD as defined by *Heath* and *Edwards* (1958). Apart from differences in interpretation concerning the nature of the lesions, especially from those of the German authors, our findings were essentially the same as those reported in the literature. The important point to be emphasized is that no differences were observed between lesions of pulmonary muscular arteries in cases of PHD related to aminorex intake and those of PHD, primary or secondary to cardiac malformations.

VII. Ultrastructure of Normal Muscular Pulmonary Arteries

1. Personal Observations

The control cases used for this study were a 46-year-old man who died of encephalitis and a 23-year-old female epileptic, treated for Hodgkin's disease and cured. Muscular arteries of between 100 and 1000μ in external diameter were studied.

The membrane of *endothelial cells* has numerous pinocytotic vesicles, especially on the side opposite the lumen. The cells have endothelial projections (*Smith* et al., 1971) on the side facing the lumen (Fig. 19). The cytoplasm contains occasional phagosomes, small mitochondria, some isolated ribosomes, and a moderately abundant smooth endoplasmic reticulum with slightly dilated cisternae. Lipid inclusions of varying density and vesiculated bodies of Weibel-Palade are sometimes observed. There are numerous microfibrils measuring about 60 Å in thickness, arranged in bundles, especially visible when the cells are contracted, a state that can be recognized by the wrinkled appearance of the nuclear membrane (*Majno* et al., 1969). Formations resembling dense bodies sometimes situated in the vicinity of the lamina elastica interna may also be observed. The intimal cells are interconnected by desmosomes.

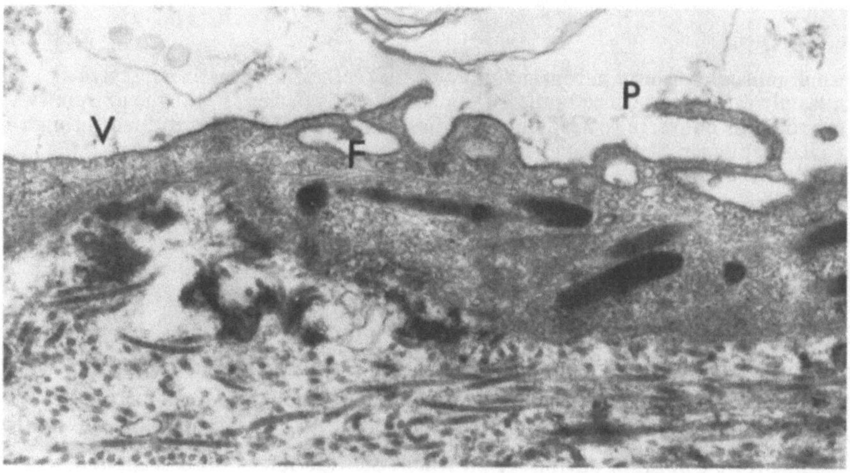

Fig. 19. Endothelial cell of normal muscular artery: endothelial projections *(P)*; pinocytotic vesicles *(V)*; microfibrils *(F)*. (Neg. 10553; reduced to 90%)

The *subintimal layer* may contain intimal muscular cells with numerous microfilaments, elongated mitochondria and a scarce rough endoplasmic reticulum. This latter feature may indicate young or poorly differentiated smooth muscle cells; it is also found in the myointimal cells observed by *Buck* (1961), in arteries of the larger circulation, and in the myofibroblasts of granulation tissue described by *Majno* et al. (1971). These intimal cells will be referred to in the present study as myointimal cells (MIC). Figure 20 shows the clearly visible dense bodies. The cells are surrounded with a delicate basal membrane and a variable quantity, usually rather small, of bundles of collagen fibers.

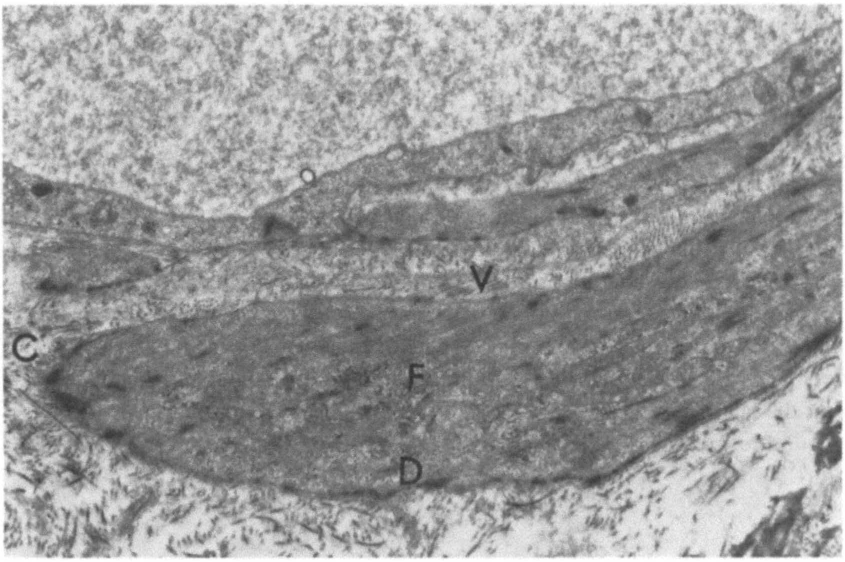

Fig. 20. Normal muscular artery: subintimal myointimal cell. Microfibrils *(F)*; scarce rough endoplasmic reticulum; dense bodies *(D)*; pinocytotic vesicles *(V)*; at the extremity of the cell microfibrils converge toward dense bodies which seem to be continued through fine fibrils toward collagen fibers *(C)*, giving a picture suggestive of a microtendon. (Neg. 6825; reduced to 90%)

The well-developed *lamina elastica interna* sometimes seems fragmented or duplicated with occasional fenestrations.

The media is made up of a varying number of smooth muscle cells (SMC), whose membrane has numerous pinocytotic vesicles (Fig. 21). They contain microfibrils that appear to run longitudinally and become dense in the dense bodies. The latter are sometimes applied to the lamina elastica interna or to bundles of collagen fibers. This arrangement suggests that this might be a point of anchorage on which the traction of SMC is applied. Occasionally the cytoplasm contains clumps of glycogen and some rough endoplasmic reticulum. There are few mitochondria which are of medium size. A thin basal membrane usually surrounds the periphery of the cells. Bundles of collagen fibers of varying density separate the external part of the media from the lamina elastica externa, which may be fragmented.

The *adventitia* is essentially composed of coarse entangled bundles of collagen fibers.

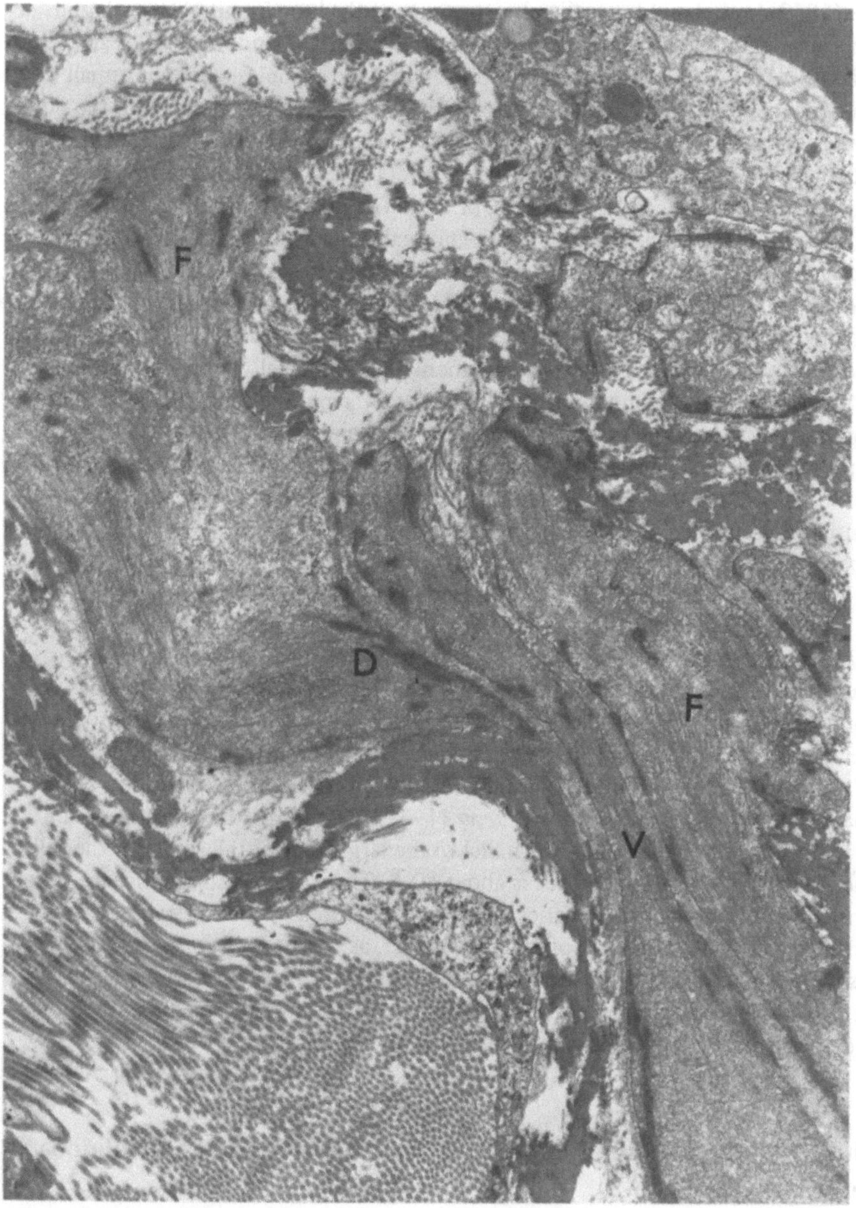

Fig. 21. Normal muscular artery: smooth muscle cells. Numerous microfibrils *(F)* condensing in dense bodies *(D)*; some pinocytotic vesicles *(V)*. (Neg. 12196; reduced to 90%)

2. Review of the Literature and Discussion

The normal ultrastructure of muscular pulmonary arteries, either human or animal, has seldom been described in the literature. The essential features have been described by

Hatt et al. (1959) in their study on the ultrastructure of the lung in mitral stenosis and congenital cardiopathies. They are briefly summarized here. The endothelium of arteries and capillaries has a membrane with pinocytotic "villi"; the cytoplasm contains small mitochondria and a poorly developed ergastoplasm. These "villi" are identical to the pinocytotic vesicles observed in the present study. They should be distinguished, however, from the endothelial projections described by *Smith* et al. (1971), which we observed as well, but which were not mentioned by *Hatt* et al. (1959). The microfibrils, observed under pathologic conditions, were not denoted in the normal state. Their presence in man and animals (guinea pig) was described by *Bensch* et al. (1964); *Caesar* (1969); *Majno* et al., (1971); *Joris* et al. (1972). The endothelial layer rests directly on the lamina elastica interna which is replaced, if absent, by the basal membrane. The media (*Hatt* et al., 1959) is composed of one or two layers of SMC arranged either circumferencially, obliquely, or longitudinally and each cell is surrounded with its own basal membrane. The cellular membrane has numerous pinocytotic "microvilli" (= vesicles). The central zone of the cytoplasm contains the usual organelles. Fibrils, concentrated in denser places (= dense bodies), to which they seem to be tied, are found at the periphery. The external elastica is discontinuous. The adventitia is composed of a layer of collagen fibers.

The appearance of pulmonary muscular arteries does not seem to be fundamentally different from that of muscular arteries of the larger circulation, either in man (*Wissler*, 1968) or in laboratory animals (*Pease* and *Molinari*, 1960): cat and monkey; *Takayanagi* et al., 1972: cat; *Gendre*, 1970a: chicken; *Wiener* and *Spiro*, 1968: various animals; *Caesar*, 1969). Our observations concur with those of *Hatt* et al. (1959) except that they did not observe bundles of microfibrils in the endothelial cells as we did and as reported by *Bensch* et al. (1969) (man and guinea pig). *Wiener* and *Spiro* (1968) pointed them out in the capillaries of the larger circulation but not in the muscular arteries. In his description of the endothelial cell of vessels of vertebrates, *Caesar* (1969) noticed that they often contain very thin filaments gathered in bundles parallel to the larger axis of the cell, near the cell membrane. The diameter of these fibrils is about 80 Å. Their functional significance is not yet known, but contractile characteristics have been attributed to them because of their resemblance to contractile elements of SMC; moreover, as will be mentioned later, the experiments of *Gabbiani* et al. (1975) seem to support this hypothesis. Although the fibrils measured by *Gabbiani* had a larger diameter than those encountered in the present study, there is no doubt that we are dealing with the identical structure.

VIII. Ultrastructure of Lesions of Muscular Pulmonary Arteries in Pulmonary Hypertensive Disease

1. Personal Observations

a) PH Related to Aminorex Intake

Electron microscopic observations were made on the following three necropsy cases from the Geneva Department of Pathology: case 8: 45-year-old male; case 26: 68-year-old female; and case 28: 68-year-old female.

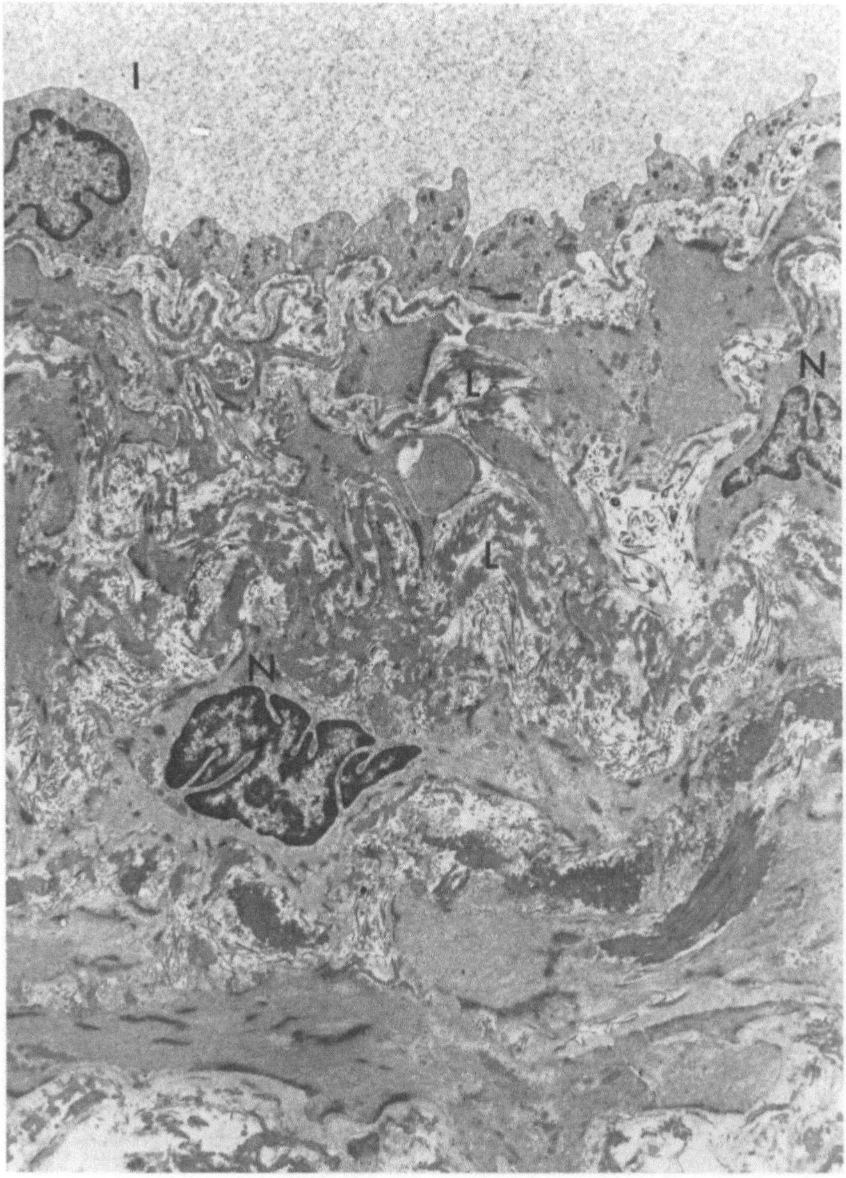

Fig. 22. Concentric intimal hyperplasia. Contraction of the artery indicated by prominent intimal cells *(I)*, folded nuclei *(N)* of myointimal cells, and smooth muscle cells; duplication of lamina elastica interna *(L)*. (Neg. 13469; reduced to 90%)

Because of the uneven distribution of the lesions, the six grades seen under the light microscope were not all seen with the electron microscope. In fact, medial hypertrophy, presumed to be an early lesion, was rarely encountered in our material. Arterial necrosis, probably characteristic of a rapid and severe evolution of PHD, was not encountered at all. We can therefore only describe the various types of intimal hyperplasia (concentric, mushroom shaped, obliterating, or fibrosing) and diffuse dilatation.

1. *Concentric intimal hyperplasia* has endothelial cells, whose appearance is similar to that of normal pulmonary arteries. It should be noted that there are many microfibrils, especially visible when the cells are contracted, a state which can be recognized by the folded appearance of the nuclear membrane (Fig. 23).

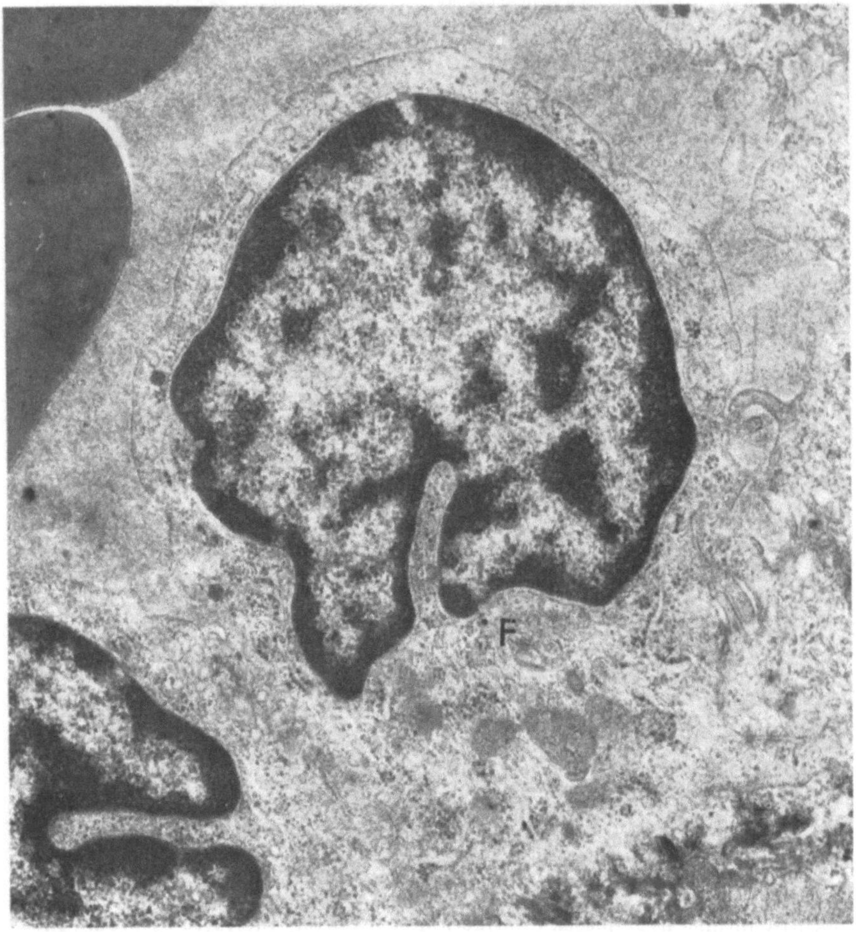

Fig. 23. Contracted intimal cell. Radially distributed microfibrils *(F)*. (Neg. 6732, reduced to 90%)

Subintimal cells are often contracted. They contain microfilaments of 60 Å mean diameter, often arranged in bundles near the cellular membrane which appear to condense in the dense bodies and to have a radial arrangement in the contracted cells. These cells contain some rough endoplasmic reticulum, and can thus be identified as myointimal cells according to *Buck* (1961) (Figs. 24 and 25). Sometimes granular

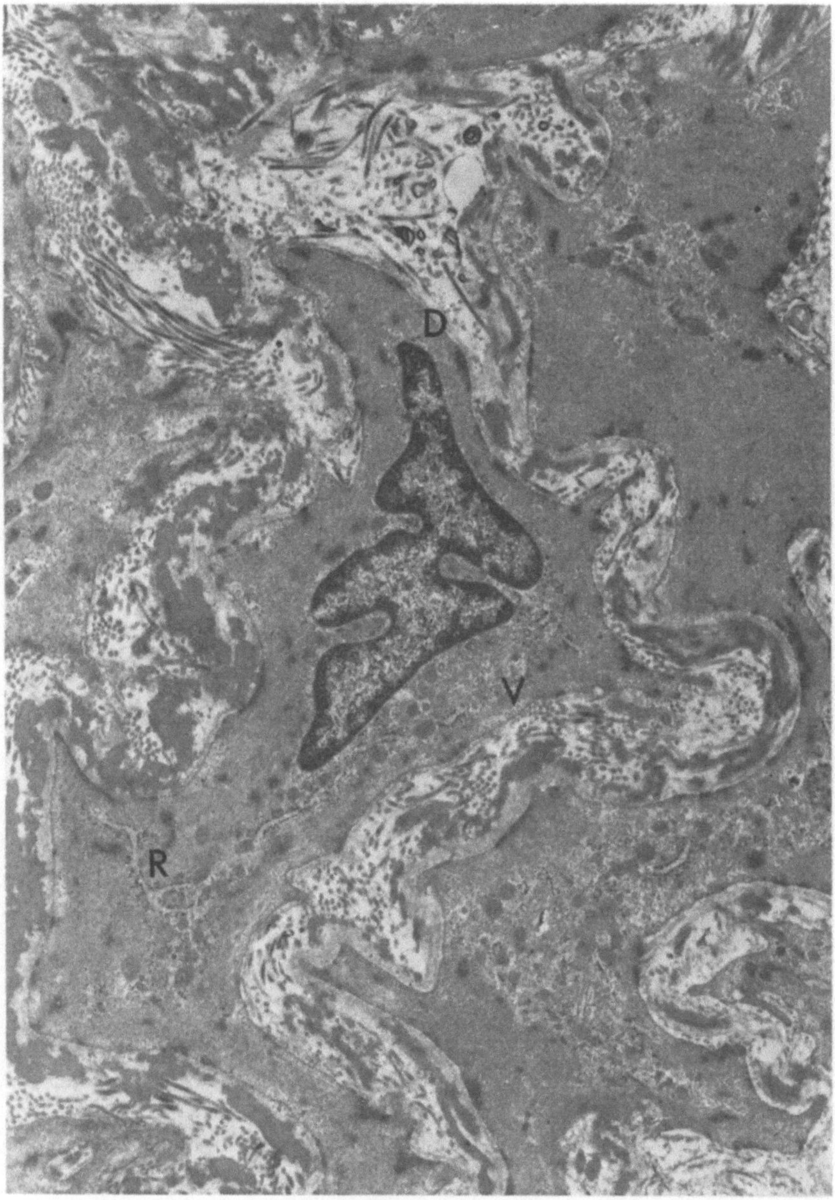

Fig. 24. Typical myointimal cell: well-developed rough endoplasmic reticulum *(R)*, numerous dense bodies *(D)*, and pinocytotic vesicles *(V)*. (Neg. 13473; reduced to 90%)

inclusions in round vesicles (lysosomes) may be found in them. Their membrane shows some pinocytotic vesicles. They are surrounded by deposits of granulo-filamentous basaloid material or fragments of material resembling elastine and irregular bundles of collagen. They are frequently seen astride the fenestrations of the lamina elastica interna.

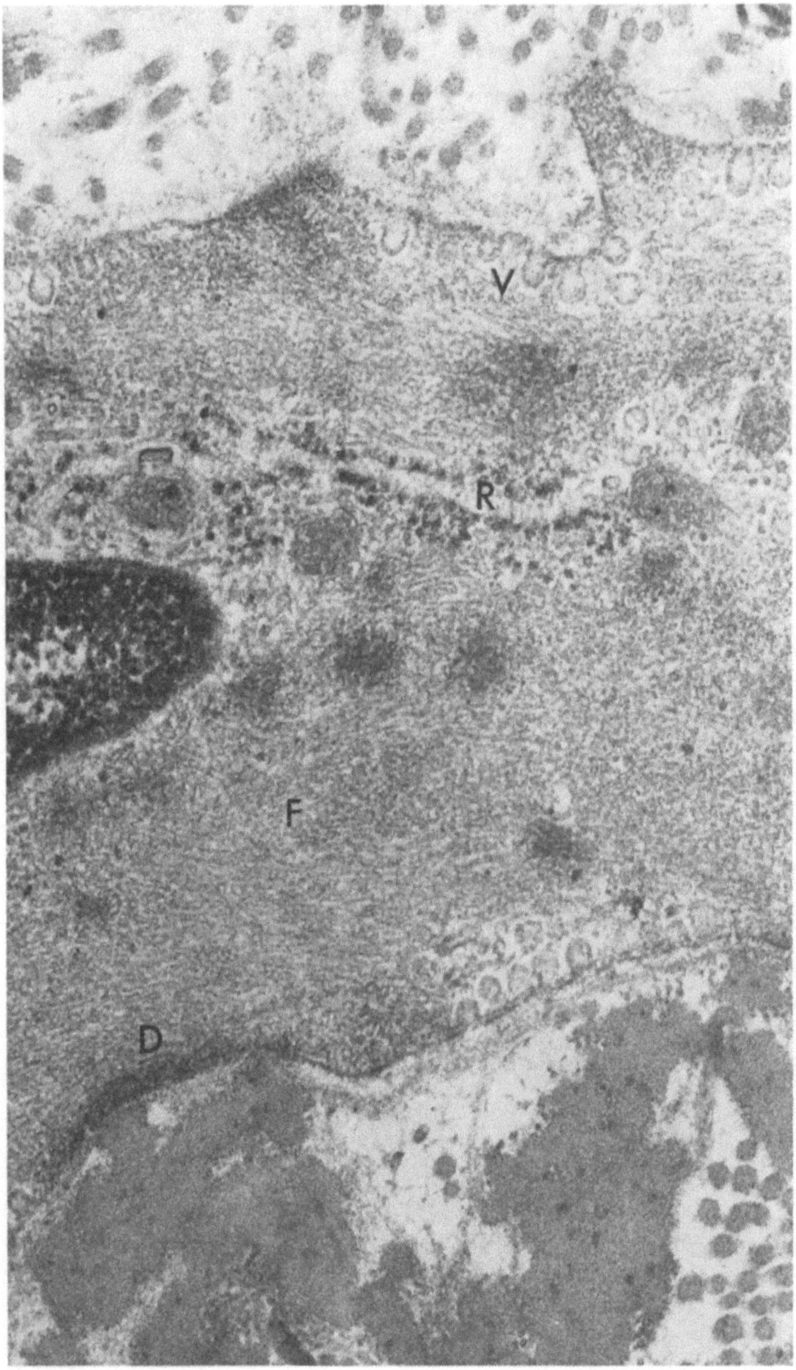

Fig. 25. Detail of Figure 24: rough endoplasmic reticulum *(R)*, dense bodies *(D)*, pinocytotic vesicles *(V)*, microfibrils *(F)*. (Neg. 13474; reduced to 90%)

The *lamina elastica interna* has fenestrations; it is sometimes duplicated.

2. *Mushroom-shaped intimal hyperplasia* has essentially the same cellular appearance as concentric intimal hyperplasia.

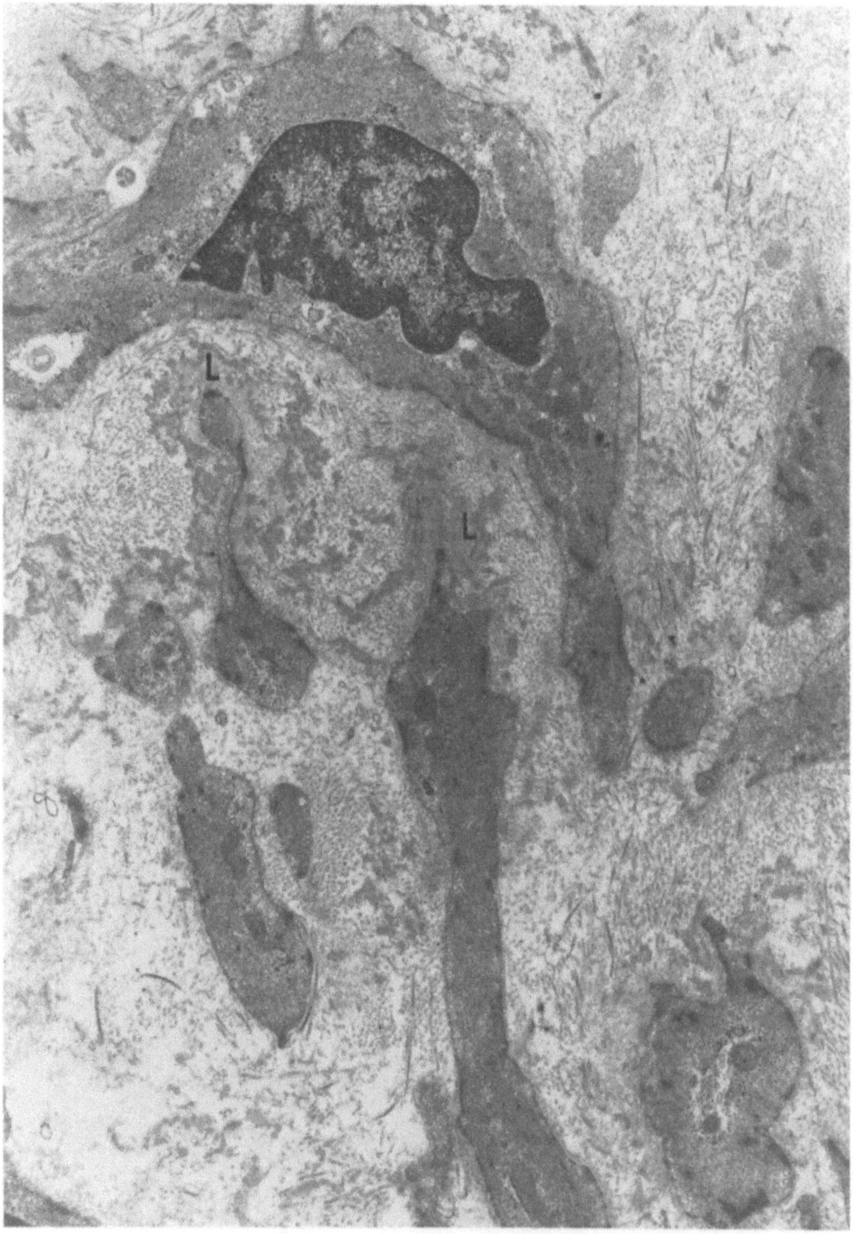

Fig. 26. Smooth muscle cell of intimal hyperplasia in close contact with newly formed appearing fragments of lamina elastica interna *(L)*. (Neg. 8666; reduced to 90%)

3. *Obliterating intimal hyperplasia* has no visible endothelium. The *subendothelial layer* contains numerous SMC of normal appearance as well as myointimal cells. Their microfibrils are condensed in the dense bodies, which appear in some places to be attached to collagen fibers or to elastic lamellae (Fig. 26). Collagen, either in the form of isolated fibers or in more or less dense bundles, and fragments of elastine and basal laminae are found between these cells. In the external part of this layer the cells are arranged radially. Here again, SMC astride the fenestration of the lamina elastica interna may be seen. The *media* is composed of elongated SMC.

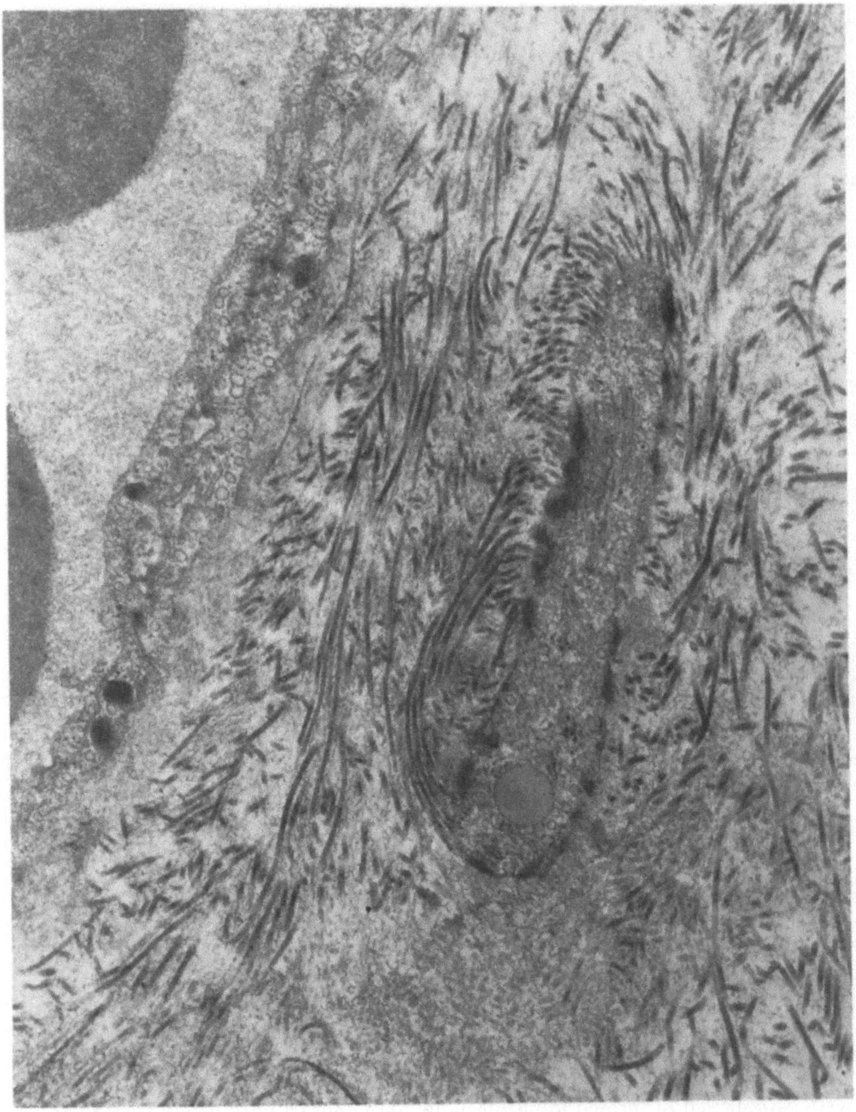

Fig. 27. A myointimal cell appearing in continuity with collagen fibers. (Neg. 4466; reduced to 90%)

4. *Intimal hyperplasia with predominating fibrosis* is similar to intimal hyperplasia, but with the MIC in a more dense connective web. Occasionally these cells appear contracted with microfibrils converging upon a dense body situated in the proximity of collagen bundles (Fig. 27).
5. *Dilatation lesions* have very flattened endothelial cells, containing numerous microfibrils. SMC and MIC of the intima are dissociated by collagen bundles.
6. *Dilatation lesions with intimal proliferation* resemble those described above but have a more composite appearance.

b) Primary Pulmonary Hypertensive Disease

The ultrastructural aspect of these three cases has been compared with that of a 35-year-old woman suffering from PPHD with no history of drug intake. Intimal hyperplasia with contracted endothelial cells, contracted intimal cells, and sometimes with an abundant rough endoplasmic reticulum were noted. All of these cells were surrounded by clearly visible, often duplicated, basal membrane. Intimal fibrosis containing MIC, numerous collagen bundles and cells suggesting fibroblasts were also observed; these cells were characterized by their rough endoplasmic reticulum, their microfilaments, the almost complete absence of pinocytotic vesicles, and by the absence of a basal membrane. The lamina elastica interna was sometimes duplicated and the SMC were occasionally visible in the fenestrations of the lamina elastica interna.

c) Secondary Pulmonary Hypertension

Two lung biopsies from children with pulmonary hypertension due to cardiac malformations were studied for comparison.

In a three-month-old girl with persistent ductus arteriosus we observed intimal hyperplasia characterized by contracted endothelial cells with a radial arrangement, which in some places gave the impression of undergoing desquamation. MIC were clearly visible and contained lipid inclusions and multivesicular bodies of Weibel and Palade. One of them contained oriented dense bodies with a "microtendon." Macrophages containing phagocytized cell remnants, probably of SMC, were observed in the wall.

Another girl, five years of age, with a situs inversus and grade IV histologic lesions of the pulmonary arteries, provided the opportunity to study *medial hypertrophy*. This is characterized by contracted SMC containing myofibrils and some rough endoplasmic reticulum; their membrane has numerous pinocytotic vesicles. Occasionally they contain lipid inclusions, myeline bodies or Weibel-Palade bodies. Contracted endothelial cells are also found. Here again, SMC penetrating into fenestrations of the lamina elastica interna are sometimes observed.

In summary, the lesions described for the various types of PHD are very similar, both ultrastructurally and optically. The two points essential for the understanding of the physiopathologic mechanism of PHD that have emerged from our study with the electron microscope are the following:

1. The presence of microfibrils in endothelial cells;

2. The existence of MIC, in varying number, in the lesions of intimal proliferation, realizing with SMC the appearance of muscularization of the intima.

2. Review of the Literature and Discussion

Although numerous studies with the light microscope have been made of pulmonary arteriosclerosis and of vascular damage in PHD, there have been very few studies devoted to the ultrastructural aspects of these two diseases.

The basic studies remain those of *Hatt* and *Rouiller* (1958) and *Hatt* et al. (1959) which are now rather old but still valid. In their first paper on the lesions observed during a severe mitral stenosis, they described the modifications of the various vascular levels, of which only the level they called broncho-arterial will be retained. In the case of moderate PHD, the parabronchial pulmonary artery reveals a subintimal hyaline infiltration, like that seen in capillaries and veins, i.e., the basal lamina is always thickened, the interstitium is crowded with fibers (collagen, elastic and elementary fibers) and cells: histiocytes, cells rich in ergastoplasm, fibroblasts (among which some may resemble our MIC). The endothelial cells in some places had a fibrillar structure, which *Hatt* et al. (1959) did not observe in normal arteries.

In cases of unquestionable PHD, these authors described intimal thickening, broadening of the media, and a crowding of interstitial spaces:

1. The intima had swollen, vacuolated endothelial cells containing abnormally abundant fibrils, whose structure resembled those of SMC; they formed one or several cell layers with a structure very similar to that of SMC, suggesting to these authors that perhaps these were intimal cells undergoing a transformation into SMC. They were associated with hyaline material with fragmented or duplicated elastic fibers.
2. The media revealed a multiplication of layers of SMC; some had a more dense fibrillary structure with diminished microvilli; in some cases, areas of connective infiltration were also noted.
3. The adventitia was crowded with both cellular and fibrous elements.

In cases of pulmonary vascular lesions related to a congenital cardiopathy accompanied by an increased intrapulmonary output, medial hypertrophy of such a degree as to be visible with the light microscope was observed. Two aspects of pulmonary arteriosclerosis are emphasized by these authors:

1. *a simple intimal fibrosis* with a fibro-cellular thickening in discontinuous areas containing modified endothelial cells, with a thickened membrane, and a fibrillar structure similar to that of SMC but with no areas of fibrillar concentration;
2. *muscularization of the intima* constituted by newly formed SMC appearing between the endothelium and the lamina elastica interna and producing layered obstructions visible with the light microscope, which seem to correspond to our obliterating type of intimal hyperplasia. This muscularization, according to *Hatt* et al. (1959), may be considered a factor that can sustain or aggravate a preexisting PHD.

Meyrick et al. (1974) studied a lung biopsy from a case of PPHD. They described a swelling of endothelial cells of nonmuscular small arteries. Since the appearance of the nuclei was not described, however, it is not certain whether these cells were truly swollen or contracted. Their description of the blocking of the arterial lumen by endothelial cells, assuming an onion-skin appearance, seems to correspond to intimal hyperplasia as we conceive it. Moreover, these authors have described medial hypertrophy of muscular arteries of larger size.

Although papers on ultrastructural modifications of pulmonary vessels in cases of PHD are practically nonexistent, experimentally induced lesions have been rather accurately described, especially those obtained with the experimental model represented by monocrotaline intoxication, a pyrrol derivate extracted from *Crotalaria spectabilis*, a plant of the Papilionaceae family. The initial effect of this substance is an endothelial lesion of capillaries, arterioles, and small muscular arteries (*Allen* and *Carstens*, 1970; *Butler*, 1970). Muscular hypertrophy is also noticed (*Allen* and *Carstens*, 1970). *Merkow* and *Klinerman*, (1966), *Kay* et al. (1969) and *Valdivia* et al. (1967) emphasized the lesions of septa and alveolar capillaries. The point of impact of this toxin is thus essentially endothelial and presumably different from that of aminorex which, according to the pharmacologic data, probably has a vasospastic effect.

Under other experimental conditions, which were possibly too artificial, e.g., anastomosis of a subclavian artery to a branch of the pulmonary artery in dogs, *Esterly* et al. (1968) described intimal proliferations, first focal, later on diffuse, tending to obliterate the arterial lumen; these proliferations were composed of poorly differentiated cells, whose structure is similar to that of SMC; the SMC in the media were described as resembling these cells, but as being less differentiated with fewer myofibrils and dense bodies and more abundant rough endoplasmic reticulum. They suggest that the endothelial cells might be the result of an endothelial hyperplastic reaction and an invasion of the intima by modified SMC coming from the media, and which subsequently might lose their morphologic characteristics.

In the field with which we are particularly concerned, that of PHD related to aminorex intake, there has been no experimental work in which ultrastructural studies are reported. This is probably due to the fact that, until now, no histologic modifications have been observed.

A comparison of the lesions of pulmonary arteries in PHD with those of arteries of the larger circulation in cases of arteriosclerosis or hypertension, reveals that they are very similar, both in spontaneous human modifications and in experimental ones (*McGee* and *Ashworth*, 1963; *Takebayashi*, 1970; *Gendre*, 1970a, b; *Stemerman* and *Ross*, 1972; *Moss* and *Benditt*, 1970; *Spiro* et al., 1965; *Backwinkel* et al., 1970; *Todd* and *Friedman*, 1972; *Scott* et al., 1972; *Wiener* and *Spiro*, 1968; *Caesar*, 1969; *Thomas* et al., 1963; *Parker* and *Odland*, 1966).

In these lesions intimal plaques have been described whose cellular population consisted mostly of SMC, but also of cells with a more or less developed rough endoplasmic reticulum, not seen in normal SMC. The presence of such organelles is interpreted in various ways: for example, it indicates that these cells are either poorly differentiated or young SMC (*Spiro* et al., 1965; *Moss* and *Benditt*, 1970), cells resembling fibroblasts (*Scott* et al., 1972) or that we are dealing with a stock cellular population (*Moss* and *Benditt*, 1970). According to *Buck* (1961), these cells are myointimal cells which differ from endothelial cells and are not immature SMC, since they retain their characteristic appearance throughout the experiment (production of intimal thickening by ligature of a carotid artery in rats). This term, adopted by many authors, seems to us the most appropriate to describe the cells we have observed and which differ from contractile interstitial alveolar cells (*Kapanci*, 1974), the latter containing a larger amount of rough endoplasmic reticulum.

The fact that these cells are sometimes visible in fenestrations of the lamina elastica interna suggested to some authors, *Backwinkel* et al. (1970);*Buck* (1961);*Spiro* et al. (1965) for example, that they originate in the media. *Stemerman* and *Ross* (1971) suggested that accessory vessels of the main artery, or patches of endothelium remaining after injury might be their source. In his discussion of the problem, *Caesar* (1969) feels that the origin of SMC of intimal plaques remains an open question; without taking a position, he suggested the following three possibilities: media, endothelium, and circulating monocytes. In the absence of positive experimental results on the effects of aminorex, we are unable to give an accurate response to this question but are inclined to believe that these cells develop on the spot.

A precise determination of the functions of MIC seems to us of utmost importance. In the opinion of *Buck* (1961) they are involved in the production of the extracellular material of the intimal thickening obtained experimentally. *Moss* and *Benditt* (1970) studied MIC in the spontaneous intimal atherosclerotic plaque of the chicken and considered them to be different from SMC, but possibly derived from them; they might also be responsible for the intimal thickening. More concerned with the relationship between the experimental data and the picture observed in human pathology, these authors placed little emphasis on the functional role of these cells. The cells demonstrated in their pictures are strikingly similar to those of our MIC. Other authors (*Parker* and *Odland*, 1966; *Thomas* et al., 1963, among others) have suggested a possible role of these cells in the genesis of the lipidic plaque, but we found no publication suggesting that they might intervene in hemodynamics.

In the field of vasomotricity, two papers seem particularly relevant in the context of MIC. *Gabbiani* et al. (1972) have shown that granulation tissue should be considered a contractile organ; it contains cells that they have named myofibroblasts, to which they attributed contractile capacities as suggested by immunofluorescent studies and demonstrated by their behavior in vitro which was similar to that of SMC. Cells with characteristics nearly identical to those of myofibroblasts have recently been observed in lung alveolar septa by *Kapanci* et al. (1974) and named interstitial contractile cells. Ultrastructurally MIC are almost identical to these two types of cells and it is entirely possible that they may have very similar functional properties. It might therefore be assumed that, once they have appeared in the intimal hyperplasia, they play an important role in the maintenance of PHD, especially in the phase of muscularization and obliteration.

The role of endothelial cells and of the increased number of microfibrils remains to be defined. *Gabbiani* et al. (1975) have recently described a numerical increase of endothelial filaments, comparable to that we have observed, in aortic endothelium of hypertensive rats; these filaments are in part composed of actin. These modified cells are thought to play a role in the increase of endothelial permeability by contracting and widening the intracellular junctions.

IX. Numerical Evaluation of Arterial Lesions

1. Personal Observations

The mean distribution of the various types of lesions, evaluated numerically in 29 cases (Table 3) shows that normal arteries, contrary to what we suggested in our previous papers, are rare. In our earlier studies normal arteries averaged 30-36% of all arteries counted. In the present study, we found that less than 1% of the arteries were normal. This discrepancy is probably due to the following facts:

1. Our first papers were based on a relatively small number of specimens per case; for the present study the sampling was much larger (10 random specimens per lung), thus giving a more representative sample.
2. Our subjective criteria for evaluation have been made more stringent, after a histologic control of all cases.
3. The present study is based on a larger number of cases (29) compared with 8 and 22 in our preliminary studies; this might improve the statistical validity.

Table 3. Mean distribution of the various types of arterial lesions

	Ax (= 29) %	PRIM. (= 4) %
N	0.91 ± 0.28	5.0 ± 1.68
I	1.52 ± 0.30	3.1 ± 1.14
II	0.65 ± 0.24	2.6 ± 1.51
III A	39.8 ± 3.65	59.8 ± 13.9
III B	46.5 ± 4.29	21.1 ± 14.8
III A + B	86.4 ± 1.16	80.9 ± 3.39
IV	8.5 ± 0.9	8.1 ± 3.18
V	1.3 ± 0.24	0.38 ± 0.38
VI	0.28 ± 0.12	0

Low grade lesions (I: medial hypertrophy; II: intimal hyperplasia) are few, since they total only 2%, compared to 17 and 13% obtained previously. This discrepancy can also probably be explained by the reasons given above.

The large majority of lesions were those of grade III (intimal fibrosis and dilatation), which represented 86.4% of the total. This percentage is much higher than 45 and 55% obtained previously. In view of this massive predominance of grade III lesions, which became difficult to interpret, we decided to revise the classification of *Heath* and *Edwards* (1958), introducing two subgroups for grade III, i.e., III A: fibrosis and III B: dilatation.

Actually the two graphs (Figs. 28 and 29) show a more equilibrated distribution, with about 40% of fibrotic lesions and 46.5% of dilatation lesions. This new distribution allows for a more accurate comparison with PPHD. Figure 29 shows, in fact, that in the latter, there is an unquestionable predominance of fibrotic lesions (about 60%) over dilatation lesions (21%).

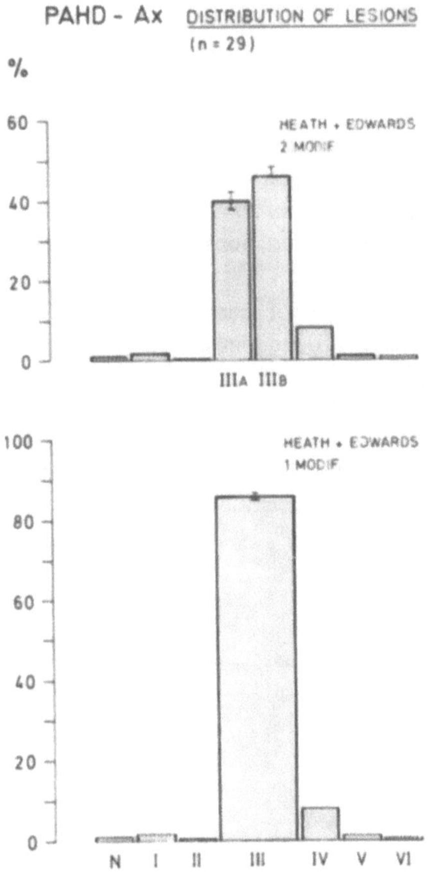

Fig. 28

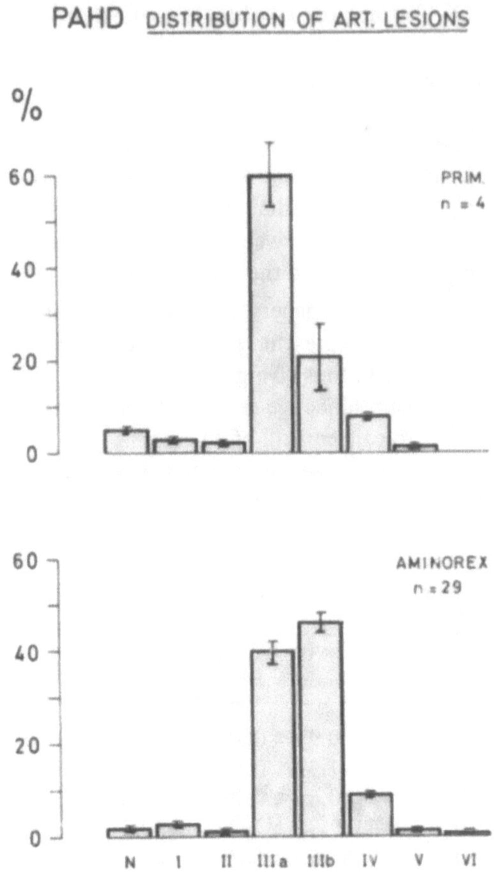

Fig. 29

Fig. 28. Distribution of lesions according to a modification of the *Heath* and *Edwards* grading. *Lower part:* global grade III; *upper part:* subdivision of grade III (*A* = fibrosis; *B* = dilatation)

Fig. 29. Distribution of lesions: comparison between primary pulmonary hypertension *(PRIM.)* and hypertension in connection with aminorex intake *(AMINOREX)*, (*n* = number of cases)

The rate of plexiform lesions (8.5%) is not significantly different from that previously observed (about 6%).

The advantage of dividing grade III into two subgroups becomes obvious when the distribution of lesions according to yearly clusters of death and age clusters at time of death is considered.

1. *Yearly clusters* (Table 4 and Fig. 30). The cases have been grouped according to the years when death occurred. It can be seen at once that fibrotic lesions increased steadily from 1968-1970. The data for the year 1971 were omitted since they were based on only two cases and therefore of no statistical value. It can also be seen that during the same period there was a marked regression in the number of dilatation

lesions. It can therefore be concluded that fibrotic lesions augment in number as a function of time between the presumed termination of drug intake and the moment of death, while dilatation lesions on the contrary diminish in number with the passing time.

Table 4. Distribution of lesions according to yearly clusters

	1968 (n = 7)	1969 (n = 13)	1970 (n = 7)	1971 (n = 2)
N	1.26 ± 0.38	0.77 ± 0.55	0.97 ± 0.47	0.45 ± 0.45
I	0.63 ± 0.34	1.6 ± 0.53	2.3 ± 0.46	1.3 ± 1.3
II	0.8 ± 0.3	0.4 ± 0.42	0.7 ± 0.47	1.3 ± 1.3
III A	27.4 ± 4.67	34.4 ± 4.48	59.6 ± 5.8	49.8 ± 21.6
III B	62.4 ± 6.35	53.9 ± 4.17	20.7 ± 6.34	33.8 ± 26.8
IIIA + B	89.8 ± 2.12	88.3 ± 1.26	80.3 ± 2.17	83.5 ± 5.2
IV	4.4 ± 1.07	7.5 ± 1.01	13.7 ± 1.66	11.0 ± 20.5
V	0.7 ± 0.34	1.1 ± 0.36	1.8 ± 0.58	2.5 ± 0.09
VI	0.5 ± 0.36	0.27 ± 0.19	0.14 ± 0.14	0

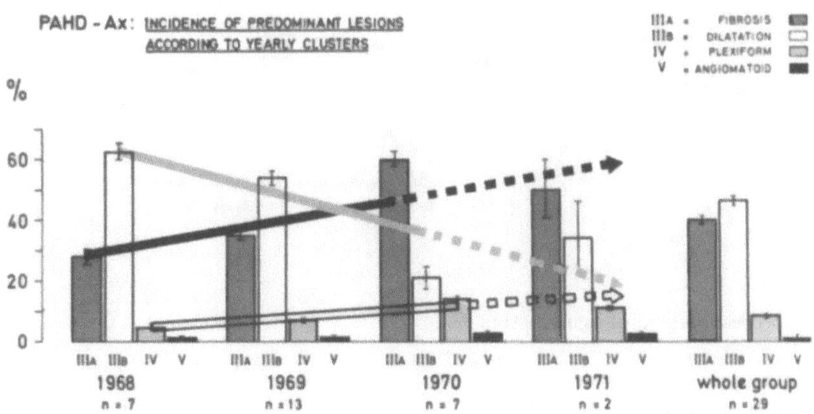

Fig. 30. Graphic representation of yearly clusters

2. *Age clusters* (Table 5 and Fig. 31). Our cases have been classified according to age in decades at time of death. This distribution shows that the number of fibrotic lesions increases by 10% during each ten-year period from the age of 50 onward. Dilatation lesions on the other hand, infrequent in younger age groups, diminish in number with increasing age.

We have tried to establish a relationship between the distribution of the lesions and the hemodynamic modifications. As we were unable to find any statistically valid relationship between the severity of the vascular injury and the magnitude of the mean PAP or PAR, we have not tried to establish any graphic demonstration.

Table 5. Distribution of lesions according to age clusters

	20-29 (n= 2)	30-39 (n = 6)	40-49 (n = 7)	50-59 (n = 5)	60-69 (n = 9)
N	2.25 ± 0.05	0.6 ± 0.47	1.6 ± 1.0	0.2 ± 0.2	0.66 ± 0.29
I	1.15 ± 1.15	1.95 ± 0.5	1.2 ± 0.8	1.8 ± 0.6	1.4 ± 0.5
II	7.0 ± 7.0	0.3 ± 0.28	0.8 ± 0.8	0.5 ± 0.3	0.8 ± 0.4
III A	27.1 ± 12.4	34.8 ± 4.8	33.1 ± 7.8	41.0 ± 9.4	50.6 ± 7.1
III B	58.7 ± 19.2	52.9 ± 6.56	57.3 ± 6.6	47.4 ±10.9	30.7 ± 8.4
III A + B	85.9 ± 6.75	87.7 ± 2.03	90.5 ± 1.87	88.4 ± 1.8	81.3 ± 2.0
IV	2.6 ± 2.6	8.3 ± 0.9	4.6 ± 0.5	7.2 ± 1.3	13.8 ± 1.4
V	0	0.8 ± 0.8	1.1 ± 0.46	1.6 ± 0.45	1.8 ± 0.3
VI	1.15 ± 1.15	0.4 ± 0.36	0.2 ± 0.2	0.26± 0.26	0.1 ± 0.1

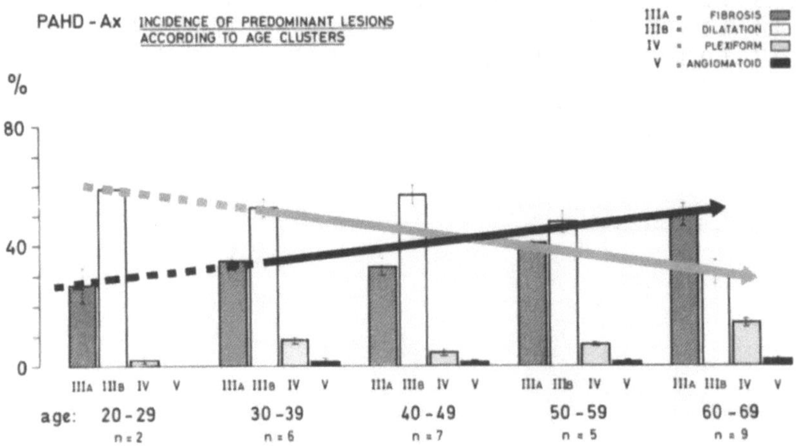

Fig. 31. Graphic representation of age clusters

2. Review of the Literature and Discussion

There has been no paper, apart from that of *Laissue* (1970), devoted to the numerical evaluation of lesions of pulmonary arteries in cases of PHD, either primary or related to aminorex intake. *Laissue* (1970) established the relative frequency of lesions of medium to marked severity, using the criteria of *Wagenvoort* et al. (1964) which are quite similar to our own. He observed mainly grade III lesions, i.e., fibrosis and dilatation. Although his evaluation is based on a grading of crosses (from + to +++), it concurs fairly well with ours, since we too found a marked predominance of fibrotic and dilatation lesions (more than 80%). *Laissue* (1970) then tried to establish a correlation between the severity of the lesions and the severity of the pulmonary arteriolar resistance; he observed that the most severe lesions were usually encountered in cases where this resistance was highest. In the present study, no such correlation was found between the distribution of lesions and the severity of MPAP or that of PAR.

Although, in the present study, no relationship between the severity of the histologic lesions and that of hemodynamic modifications could be established, some interesting observations concerning the evolution of lesions were made. With the subdivision of grade III into two subgroups (III A = fibrosis, III B = dilatation) certain problems became easier to understand. It was observed that fibrotic lesions tend to increase in number as a function of time between the presumed termination of drug intake and the date of death, as well as with the age at death. Conversely, the rate of dilatation lesions evolves in the opposite direction. The question naturally arose therefore as to whether, once the effect of the drug had worn off, it might not have been the effects of the evolution of the histologic lesions that caused the severe consequences for the patients. It can be assumed that the first effect of aminorex is a vasospastic one, as demonstrated by the pharmacologic experiments of *Kraupp* (1969). This would be followed by an intimal lesion, of which hardly any measurable morphologic signs remained by the time our observations were made. The intimal lesions (cellular proliferations, muscularization of the intima) contribute to the maintenance and even possibly the aggravation of the PHD; later on they tend to undergo fibrosis while more or less obliterating the lumen, rather than dilating it. Thus the muscular arteries undergo a transformation into rigid tubes, segmentally obliterated, which maintain the PAP and the PAR at a high rate. The further obstacle, represented by the plexiform lesions, even though they are infrequent, must be added to this.

Without entering into a discussion of the possible initial repercussions of the level of the alveolar septa due to hypoxia, as suggested by the data of *Kapanci* et al. (1974), or of experimental modifications reported by *Valdivia* (1967) (monocrotalin in rats) it would seem that the hypothetical prolonged vasospasm does, in fact, induce a chain reaction of morphologic lesions which contribute to worsen and then fix this pulmonary hypertension.

X. Morphometric Examination

1. Personal Results

The values indicated in Section IV (Material and Methods) were obtained from points and intersections registered in 14 cases of PHD related to aminorex intake and in 3 cases of PPHD and compared with those of 3 control cases.

The results are shown in Table 6.

The mean thickness of the media (τm), expressed in μ was significantly greater in cases of PHD, both primary and related to aminorex intake.

The mean thickness of the intima (τi) in cases of PHD was more than twice that observed in control cases.

The mean total thickness of the wall ($\tau t = \tau m + \tau i$) was also twice that of controls. The increase in the thickness of the wall and in that of its constituants is shown in Figure 32.

The volume density of the lumen compared to the total volume of muscular arteries is expressed by the index Vv_1^a. Table 6 demonstrates that this is reduced to half of the normal in cases of PHD. In Figure 33, where this index is expressed graphically, it can be seen that in cases of PHD, the lumen is reduced by about 50% compared with that of controls.

Table 6. Results of morphometry

	C (= 3)	A (= 14)	PRIM. (= 3)
τm (μ)	4.4 ± 0.72	7.2 ± 0.44	6.5 ± 0.26
τi (μ)	6.4 ± 0.19	13.7 ± 1.19	20.0 ± 0.55
τT (μ)	10.8 ± 0.7	20.9 ± 1.5	26.5 ± 0.81
VL %	69.7 ± 3.9	38.6 ± 2.36	32.2 ± 3.6
Ce (mm)	0.4 ± 0.15	5.46 ± 0.35	3.07 ± 1.77
MPAP (mm Hg)	---	51.8 ± 3.2 (12 cases)	53.7 ± 4.2
PAR (dyn/sec/cm^{-5})	---	1182.4 ± 139.9 (10 cases)	1490 ± 182.5

Main parameters measured: mean thickness of media (τm), of intima (τi), of whole wall (τT); volume density of lumen (VL %); mean circumference of lamina elastica interna (Ce).

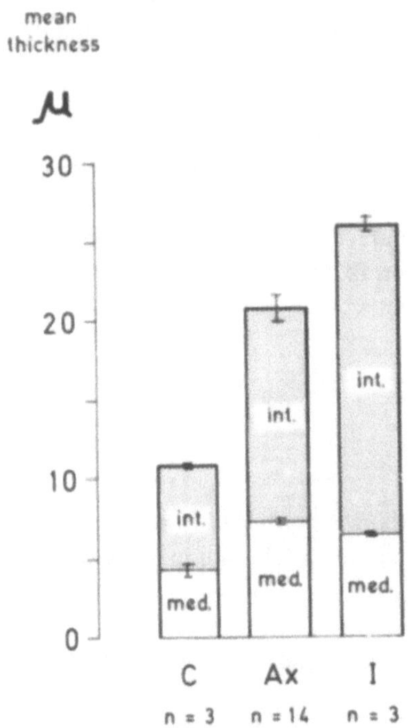

Fig. 32. Mean thickness of arterial wall and its constituents (*int.* = intima; *med.* = media; *C* = controls; *Ax* = patients with aminorex intake; *I* = idiopathic pulmonary hypertension; *n* = number of cases)

In an attempt to establish a statistical relationship between the histologic alterations and the importance of hemodynamic modifications, values for the two main constituents of the wall (intima and media) as well as their sum in function of the mean pulmonary arterial pressure have been tested. The regression line and the coefficient of correlation *r* have been calculated (Geigy tables Nos. 632 and 704) and the functional dependence has been tested with the *t* test (Geigy table No. 76). This statistical analysis revealed no correlation for any of the parameters tested. The *r* coefficients (- 0.32 for τi, - 0.615 for τm, and - 0.254 for τt) never diverged significantly from zero.

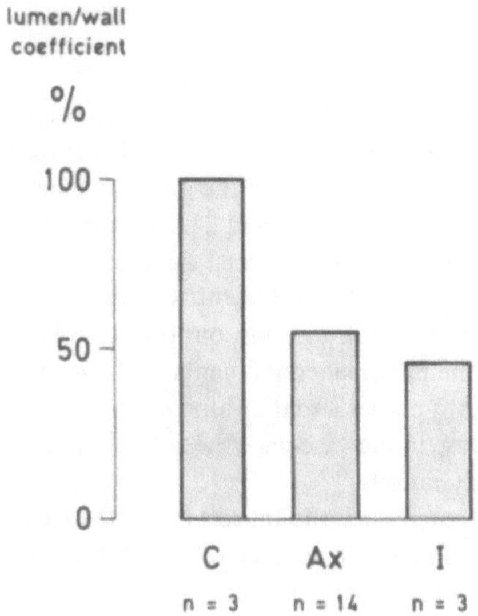

Fig. 33. Reduction of volume density of the lumen (lumen/wall coefficient) by about 50% in cases with aminorex intake *(Ax)* and primary PHD *(I)*, compared with controls *(C)*

Of the other parameters calculated, we shall retain only *the mean circumference of the lamina elastica Ce.* It can be seen that this is highly increased in cases of PHD, (between 13.6 for aminorex and 7.6 for PPHD, that of control arteries being equal to 1 as a reference). This increase must, in our opinion, be attributed to the great number of dilatation lesions, as has been stated in Section IX.

2. Review of the Literature and Discussion

In a preliminary paper (*Kapanci* and *Widgren*, 1970) we reported the thickening of the wall of muscular arteries and its constituents, which caused a decrease in the volume of the arterial bed. *Laissue* (1970) also observed, by means of a semiquantitative method of evaluation, an increase in the thickness of the intima and a reduction in the size of the vascular lumen. *Ostertag* (1969), using the method described by *Wagenvoort* (1960) (see below) tried to establish a relationship between the area of the media and the total area of parenchyma per cm^2 of histologic section. In one case he observed a very distinct increase in the area of the media of arteries of less than 100μ in diameter, whereas in the second case he found that it was diminished, at least for arteries larger than 100μ.

Apart from the studies of *Weibel* (1963) on the morphometry of the normal lung, all of the morphometric evaluations of pulmonary arteries in cases of PHD have been made by techniques which were apparently simple but which involved a high degree of subjectivity (semiquantitative methods; measurements with graduated eye-pieces or on planimetric projections; deliberate choice of arteries, as in the study of *Wagenvoort*, 1960).

Zwahlen shares our scepticism as to the validity of conclusions based on these methods. A stereologic method, based on the data of *Weibel* (1963), allows for the elimination, to a great extent, of the above-mentioned disadvantages of "handicraft" methods which are statistically inadequate; the serious disadvantage represented by deliberate choice of lesions is avoided by using random sampling. The objective unchosen sample is, in fact, indispensable for the application of statistical methods (*Zwahlen*, 1974).

Wagenvoort and *Wagenvoort* (1970) published a study on pulmonary vessels in 156 cases of PHD, of which 110 were primary PHD. They used a method of measurement expressing the thickness of the media or the intima as a percentage of the external diameter of the artery on a transverse section. They thus obtained a thickness of the media ranging between 10.8 and 13.9% and a thickness of the intima between 49.5 and 24.9%. It is necessary to go back to *Wagenvoort*'s paper of 1960 to find the mean normal value of 4.64 for the media; the value for the intima has never been given. Due to the entirely different methods used, a comparison of the numerical data of these authors with our own is not possible. They did, however, report a thickening of the principal constituents of the wall inducing a stenosis of the lumen. We agree with their hypothesis that a vasoconstrictor mechanism associated with, or determining, an increase in the vascular tone is the initiating factor inducing an entire series of morphologic alterations.

Although the mean circumference of the vessels is increased, the thickening of the wall causes a decrease of the mean volume of the arterial lumen. The absence of any statistical correlations between the measured alterations and the severity of the hemodynamic modifications suggests that factors other than those that we have measured contribute to the development of PHD as well as to its persistence.

XI. General Discussion

Following a sudden increase in the number of cases of primary pulmonary hypertensive disease in Switzerland, the Federal Republic of Germany and Austria in 1966, epidemiologic investigations showed that persons suffering from this disease had taken an anorexigenic drug, aminorex fumarate (Menocil, Cilag). A study was made of the lungs of 37 cases deceased in Switzerland, the results of which are summarized below.

Clinical data indicate that the great majority of patients were women whose average age ranged from 30-50 years, and who were affected by PHD either during the course of or after a weight-reducing treatment. No correlation has been found between the dose taken, the duration of the treatment, and the severity of hemodynamic modifications. The only laboratory finding of note was a distinct elevation of the hematocrit, and possibly, a pathologic platelet adhesivity.

Histologic examination of muscular pulmonary arteries from 100-1000μ diameter showed lesions entirely comparable to those described by *Heath* and *Edwards* (1958) in cases of pulmonary hypertension secondary to cardiac malformation. They were in no significant way different from the lesions we observed in cases of primary or secondary pulmonary hypertension.

The ultrastructure of intimal hyperplasia, whether concentric, obliterating, or fibrosed, is composed of myointimal cells (MIC) and smooth muscle cells (SMC), characteristic of muscularization of the intima as seen with the light microscope.

Numerical evaluation of the various types of arterial lesions has demonstrated the predominance of the grades of intimal fibrosis and diffuse dilatation. The fibrotic lesions increase with time between the end of drug intake and death as well as with age at time of death. However, no statistically valid correlation between the severity of vascular lesions and that of hemodynamic modifications was found.

Morphometric observations have revealed an increase in the mean thickness of the arterial wall, which was found to attain a thickness twice that found in control cases. This corresponds to a 50% reduction of the vascular bed. No statistically significant correlation could be demonstrated, however, between these results and the severity of hemodynamic changes.

The first pharmacologic studies published either did not mention any significant modifications of the pulmonary circulation in animals which could be attributed to aminorex or had not investigated them (*Yelnosky*, 1966). It was only after the sudden increase in the number of cases of PHD, first in Switzerland, then in Germany and Austria, was noticed and the possible relationship between this disease and aminorex intake was suspected, that an entire series of experimental studies were undertaken by various groups of pharmacologists and pathologists. Although a certain number of publications denied any influence of this drug on the pulmonary hemodynamics, others showed that aminorex intake caused an increase of PAP, either of short duration following a perfusion (*Kraupp* et al., 1969, 1970). or prolonged after ingestion (*Lüllmann* et al., 1972). Histologic investigations, however, revealed no pulmonary vascular modifications. It should be borne in mind that none of these studies included electron-microscopic observations of muscular arteries or alveolar septa.

In spite of the weaknesses of some of these pharmacologic tests, it may be assumed that the effect of aminorex is to cause an increase in the arterial tone which could result in a spasm. It is possible that this is the point of departure in a chain reaction inducing endothelial lesions followed by an intimal reaction. Since only a low proportion (1-2‰) of the persons who took this drug developed PHD, it must be postulated that other as yet unknown mechanisms have an adjuvant effect, synergic or potentiating in the pathogenesis of the disease. One experimental finding of importance here is that of *Peters* and *Gourzis* (1970) who observed that the increase of the pulmonary perfusion pressure due to microemboli of diatomaceous earth was accentuated by the injection of aminorex (or amphetamine). This strongly suggests that, although aminorex injected alone has only a temporary effect, it must have a potentiating effect when administered with another agent. Unfortunately, it was not possible to identify this agent in human beings, in spite of thorough epidemiologic investigations. One fact remains certain: at least 593 patients affected with PHD have been registered by various cardiology centers in the three countries mentioned, and of these, there have been at least 85 who died as a result of the disease.

None of the light-microscopic data that we have assembled provided any new light on the nature of the lesions of muscular arteries in pulmonary hypertension, whether related to aminorex intake, idiopathic, i.e., without known drug intake, or secondary to a cardiac malformation with a shunt.

Some clinical studies have suggested the existence of intrapulmonary arteriovenous anastomoses. *Follath* et al. (1971) and *Gahl* et al. (1970) consider them to be the cause of a drop in pO_2. *Gurtner* et al. (1968a) and *Kaindl* (1971) think that they are responsible for the early venous phase which they observed during selective pulmonary angiography in a certain number of cases. We found no convincing evidence of this morphologic peculiarity in the slides that we examined nor of any broncho-pulmonary anastomoses connected to the plexiform and angiomatoid lesions, such as reported by *Liebow* and *Downing* (1968).

Our investigations using the electron microscope allowed a precise description of the ultrastructure of normal pulmonary arteries, about which, in human beings at least, relatively little is known. The typical appearance of endothelial cells, with an abundance of microfilaments suggests that they are endowed with a contractile capacity. But is not known whether they play an active or passive part in the supposed vasospasm, which is presently thought to be the principal effect of aminorex, as far as one makes abstraction of another possible effector organ: the alveolar septum. We would also like to stress the presence, in the normal subintimal layer, of occasional myointimal cells (MIC). Morphologically, there is no difference between the muscular pulmonary artery and the muscular artery of the larger circulation; the medium in which both are situated, however, must have functional significance.

PHD lesions of intimal hyperplasia, concentric, obliterating, or undergoing fibrosis, were closely studied in the course of the present investigation. These variants are characterized by a relatively large number of myointimal cells, which can be recognized by their bundles of microfilaments, often oriented, by their low content of rough endoplasmic reticulum, and by their membrane containing pinocytotic vesicles. In obliterating lesions, they are associated with numerous smooth muscle cells. MIC have an ultrastructure nearly identical to that of alveolar interstitial contractile cells as described recently by *Kapanci* et al. (1974). The latter cells react differently from arterial or bronchial SMC to various stimuli, particularly hypoxia and are considered responsible for the regulation of the ventilation/perfusion rate. MIC also have structural features very similar to those of myofibroblasts of granulation tissue, to which *Majno* (1971) and his team (*Gabbiani* et al., 1972) have attributed contractile capacity. Considering the great similarity of MIC to alveolar interstitial contractile cells and myofibroblasts we believe they must surely have similar functional characteristics. Their frequently contracted appearance leads one to think that MIC and SMC must play an aggravating role in PHD by complicating the vasospasm that is probably induced by the drug. Thus MIC, because of their morphologic kinship with SMC, as well as the SMC may contribute to increase the tone of the arterial wall. Morphologically, because of the hindrance represented by intimal thickening and especially because of the reduction of the vascular lumen, they contribute to fix the pulmonary hypertension.

Our morphometric investigations have revealed an apparently paradoxical feature of the lesions of muscular arteries, i.e., a marked increase of the mean circumference of the lamina elastica interna; this can be attributed to the great number of dilatation lesions as well as to complicated vascular modifications (plexiform and angiomatoid lesions).

The morphometric data also allowed for a quantitative evaluation of some of the aspects of arterial lesions. The total thickness of the intima and the media, and therefore of the wall, was found to be on average twice that of muscular arteries of control cases. This results in a reduction of about 50% of the muscular arterial bed. We are thus dealing with an anatomic modification that is irreversible, or reversible to only a slight extent. It is our opinion that these lesions may develop into fibrosis. Whether this results in an obliteration of the lumen or not, seems to make no difference, functionally. This fibrotic structure, from a functional point of view, can only be viewed as a rigid, rather narrow tube with very little motivity.

As has been mentioned in the discussion on the numerical evaluation of arterial walls, intimal fibrosis was found in 40% of the vessels. As this type of lesion tends to increase with time between the presumed termination of drug intake and the time of death, as well as with the age of the patient, it may be presumed that the pulmonary hypertension becomes fixed. Even though plexiform lesions constitute only a relatively small percentage of all the lesions, they are nevertheless significant in number and must certainly play a role in the development as well as the aggravation of PHD.

It would be of interest to know why, in patients deceased long after the last absorption of the drug, one may still observe grade II lesions, i.e., intimal cellular hyperplasia, and particularly intimal proliferation in the vicinity of fibrinoid material. It is our opinion that, once the pulmonary hypertension is established, it continues, even in the absence of the triggering or potentiating agent, to produce endothelial lesions on which degraded fibrin is deposited; this becomes phagocytized, thus inducing a new proliferation of intimal cells, as suggested by *Kapanci* (1965). The same mechanism could also explain the development of plexiform lesions. (Their morphologic and functional significance has already been discussed.)

It is evident, therefore, that we are confronted with a vicious cycle consisting of "morphologic healing" (fibrosis and dilatation), "histologic activity" (intimal proliferation and plexiform lesions), "functional activity " (muscular spasm and muscularization of the intima), and "functional inactivity" (angiomatoid lesions). Any treatment, therefore, whether spasmolytic or fibrinolytic, can at best result in only a transitory improvement of the patient's condition.

It is our opinion that a direct or indirect action of aminorex, probably potentiated, causes an increase in the tone of the muscular arteries, which is followed by a vasospasm. This induces an endothelial lesion on which fibrinoid material is deposited, producing intimal proliferation. This proliferation, in which MIC, then SMC are abundant, contributes to the aggravation of the PHD. This histologic lesion, already irreversible, may obliterate the lumen or undergo fibrosis. The thickening of the arterial wall, which induces a reduction of the vascular bed, leads to a new aggravation of the pulmonary hypertension. Considered from a dynamic point of view, this series of morphologic modifications constitutes a vicious circle of lesions which contribute to maintain and aggravate the PHD, and which resist any etiologic therapy.

In conclusion, it can be said that although the pathogenesis of PHD related to aminorex intake has not been definitely established, and although the responsibility of this drug has been questioned by *Heath* et al. (1971) because of their own negative experimental results and the equivocal results of others, the following epidemiologic findings lend undeniable support to the existence of such a relationship:

1. The frequency of cases of PHD increased suddenly in 1966; it remained high until 1969, reaching a rate 10-20 times higher than that of previous years (*Rivier*, 1972). With a slight delay, this increase is concomitant with the commercialization of aminorex at the end of 1965.
2. The frequency of PHD diminished suddenly from 1969, also with a certain delay after the drug was withdrawn from the market in 1968. This disease is now as infrequent as it was before 1966 (*Rivier*, 1972, personal communication).
3. PHD showed a sudden and marked rise in frequency in the three countries only where the drug was on sale (Switzerland, the Federal Republic of Germany, and

Austria). On the other hand, in the countries where aminorex was not sold, there was no change in the frequency of PHD during the years in question (*Ferrero*, 1970).

Another important fact must be kept in mind. Not all of the subjects who had taken aminorex (probably several hundredthousand) developed PHD; only one to two per thousand became ill. The reason for this low morbidity rate is still not clear. Two possibilities that come to mind are:

1. A particular susceptibility of some subjects to the drug, which had no secondary effect in 998 to 999‰ of persons who took it.
2. The drug acts as a potentiating agent in the presence of other drugs or metabolic disorders as yet unexplained.

Unfortunately none of the investigations have provided any answers to these questions so the "aminorex matter" has become a tragic and mysterious chapter in the history of Medicine.

XII. Summary

Following reports in 1967, of an increasing number of cases of PHD, various clinical investigations undertaken in Switzerland, the Federal Republic of Germany, and Austria, demonstrated a possible causal relationship between this disease and the absorption of an anorexigenic drug, aminorex fumarate (Menocil, Čilag). The present paper is based on the study of the lungs of 37 cases deceased in Switzerland. The results of light and electron-microscopic studies, of numerical evaluation and of morphometry of arterial lesions are presented.

The modifications of pulmonary muscular arteries examined under the light microscope are in every way similar to those described in pulmonary hypertension secondary to cardiac malformations. The arteries have ultrastructural features resembling those of atherosclerosis of the larger circulation, either spontaneous or experimental: the subintima is the seat of muscularization composed of smooth muscle cells and myointimal cells which are thought to worsen the pulmonary hypertension induced by the drug or another mediator. The counting of arterial lesions has demonstrated a marked predominance of dilatation and fibrotic lesions. The latter increase in frequency with age and as a function of time between the beginning of drug intake and death. Morphometric studies showed a marked thickening of the arterial wall and a reduction of the vascular bed by about 50%. No statistical relationship was found between morphometric and numerical results and hemodynamic data.

The findings of our study do not allow for a precise explanation of the action of aminorex. It is possible, however, to suggest the hypothesis that some triggering agent (aminorex alone or combined with a potentiating agent) causes a vasospasm which induces a series of morphologic and functional lesions which tend to be self-perpetuating. This is based on morphologic alterations which rapidly become irreversible and resistant to any etiologically directed therapy. The reason why only 1-2‰ of those persons who had taken aminorex were affected by pulmonary hypertension remains a mystery.

XIII. Zusammenfassung

Nach der plötzlichen Zunahme, im Jahre 1967, der Fälle von pulmonaler Hypertonie, haben verschiedene klinische Untersuchungen in der Schweiz, der Bundesrepublik Deutschland und Oesterreich, einen möglichen kausalen Zusammenhang mit der Einnahme eines Abmagerungsmittels, Aminorexfumarat (Menocil, Cilag) hervorgebracht. Die vorliegende Arbeit ist auf der Untersuchung der Lungen von 37 in der Schweiz verstorbenen Patienten begründet. Die Ergebnisse der licht- und elektronenmikroskopischen Untersuchungen, der Zählung und der Morphometrie der arteriellen Veränderungen werden dargestellt. Die Läsionen der muskulären Lungenarterien sind lichtmikroskopisch denen bei einem Vitium cordis sekundären pulmonalen Hypertonie weitaus vergleichbar. Diese Arterien zeigen ultrastrukturelle Veränderungen, die vergleichbar mit denen der Atherosklerose der großen Zirkulation sind: die Subintima ist der Sitz einer Muskularisation, die aus glatten Muskelzellen und myointimalen Zellen besteht; diese dürften die durch ein Medikament oder einen anderen Mediator induzierte pulmonale Hypertonie verschlimmern.

Die Zählung der Arterienveränderungen hat das deutliche Vorherrschen der Fibrose und Dilatation erbracht. Diese letzte Veränderung nimmt numerisch mit dem Alter und der zwischen dem Anfang der Medikation und dem Tod verflossenen Zeit zu. Die morphometrischen Untersuchungen ergeben eine deutliche Verdickung der Arterienwand und eine Abnahme um etwa 50% des vaskulären Bettes. Ein statistischer Zusammenhang zwischen den morphometrischen und numerischen Ergebnissen und den haemodynamischen Daten konnte nicht erwiesen werden.

Unsere Untersuchungen konnten die Wirkungsweise des Aminorexfumarats nicht erklären. Jedoch scheint es möglich, die Hypothese vorzuschlagen, daß ein "Trigger-Agens" (Aminorex allein oder mit einem potentialisierenden Agens assoziiert) einen Vasospasmus induziert, der eine ganze Kette von morphologischen und funktionellen Veränderungen, die sich in einem Circulus viciosus fortsetzen, zur Folge hat. Dieser beruht auf rasch irreversibel gewordenen morphologischen Läsionen, die jeglichen ätiologisch-gezielten therapeutischen Versuchen im Wege stehen. Der Grund, warum nur 1-2‰ der Personen, die Aminorex eingenommen haben, an einer pulmonalen Hypertonie erkrankt sind, bleibt unerklärt.

Acknowledgments: We wish to thank Prof. Y. Kapanci for his help in the preparation of this study, for having authorized the use of previous joint publications, and for having made his morphometric equipment available to us.

We are also deeply indebted to:

The pathologists who have provided autopsy material:

in Lausanne: Prof. Ch. Hedinger, Prof. D. Gardiol, and Dr. T. Hardmeier;

in Neuchâtel: Dr. R.P. Baumann and Dr. O. Fryc;

in Bern: Prof. H. Cottier and Dr. J. Laissue;

in Basle: Prof. H.U. Zollinger;

in Zürich: Prof. E. Uehlinger and Prof. J.R. Rüttner;

in Winterthur: Dr. B. Egloff;

in St. Gallen: Dr. Dirnhofer;

the cardiologists and internists who answered our questionaires and made their clinical and hemodynamic data available to us:

in Geneva: Prof. P. Duchosal and Prof. P. Moret, Division of Cardiology;

in Lausanne: Prof. J.L. Rivier;

in Neuchâtel: Dr. J. Jornod, PD;

in Bern: Prof. H.P. Gurtner;

in Basle: Prof. W. Schweizer; the Direction of the Department of Medicine, St Claraspital;

in Zürich: Dr. M. Rothlin:

in Winterthur: Prof. F. Wuhrmann;

in Zug: Dr. G. Keiser, PD;

in Münsterlingen: Direction of the Department of Medicine;

in Mammern: Dr. A.O. Fleisch;

Dr. A. Assimacopoulos for his help with morphometric methods and statistical calculations;

Dr. K. Hanisch, medical director of Cilag-Chemie AG (Schaffhausen) which provided financial assistance;

the Swiss Society of Cardiology for financial help;

Mrs. Judith Noebels for her editorial help with the English version of this paper;

the laboratory technicians of the Department of Pathology:

Mrs. Danielle Berclaz-Coudray, Mrs. Elisabeth Bouchardy-Halter, Miss Marie-Claire Clottu, Mrs. Kerstin Tamm-Janvert, Miss Geneviève Leyvraz, Miss Unn Lutzen;

the secretaries of the Department of Pathology: Mrs. Nicole Duruz-Mayet, Carole Reiss, Béatrice Roche, Jacqueline Schaffter, Claire-Lise Seignemartin;

Miss Lucienne Ducret, librarian at the Department of Pathology;

Mr. Louis Bodoira, archivist;

Mr. Etienne Denkinger and Mr. Jean-Claude Rumbeli for photographic work;

Mr. Charles Ryser and his team for help with the tables and drawings.

References

A. Symposia and round tables

1. VIENNA: Primäre pulmonale Hypertension. 11. Kardioangiologische Diskussion der Österreichischen Kardiologischen Gesellschaft in Wien, 23. November 1968; Reports in Wien. Z. inn. Med. **50**, 449-511 (1969).
2. HANNOVER: Arbeitstagung: Pulmonale Hypertonie; 20-21 Februar 1970; Reports in Z. Kreislaufforsch. **59**, 865-933 (1970).
3. BÜRGENSTOCK: Symposium: Adipositas, Kreislauf und Anorektika, 15.-16.5.1970. Blankart, R. (ed.). Bern-Stuttgart-Vienna: Hans Huber 1974.
4. MONTREUX: Diskussion zum Hauptthema: Die pulmonale Hypertonie. Schweiz. med. Wschr. **100**, 154-158 (1970).
5. BASEL: Roundtable: Häufung der primär vaskulären pulmonalen Hypertonie in der Schweiz 1967-1970, 11.6.1970. In: Schweiz. med. Wschr. **100**, 2146-2159 (1970).

B. Isolated papers

Allen, J.R., Carstens, L.A.: Clinical signs and pathologic changes in Crotalaria spectabilis intoxicated rats. Amer. J. vet. Res. **31**, 1059-1070 (1970).
Backmann, R.: Pathologische Anatomie der primären pulmonalen Hypertonie. Biopsiebefunde bei Patienten und bei experimenteller pulmonaler Hypertonie. Z. Kreislaufforsch. **59**, 931-933 (1970)
Backmann, R., Ebbinghaus, H.: Zellproliferation in Rattenlungen nach Anorektikagaben. Eine autoradiographische Untersuchung. Verh. dtsch. Ges. Path. **56**, 517-521 (1972).
Backmann, R. Haschemian, A., Kemper, F., Morgenroth, K., Jr.: Einfluß von Anorektika auf die Lunge von Ratten; (see Symposia: BÜRGENSTOCK, 123-129).
Backwinkel, K.P., Schmitt, G., Themann, H., Hauss, H.W.: Elektronenmikroskopische Untersuchungen über Frühveränderungen der Koronararterien bei experimenteller Hypertonie. Beitr. Path. **141**, 374-391 (1970).
Baghirzade, M.F., Donath, K.: Menocil-Medikation und pulmonaler Hochdruck. Verh. dtsch. Ges. Path. **53**, 473-481 (1969).
Bass, O., Gurtner, H.P.: Verlauf der primärvaskulären pulmonalen Hypertonie nach Einnahme von Aminorex fumarat (Menocil). Vorläufige Mitteilung. Schweiz. med. Wschr. **103**, 1794 (1973).
Bensch, K.G., Gordon, G.B., Miller, L.: Fibrillar structures resembling leiomyofibrils in endothelial cells of mammalian pulmonary blood vessels. Z. Zellforsch. **63**, 759-766 (1964).
Berthou, J.D., Delahaye, J.P., Brune, J., Dorsit, G., Delaye, J., Roques, J.C., Gonin, A.: Hypertension artérielle pulmonaire apparemment primitive chez une femme ayant pris des anorexigènes (Aminorex). Rev. Tuberc. Pneumol. **36**, 944-951 (1972).
Blankart, R.: Der Zeitfaktor im Ablauf der primär vaskulären pulmonalen Hypertonie. Schweiz. med. Wschr. **100**, 2157-2158 (1970).
Blankart, R.: Epidemiologische Untersuchungen zur Häufung der pulmonalen Hypertonie in den Jahren 1967-1969; (see Symposia: BÜRGENSTOCK, 187-201).
Borbely, F., Pasi, A., Velvart, J.: Die akute perorale Vergiftung durch 2-amino-5-phenyloxazolin-fumarat beim Menschen anhand von 30 Beobachtungsfällen. Arch. Toxikol. **26**, 117-124 (1970).
Bouvier, C.A., Ritschard, J., Meykadeh, F., Fournet, P., Duchosal, P.W.: Traitement fibrinolytique de l'hypertension pulmonaire sévère. Justification, modalités et critique. Schweiz. med. Wschr. **100**, 148-151 (1970).
Brandt, H.J., Preussler, H.: Die "primäre" vaskuläre pulmonale Hypertonie. I. Bioptische Untersuchungen bei vaskulärer pulmonaler Hypertonie. Z. ärztl. Fortbild. **19** (1969).
Bredt, H.: Die primäre Erkrankung der Lungenschlagader in ihren verschiedenen Formen (Arteriopathia pulmonalis idiogenica). Virchows Arch. path. Anat. **284**, 126-153 (1932).
Brunner, H., Stepanek, J.: Drucksteigerungen im kleinen Kreislauf des Hundes durch Aminorex. Naunyn-Schmiedebergs Arch. Pharmakol.**266**, 304-305 (1970).
Buck, R.C.: Intimal thickening after ligature of arteries; an electron microscopic study. Circulat. Res. **9**, 418-426 (1961)
Butler, W.H.: An ultrastructural study of the pulmonary lesion induced by pyrrole derivatives of the pyrrolizidine alkaloids. J. Path. **102**, 15-19 (1970).
Byrne-Quinn, E., Grover, R.F.: Aminorex (Menocil) and amphetamine: acute and chronic effects on pulmonary and systemic haemodynamics in the calf. Thorax **27**, 127-131 (1972).
Caesar, R.: Gefäße und Herz im elektronenmikroskopischen Bild. In: Kaufmann, E., Staemmler, M. (eds.): Lehrbuch der speziellen pathologischen Anatomie. Berlin: de Gruyter 1969, Erg. Bd. I/1/4, pp. 700-812.
Chardonnens, J.D.: Une entité clinique nouvelle et éphémère: l'hypertension artérielle pulmonaire consécutive à la prise d'un anorexigène. (Etude de 21 observations.) Rev. méd. Suisse rom. **92**, 845-878 (1972).
Chardonnens, J.D., Mahaim, Ch.: Hypertension pulmonaire après cure d'amaigrissement. Praxis **58**, 236-242 (1969).
Corrodi, P., Bühlmann, A.A.: Zur Prognose des Cor pulmonale bei multipler Gefäßobstruktion unter Dauerantikoagulation. Schweiz. med. Wschr. **103**, 96-100 (1973).

Dienstl., F.: Primäre pulmonale Hypertension. Klinische Beobachtungen an der Medizinischen Universitätsklinik in Innsbruck. Wien. Z. inn. Med. **50**, 462-464 (1969).
Downing, S.E., Vidone, R.A., Brandt, H.M., Liebow, A.A.: The pathogenesis of vascular lesions in experimental hyperkinetic pulmonary hypertension. Amer. J. Path. **43**, 739-765 (1963).
Engelhardt, A., Kroneberg, G., Stoepel, K., Stötzer, H.: The effect of acute and chronic administration of sympathicomimetic substances on the systemic and pulmonary circulation. Proc. Europ. Soc. Study Drug Toxicity. Amsterdam: Excerpta Medica 1971, Vol. XII, pp. 110-117.
Engelhardt, R., Hort, W.: Über die Wirkung von Aminorex am Kreislauf der Ratte im Langzeitversuch. Naunyn-Schmiedebergs Arch. Pharmacol. **266**, 318-319 (1970).
Esterly, J.A., Glagov, S., Ferguson, D.J.: Morphogenesis of intimal obliterative hyperplasia of small arteries in experimental pulmonary hypertension. An ultrastructural study of the role of smooth-muscle cells. Amer. J. Path. **52**, 325-347 (1968).
Ferrero, C.: Rapport à la table ronde: "Die pulmonale Hypertonie". Schweiz. med. Wschr. **100**, 155 (1970).
Feuardent, R.: Une cause rare de coeur pulmonaire: "L'endofibrose oblitérante idiopathique des artérioles du poumon". Presse méd. **61**, 594-597 (1953). Thèse, Genève, No. 2175.
Follath, F., Burkart, F., Schweizer, W.: Drug-induced pulmonary hypertension? Brit. med. J. **1971 I**, 265-266.
Gabbiani, G., Badonnel, M.-C., Rona, G.: Cytoplasmic contractile apparatus in aortic endothelial cells of hypertensive rats. Lab. Invest. **32**, 227-234 (1975).
Gabbiani, G., Hirschel, B.J., Ryan, G.B., Statkov, P.R., Majno, G.: Granulation tissue as a contractile organ. A study of structure and function. J. exp. Med. **135**, 719-734 (1972).
Gadd, K., Gunne, L.M.: Klinisk värdering av de centralstimulerande egenskaperna hos ett aptitreducerande medel. Läkartidningen **67**, Suppl. II, 100-102 (1970).
Gahl, K., Fabel, H., Greiser, E., Harmjanz, D., Ostertag, H., Stender, H.St.: Primär vaskuläre pulmonale Hypertonie. Bericht über 21 Patienten. Z. Kreislaufforsch. **59**, 868-883 (1970).
GEIGY Scientific Tables, 7th Edition, Basle 1972.
Gendre, P.: Observations sur l'infrastructure de la paroi artérielle chez le poulet, en vue de l'athérome cholestérolique expérimental. C.R. Acad. Sci. (Paris) **270**, 1356-1358 (1970a).
Gendre, P.: Ultrastructure de la paroi artérielle humaine saine et athéroscléreuse. Path. europ. **5**, 283-306 (1970b).
Georgii, A.: Discussion in Baghirzade, M.F., and Donath, K. (1969).
Gertsch, M., Stucki, P.: Weitgehend reversible primär vaskuläre pulmonale Hypertonie bei einem Patienten mit Menocil-Einnahme. Z. Kreislaufforsch. **59**, 902-908 (1970).
Gurtner, H.P., Gertsch, M., Salzmann, C., Scherrer, M., Stucki, P., Wyss, F.: Häufen sich die primär vaskulären Formen des chronischen Cor pulmonale? Schweiz. med. Wschr. **98**, 1579-1589; 1695-1707 (1968a).
Gurtner, H.P., Scherrer, M., Stucki, P.,Wyss, F.: Häufen sich die primär vaskulären Formen des Cor pulmonale? Helv. med. Acta Suppl. **48**, 108-109 (1968b).
Hadler, A.J.: Studies of Aminorex, a new anorexigenic agent. J. clin. Pharmacol. **7**, 296-302 (1967).
Hadler, A.J.: Further studies on Aminorex, a new anorexigenic agent. Curr. ther. Res. **12**, 639-644 (1970).
Harris, P.: Pathogenesis of pulmonary hypertension. Schweiz. med. Wschr. **100**, 130-133 (1970).
Hatt, P.Y., Rouiller, Ch.: Les ultrastructures pulmonaires et le régime de la petite circulation. I. Au cours du rétrécissement mitral serré. Path. Biol. **6**, 1371-1397 (1958).
Hatt, P.Y., Rouiller, Ch., Grosgogeat, Y.: Les ultrastructures pulmonaires et le régime de la petite circulation. II. Au cours des cardiopathies congénitales comportant une augmentation du débit sanguin intrapulmonaire. Path. Biol. **7**, 515-544 (1959).
Heath, D., Edwards, J.E.: The pathology of hypertensive pulmonary vascular disease. Circulation **18**, 533-547 (1958).
Heath, D., Kay, J.M., Smith, P.: A quantitative pathological study of the heart and pulmonary arteries of rats and dogs after prolonged feeding with aminorex fumarate (Menocil). Proc. 12th Meeting Europ. Soc. Study Drug Toxicity. Amsterdam: Excerpta Medica 1971, pp. 96-100.

Höfer, W., Paeprer, H.: Die "primäre" vaskuläre pulmonale Hypertonie. Z. ärtzl. Fortbild. **19** (1969).
Hotz, H.W.: Untersuchungen mit dem neuen Appetitzügler Aminorex-Fumarat (Menocil®) bei Patienten mit inneren Erkrankungen. Praxis **56**, 1538-1540 (1967).
Inglesby, T.V., Singer, J.W., Gordon, D.S.: Abnormal fibrinolysis in familial pulmonary hypertension. Amer. J. Med. **55**, 5-14 (1973).
Jahrmärker, H., Grohmann, H., Seibel, K., Frey, K.W., Fruhmann, G., Finsterer, H.: Cor pulmonale chronicum durch primär vaskuläre Hypertonie bei jüngeren Menschen. Verh. dtsch. Ges. inn. Med. **75**, 432-436 (1969).
Joris, I., Majno, G., Ryan, G.B.: Endothelial contraction in vivo: a study of the rat mesentery. Virchows Arch. (Zellpath.) **12**, 73-83 (1972).
Jornod, J., Widgren, S., Fischer, G.: Etude anatomo-clinique d'un cas d'hypertension artérielle pulmonaire chez une jeune femme. Schweiz. med. Wschr. **100**, 151-154 (1970).
Kaindl, F.: Primäre pulmonale Hypertension. Wien. Z. inn. Med. **50**, 451-453 (1969).
Kaindl, F.: Medikamentös bedingte pulmonale Hypertonie. Erfahrungen in Österreich. Wien. klin. Wschr. **83**, 373-375 (1971).
Kapanci, Y.: Thrombopathie des petites artères pulmonaires chez un nouveau-né sans malformations cardiaques. Press méd. **65**, 2150-2153 (1957).
Kapanci, Y.: Hypertensive pulmonary vascular disease. Endothelial hyperplasia and its relations to intravascular fibrin precipitation. Amer. J. Path. **47**, 665-676 (1965).
Kapanci, Y., Assimacopoulos, A., Zwahlen, A., Irle, C., Gabbiani, G.: "Contractile interstitial cells" in pulmonary alveolar septa. A possible regulator of ventilation/perfusion ratio? Ultrastructural and immunofluorescent studies. J. Cell Biol. **60**, 375-392 (1974).
Kapanci, Y., Widgren, S.: Hypertension pulmonaire en relation avec l'absorption de Menocil. Etude histologique et morphométrique préliminaire des artères pulmonaires. Schweiz. med. Wschr. **100**, 2150-2152 (1970).
Kay, J.M., Smith, P., Heath, D.: Electron microscopy of Crotalaria pulmonary hypertension. Thorax **24**, 511-526 (1969).
Kay, J.M., Smith, P., Heath, D.: Aminorex and the pulmonary circulation. Thorax **26**, 262-270 (1971).
Kew, M.C.: Aminorex fumarate: a double-blind trial and examination for signs of pulmonary arterial hypertension. S. Afr. med. J. **44**, 421-423 (1970).
Kotscher, E., Lobenwein, E.: Radiologischer Beitrag zur primären pulmonalen Hypertension nach Appetitzüglern. Fortschr. Röntgenstr. **119**, 141-146 (1973).
Krähenbühl, B., Laperrouza, C., Moret, P.: L'hypertension pulmonaire "essentielle" ou "thrombo-embolique". Schweiz. med. Wschr. **98**, 1311-1320 (1968).
Kraupp, O.: Tierexperimentelle Untersuchungen zur Ätiologie der primären Pulmonalhypertension. Wien. Z. inn. Med. **50**, 493-496 (1969).
Kraupp, O., Stühlinger, W., Raberger, G., Turnheim, K.: Die Wirkung von Aminorex (Menocil®) auf die Hämodynamik des kleinen und großen Kreislaufs bei i.v. Darreichung am Hund. Naunyn-Schmiedebergs Arch. Pharmacol. **264**, 389-405 (1969).
Kraupp, O., Stühlinger, W., Turnheim, K.: Vergleichende Untersuchungen über die Wirkung von Aminorex, Ephedrin, Noradrenalin, Amphetamin und N-Chlorpropylamphetamin auf die Hämodynamik des großen und kleinen Kreislaufs an Hunden. Schweiz. med. Wschr. **100**, 2154 (1970).
Krueger, G.A.W.: Die Behandlung der komplizierten Adipositas mit Menocil in der Allgemeinpraxis. Landarzt **44**, 896-900 (1968).
Laissue, J.: Ergebnisse pathologisch-anatomischer Untersuchungen bei Patienten mit primär vaskulärer pulmonaler Hypertonie. Schweiz. med. Wschr. **100**, 2152-2153 (1970).
Lang, E., Haupt, J., Köhler, J.A., Schmidt, J.: Cor pulmonale durch Appetitzügler? Münch. med. Wschr. **111**, 405-412 (1969).
Leuschner, F., Otto, H., Wagener, H.H.: Die Verträglichkeit von Aminorex und einigen weiteren Verbindungen bei länger dauernder Verabreichung an Beagle-Hunde. Naunyn-Schmiedebergs Arch. Pharmacol. **266**, 391-392 (1970).
Liebow, A.A., Downing, S.E.: Cardiopulmonary disease. Chap. 21. In: Pathology of the heart and blood vessels, 3rd ed. Gould, S.E. (ed.). Springfield: Charles C Thomas 1968.

Loogen, F.: Primäre pulmonale Hypertonie. Ber. Kommission dtsch. Ges. Kreislaufforsch. (1972).
Loogen, F., Kübler, W.: Zum Thema primäre pulmonale Hypertonie. Z. Kreislaufforsch. **59**, 865-867 (1970).
Lüllmann, N.H., Parwaresch, M.R., Sattler, M., Seiler, K.U., Siegfriedt, A.: The effect of anorectic agents on the pulmonary pressure and morphology of rat lungs after chronic administration. Arzneimittel-Forsch. **22**, 2096-2099 (1972).
Majno, G., Gabbiani, G., Hirschel, B.J., Ryan, G.B., Statkov, P.R.: Contraction of granulation tissue in vitro: similarity to smooth muscle. Science **173**, 543-550 (1971).
Majno, G., Shea, S.M., Leventhal, M.: Endothelial contraction induced by histamine-type mediators. An electron microscopic study. J. Cell Biol. **42**, 647-672 (1969).
Masshoff, W., Röhrer, H.D.: Pulmonaler Hochdruck bei Glomangiose der Lungen. Klin. Wschr. **42**, 655-663 (1964).
McGee, W.G., Ashworth, C.T.: Fine structure of chronic hypertensive arteriopathy in the human kidney. Amer. J. Path. **43**, 273-299 (1963).
Meessen, H.: Discussion in Baghirzade, M.F., and Donath, K. (1969).
Merkow, L., Kleinerman, J.: An electron microscopic study of pulmonary vasculitis induced by monocrotaline. Lab. Invest. **15**, 547-564 (1966).
Meyrick, B., Clarke, S.W., Symons, C., Woodgate, D.J., Reid, L.: Primary pulmonary hypertension. A case report including electronmicroscopic study. Brit. J. Dis. Chest **68**, 11-20 (1974).
Mielke, H., Seiler, K.U., Stumpf, U., Wassermann, O.: Über die Beziehung zwischen dem Serotoninstoffwechsel und der pulmonalen Hypertonie bei Ratten nach Gabe verschiedener Anorektika. Z. Kardiol. **62**, 1090-1098 (1973).
Mönckeberg, J.G.: Über die genuine Arteriosklerose der Lungenarterie. Dtsch. med. Wschr. **33**, 1243-1246 (1907).
Moss, N.S., Benditt, E.P.: The ultrastructure of spontaneous and experimentally induced arterial lesions. II. The spontaneous plaque in the chicken. Lab. Invest. **23**, 231-245 (1970).
Naeye, R.L.: Pathology of the pulmonary circulation. Ch. 12. In: The Lung. International Academy of Pathology Monograph No. 12. Liebow, A.A., Smith, D.E. (eds.). Baltimore: Williams and Wilkins 1968.
Nager, F.: Diskussionsbeitrag at round table: "Die pulmonale Hypertonie". Schweiz. med. Wschr. **100**, 155-156 (1970).
Norderø, E., Müller, B.: Anoreksikum som mulig årsag til hypertension arteriae pulmonalis. Ugeskr. Laeg. **132**, 382-386 (1970).
Obiditsch-Mayer, I.: Lingulabiopsien bei Patienten mit primärer pulmonaler Hypertonie. Wien. Z. inn. Med. **50**, 486-489 (1969).
Obiditsch-Mayer, I., Kletter, G.: Morphologische Untersuchungen an bioptischem und Sektionsmaterial bei pulmonaler Hypertonie nach "Menocil"-Medikation. Verh. dtsch. Ges. Path. **56**, 509-512 (1972).
Ostertag, H.: Lungenveränderungen nach Menocil-Einnahme. Vortrag auf der Herbsttagung Deutscher Pathologen in Wiesbaden (1969).
Parker, F., Odland, G.F.: A correlative histochemical, biochemical and electron microscopic study of experimental atherosclerosis in the rabbit aorta with special reference to the myo-intimal cell. Amer. J. Path. **48**, 197-239 (1966).
Pasi, A.: Zur toxischen Gefährdung durch Appetitzügler. Schweiz. Apoth.-Ztg. **105**, 890-897 (1967).
Pease, D.C., Molinari, S.: Electron microscopy of muscular arteries; pial vessels of the cat and monkey. J. Ultrastruct. Res. 3, 447-468 (1960).
Pendl, O.: Diskussionsbeitrag, 11. Kardioangiologische Diskussion der Österreichischen Kardiologischen Gesellschaft in Wien; Wien. Z. inn. Med. **50**, 508-509 (1969).
Peters, L., Gourzis, J.T.: Pharmacological, toxicological and clinical observations with Aminorex in the United States. Symposium on Obesity, Circulation and Anorexigens, Bürgenstock 15.-16.5.1970 (see Symposia: BÜRGENSTOCK).
Poos, G.J., Carson, J.R., Roseman, J.D., Roszkowski, A.P., Kelley, N.M., McGowin, J.: 2-Amino-5-aryl-2-oxazolines. Potent new anorexigens. J. med. Chem. **6**, 266-272 (1963).
Rivier, J.L.: Hypertension artérielle pulmonaire primitive. Schweiz. med. Wschr. **100**, 143-145 (1970).

Rivier, J.L., Jaeger, M., Reymond, Cl., Desbaillets, P.: Hypertension artérielle pulmonaire primitive et anorexigène. Arch. Mal. Coeur **65**, 787-796 (1972).
Romberg, E.: Über die Sklerose der Lungenarterie. Dtsch. Arch. klin. Med. **48**, 197-206 (1891).
Roos, B., Steen, B., Svanborg, A.: Prövning av ett aptitnedsättande preparat (aminorex). Läkartidningen **67**, Suppl. II, 96-99 (1970).
Rutishauser, E., Blanc, W.: Anastomoses artério-veineuses glomiques du poumon avec syndrome d'insuffisance droite et cyanose. Schweiz. Z. Path. **13**, 61-65 (1950).
Rutishauser, E., Feuardent, R.: Schnittrekonstruktion glomusartiger Bildungen (Anastomositis) der Lunge. Schweiz. Z. Path. **15**, 641-644 (1952).
Sandoval, R.C., Wang, R.J.H., Rimm, A.A.: Body weight changes in overweight patients following an appetite suppressant in a controlled environment. J. clin. Pharmacol. **11**, 120-124 (1971).
Schulz, H.: Diskussionsbeitrag in Baghirzade, M.F., and Donath, K. (1969).
Schuster, H.P., Lang, K., Baum, P., Just, H.G.: Akute Intoxikation mit Aminorexfumarat. Bericht über den Verlauf eines Falles mit schwerer suizidaler Vergiftung ohne Anzeichen einer pulmonalen Hypertonie. Med. Welt **40**, 2212-2215 (1969).
Scott, R.F., Daoud, A.S., Florentin, R.A.: Animal models in atherosclerosis. In: The Pathogenesis of Atherosclerosis. Wissler, R.W., Geer, J.C., Kaufman, N. (eds.). Baltimore: Williams and Wilkins 1972, Ch. 4, pp. 120-146.
Smith, P., Heath, D.: Aminorex and the pulmonary circulation of rats fed on a high fat diet. Arzneimittel-Forsch. **24**, 1277-1279 (1974).
Smith, P., Heath, D., Kay, J.M., Wright, J.S., McKendrick, C.S.: Pulmonary arterial pressure and structure in the patas monkey after prolonged administration of aminorex fumarate. Cardiovasc. Res. **7**, 31-38 (1973).
Smith, U., Ryan, J.W., Michie, D.M., Smith, D.S.: Endothelial projections as revealed by scanning electron microscopy. Science **173**, 925-927 (1971).
Spencer, H.: Pathology of the lung, 2nd ed. Oxford: Pergamon Press 1968.
Spiro, D., Lattes, R.G., Wiener, J.: The cellular pathology of experimental hypertension. I. Hyperplastic arteriolarsclerosis. Amer. J. Path. **47**, 19-49 (1965).
Steim, H.: Zur Ätiologie der primären Pulmonalhypertension. Wien. Z. inn. Med. **50**, 497-499 (1969).
Stemerman, M.B., Ross, R.: Experimental arteriosclerosis. I. Fibrous plaque formation in primates, an electron microscope study. J. exp. Med. **136**, 769-789 (1972).
Stepanek, J.: Pulmonale Hypertonie nach chronischer Aminorexgabe beim Hund. Vorläufige Mitteilung. Schweiz. med. Wschr. **103**, 1793 (1973).
Stepanek, J., Zak, F.: Zweijährige perorale Applikation von Aminorex am Hund. Z. Kardiol. **64**, 749-781 (1975).
Stepanek, J., Zolliker, V.R.: Kreislaufwirkungen des Anorektikum Diphemethoxidin im Vergleich zur Wirkung von Amphetamin und Aminorex. Arch. int. Pharmacodyn. **191**, 376-399 (1971).
Stühlinger, W., Raberger, G., Kraupp, O.: Die Wirkung von Aminorex (Menocil®) auf die Hämodynamik der Lunge. Naunyn-Schmiedebergs Arch. Pharmacol. **264**, 314-315 (1969).
Stühlinger, W., Raberger, G., Kraupp, O.: Vergleichende Untersuchungen über die Wirkung von Aminorex, Amphetamin und Ephedrin auf die Hämodynamik des großen und kleinen Kreislaufes an Hunden. Z. Kreislaufforsch. **60**, 73-82 (1971).
Stühlinger, W., Turnheim, W., Raberger, G., Kraupp, O.: Die retrograde Beeinflussung des pulmonalen Gefäßwiderstandes durch Änderungen der Hämodynamik des linken Herzens als Ursache der unterschiedlichen Wirkung verschiedener Sympathomimetika auf den kleinen Kreislauf narkotisierter Hunde. Naunyn-Schmiedebergs Arch. Pharmacol. **266**, 463-464 (1970).
Takayanagi, T., Rennels, M.L., Nelson, E.: An electron microscopic study of intimal cushions in intracranial arteries of the cat. Amer. J. Anat. **139**, 415-429 (1972).
Takebayashi, S.: Pathogenesis of obliterating endarteritis in stomach ulcer. Electron microscopic study. Acta path. jap. **20**, 37-46 (1970).
Takebayashi, S., Themann, H., Manitz, G., Giese, W., Backwinkel, K.P.: Ultrastructural study of arteriolar lesions in human kidney: concerning the histogenesis of hyaline necrosis and other media alterations. Beitr. Path. **144**, 1-22 (1971).

Thomas, W.A., Jones, R., Scott, R.F., Morrison, E., Goodale, F., Imai, H.: Production of early atherosclerotic lesions in rats characterized by proliferation of "modified smooth muscle cells". Exp. molec. Path. Suppl. **1**, 40-61 (1963).
Todd, M.E., Friedman, S.M.: The ultrastructure of peripheral arteries during the development of DOCA hypertension in the rat. Z. Zellforsch. **128**, 538-554 (1972).
Trell, E.: Primary and chronic thromboembolic pulmonary hypertension. Nosologic spectrum. Angiology **23**, 558-574 (1972).
Valdivia, E., Lalich, J.L., Hayashi, Y., Sonnad, J.: Alterations in pulmonary alveoli after a simple injection of monocrotaline. Arch. Path. **84**, 64-76 (1967).
Wade, G., Ball, J.: Unexplained pulmonary hypertension. Quart. J. Med. **26**, 83-119 (1957).
Wagenvoort, C.A.: Vasoconstriction and medial hypertrophy in pulmonary hypertension. Circulation **22**, 535-546 (1960).
Wagenvoort, C.A., Heath, D., Edwards, J.E.: The pathology of the pulmonary vasculature. Springfield: Charles C Thomas 1964.
Wagenvoort, C.A., Wagenvoort, N.: Primary pulmonary hypertension: a pathologic study of the lung vessels in 156 clinically diagnosed cases. Circulation **42**, 1163-1184 (1970).
Weibel, E.R.: Principles and methods for the morphometric study of the lung and other organs. Lab. Invest. **12**, 131-155 (1963).
Weibel, E.R.: Morphometry of pulmonary circulation. Internat. Symposium pulmonary Circulation, Prague, 1969. Progr. Resp. Res. Basle: Karger 1970, Vol. V, pp. 2-12.
Weibel, E.R., Gomez, D.M.: Architecture of the human lungs. Science **137**, 577-585 (1962).
Weibel, E.R., Kistler, G.S., Scherle, W.F.: Practical stereological methods for morphometric cytology. J. Cell Biol. **30**, 23-38 (1966).
Weissel, W.: Primäre pulmonale Hypertension. Diskussionsbemerkung. Wien. Z. inn. Med. **50**, 501-502 (1969).
Widgren, S., Kapanci, Y.: Hypertension pulmonaire en relation avec l'absorption d'un médicament anorexigène. Communication à l'assemblée annuelle de la Société suisse de Pathologie, Baden, 31.10.1969.
Widgren, S., Kapanci, Y.: Menocilbedingte pulmonale Hypertonie. Vorläufige morphologische Ergebnisse über 8 pathologisch-anatomisch untersuchte Fälle. Z. Kreislaufforsch. **59**, 923-929 (1970a).
Widgren, S., Kapanci, Y.: Pulmonary hypertensive disease in connection with the absorption of Menocil. Morphologic and preliminary quantitative investigations of 22 autopsied cases. Symposium on Obesity, Circulation and Anorexigens, Bürgenstock, 15.-16.5. 1970b.
Widgren, S., Kapanci, Y.: Pulmonary arterial hypertensive disease in patients treated with aminorex. Electron microscopic and morphometric studies in fatal cases observed in Switzerland. 11th Conference of the International Society of Geographical Pathology, Newcastle-upon-Tyne, August 17-22, 1972.
Wiener, J., Spiro, D.: The fine structure of experimental vascular disease. In: Gould, S.E. (ed.): Pathology of the Heart and Blood Vessels, 3rd ed. Springfield: Charles C Thomas 1968, Ch. 24.
Will, J.A., Bisgard, G.E.: Haemodynamic effects of oral aminorex and amphetamine in unanaesthetized beagle dogs. Thorax **27**, 120-126 (1972).
Wirz, P., Arbenz, U.: Primär vaskuläre pulmonale Hypertonie in der Schweiz, 1965-1970. Schweiz. med. Wschr. **100**, 2147-2150 (1970).
Wissler, R.W.: The arterial medial cell, smooth muscle or multifunctional mesenchyme? J. Atheroscler. Res. 8, 201-213 (1968).
Wood, L.C., Owen, J.A., Jr.: Clinical evaluation of a new anorexigenic drug, aminoxaphen, in obese patients. J. New Drugs 5, 181-185 (1965).
Yelnosky, J., Hewson, R.J., Mundy, J., Mitchell, J.: The cardiovascular effects of Aminorex, a new anorexigenic agent. Arch. int. Pharmacodyn. **164**, 412-418 (1966).
Zwahlen, A.: La morphométrie: méthode d'évaluation morphologique de la fonction. Thèse Genève, No. 3379 (1974).

DNA Injuries, Their Repair, and Carcinogenesis

J.L. VAN LANCKER

I. Introduction

The cancer cell is characterized by a distortion of gene expression that makes it escape normal control of at least some cellular functions, and provides it with survival advantages over many, if not most, cells of the organism. The special properties of the cancer cells are transferred from one generation of cells to the next. The conversion of a normal into a cancer cell is achieved by DNA and RNA viruses, ultraviolet light, ionizing radiation, and chemical carcinogens. This process is usually believed to occur in two stages: initiation and promotion. Although the molecular events responsible for initiation and promotion are not known, it seems logical to assume that the initiation process results from the interaction between the physical, biological, or chemical agent and the genome (for review [316]).

The nonviral carcinogens have at least one property in common: they all cause damage to DNA. UV light induces the formation of numerous photo products [42, 56, 269, 273, 316, 317], among them thymine dimers. In addition to single- and double-strand breaks, X-radiation causes base alterations (those involving the thymine residues have been most extensively studied) [277, 317], and interstrand cross-links. Most, if not all, chemical carcinogens are believed to be converted to electrophiles that will react with nucleophiles [95, 100, 131, 163, 210, 211, 265] (Fig. 1). For example, the conversion of acetylaminofluorene to the ultimate carcinogen has been described in detail [205, 207, 208, 209, 212], and is summarized in Figure 2.

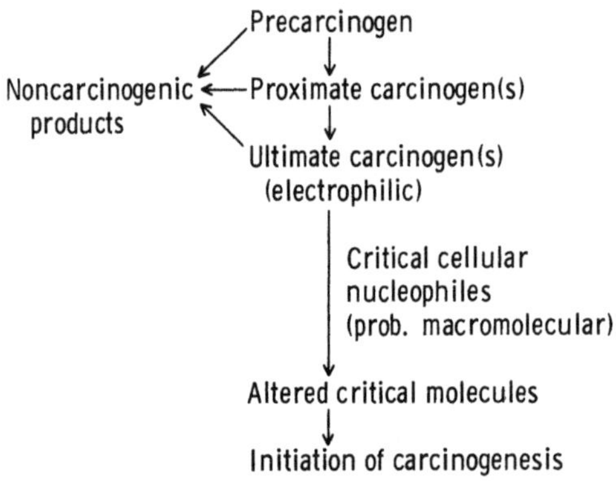

Fig. 1. Scheme for activation and detoxication of chemical carcinogens

Miller has proposed a working hypothesis for the mode of action of the amino dyes and the aminofluorenes. When efforts were made to prepare N-hydroxy methylaminobenzene (MAB), the compound was found to be too unstable to isolate. To obviate this difficulty, *Miller* synthesized a compound that presumably would yield the N-hydroxy MAB in vivo. N-Benzoyloxy MAB, the benzoester of N-hydroxy MAB, was synthesized. Somewhat to *Miller*'s surprise, this compound was a more active carcinogen than the N-hydroxy MAB. Benzoyloxy MAB produces carcinomas at the injection site and even kidney cancers in young rats. In vitro, N-benzoyloxy MAB reacts nonenzymatically at a neutral pH with DNA, RNA, guanosine, tryptophan, tyrosine, methionine, cystine, and a number of other nucleophiles. The molecular complex formed in such reactions is not always known. In methionine it is speculated that a sulfonium derivative is formed, the 3-(methione-s-yl) MAB, which decomposes to yield the 3-methyl-mercapto MAB. The sequence of metabolic changes that occur in vivo is illustrated in Figure 3.

Similarly, esters of N-hydroxy derivatives of fluorene are more carcinogenic than AAF and N-hydroxy AAF in the sense that the former produce carcinomas at the injection site. The N-acetoxy AAF has properties reminiscent of those of the N-benzoyloxy MAB and reacts with nucleophiles. With methionine, the ester yields the 0-(methionine-8-yl) AAF, which decomposes to yield 0-methyl-mercapto-AAF. On the basis of the findings made with both AAF and MAB, *Miller* and his associates [208] have proposed a working hypothesis suggesting that the carcinogen is converted in a first step to an N-hydroxy derivative of the amine or the amide, which in turn is converted to a reactive ester that is capable of

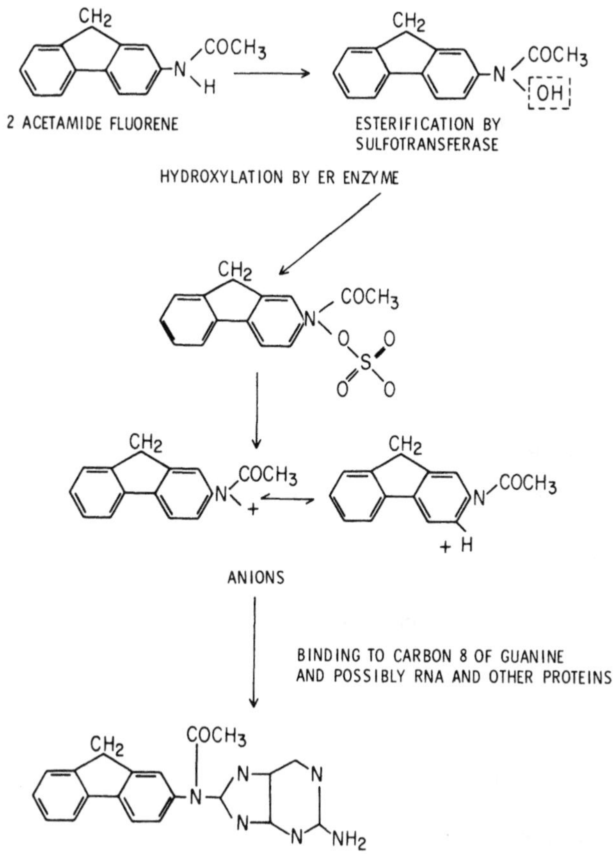

Fig. 2. Activation scheme of AAF. (*Miller* and *Miller*, 1969)

attacking nucleophiles such as DNA, RNA, and proteins. The binding then yields a cell with permanent losses of macromolecules regulating growth. A general scheme for the metabolism of aromatic compounds has also been proposed (Fig. 4). The conversion of aromatic compounds made of one or more benzene rings to the corresponding phenols involves the formation of highly reactive arene oxides in which an aromatic double bond is converted to an epoxide. The fate of the epoxide is multiple: (1) it may isomerize spontaneously and yield the phenol; (2) it may be attacked by a microsomal enzyme, "epoxide hydrase" which will convert it to the *trans*dihydrodiols; (3) it may spontaneously react with nucleophiles including the thiol group of glutathione, protein, and nucleic acids; or (4) it may be complexed enzymatically to glutathione by an enzyme found in the cytosol, glutathione-S-epoxide transferase.

The epoxide hydrase has been partially purified from rat liver, but it is likely that more than one form of the enzyme exists. Similarly, at least two and possibly three different glutathione transferases have been found in rat liver.

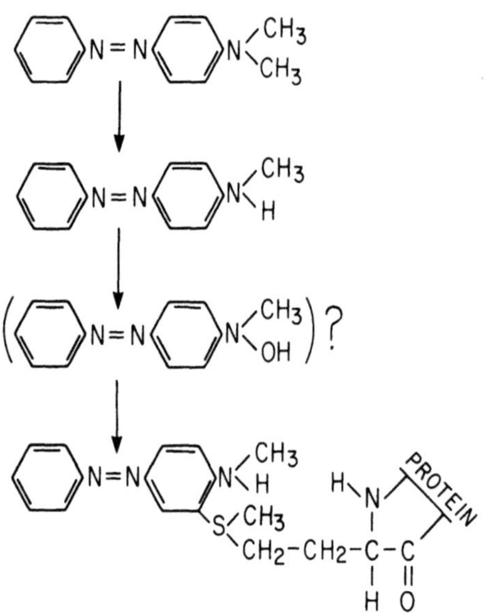

Fig. 3. Probable metabolic sequence for aromatic amino acids

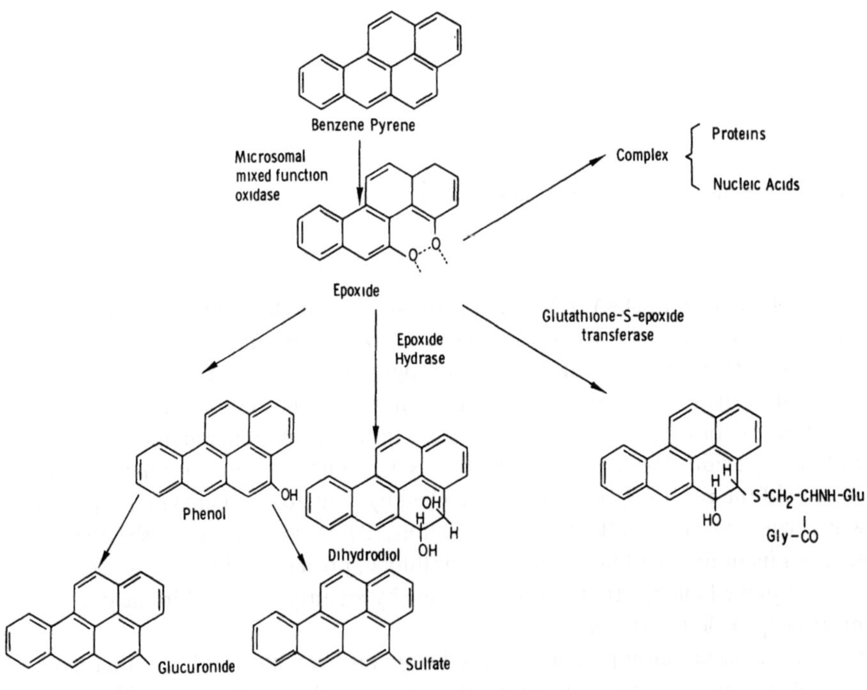

Fig. 4. Metabolism of K region epoxide

The intermediate formation of K region epoxide has been convincingly established (using liver microsomal preparations in which the epoxide hydrase was inhibited) for naphthalene, phenanthrene, benz(a)anthracene, dibenz(a)-anthracene, pyrene, 4,5 benzo-(a) pyrene, 7,12 dimethylbenz(a)anthracene, and other substituted benz(a)anthracenes.

The significance of these findings resides in the fact that the K region epoxide might constitute the proximate carcinogen. Although initial experiments in which K epoxides were administered in vivo showed that the epoxide was less active than the initial aromatic hydrocarbon, K region epoxides of benz(a)anthracene, dibenz(a)anthracene, and 3 methylcholanthrene were shown to be more active in transforming cells in vitro than the parent hydrocarbon. An important consideration in the potential carcinogenicity of a given hydrocarbon is that the steady-state concentration of the arene oxide depends not only upon the rate at which it is formed, but also upon its proximity to critical macromolecules and its rate of conversion to phenol, dihydrodiols, etc. Therefore, the potential susceptibility of a given tissue to a specific aromatic hydrocarbon will be a function of the permeability of the cell to the compound and the interreaction between the activities of the mixed-function oxidase, the epoxide hydrase, and the glutathione transferase. The situation will be further complicated in that covalent binding of the epoxide, for example to DNA, can at least in part be repaired.

Although most chemical carcinogens also bind to protein and mucopolysaccharides [205, 206, 213], binding to DNA is assumed to be of special significance because it might cause somatic mutations which might then trigger initiation of the carcinogenic event.

If there is little doubt that chemical carcinogens alter the expression of the genotype into the phenotype probably by modifying the genome, the precise molecular alteration responsible for the changes in gene expression are unknown. They could be of at least three different types: a somatic mutation resulting from direct damage of DNA; a rearrangement of the regulatory proteins controlling gene expression as a result of deletion or damage to the proteins; a permanent alteration of gene expression resulting from a disruption of yet unknown feedback loops, for example, between the cell membrane and the genome, or a combination of two or more of these events.

Thus, the initiation of cancer by nonviral carcinogens can be studied by different approaches: the cooperation between carcinogen and virus; the abolition of immunosurveillance, and the direct binding of the carcinogens to macromolecules. Each approach has its merits, and no one can at this point predict which mechanism or mechanisms will prevail in converting a normal into a cancer cell. This discussion will center on the role of the binding of carcinogens to macromolecules in the initiation of cancer.

II. Somatic Mutation Theory

Because of unlimited variability in the morphological appearance of the cancer cell and the observation that the morphological changes were transferred from one generation of cells to another, *Bauer* in 1928 [11] and *Bovery* in 1929 [28] independently proposed the somatic mutation theory for carcinogenesis. The evidence for the somatic mutation theory was circumstantial for a long time; it included the observation that the production of cancer in mice is linked to the occurrence of two consecutive random mutations [51, 143]. New interest in the somatic mutation theory was instigated by the discovery that nonviral carcinogens damage DNA (see above), and that they act as mutagens [171, 303].

Although *Burdette* concluded in 1955 [34] that there was no correlation between mutagenicity and carcinogenicity, this conclusion was premature because little was known of the molecular conversion of the carcinogen to the ultimate carcinogen, and nothing was known of the mechanisms of DNA repair.

If it cannot be excluded that chemical carcinogens act indirectly by facilitating the incorporation of a virus (DNA virus or an RNA provirus into the genome), or by abolishing immunosurveillance of cells which are already initiated, neither can it be excluded that carcinogens act directly.

At least three types of findings support this view: (1) the demonstration that many carcinogens need to be metabolically converted to exert their nefarious effects, which include binding to macromolecules, (2) the demonstration that many known carcinogens are also mutagens, (3) the demonstration that carcinogens bound to DNA can be removed by at least two different repair mechanisms, excision and postreplication repair, and that in absence of repair, the incidence of cancer is greater and the latent period for the appearance of cancer is reduced.

III. Mutation and Carcinogenesis

Mutations can be grouped into two major categories: intragenic (affecting the DNA) and chromosomal (affecting the entire chromosome). There are two classes of intragenic mutations: frameshift mutation and base pair substitutions. In frameshift mutation a small number of base pairs are either deleted or added. The addition or the deletion of one single base pair makes the reading of each codon incorrect beyond the point of mutation, because the reading is shifted forward one base in the case of base pair deletion, and backward one base in the case of base addition. As a result of such mutation a large number of greatly altered proteins are coded for.

Base pair mutations have been grouped into transition mutations (replacement of one purine or pyrimidine base by another purine or pyrimidine base respectively) and transversion mutations (in replacement of a purine base by a pyrimidine base, or vice versa) (for review, see [171, 303]).

Changes in chromosomes will not be discussed here. It suffices to point out that they include changes in groups of genes on a chromosome, deletion, duplication, or rearrangement of genes in chromosomes, loss of chromosomes (monosomy), or gain of a chromosome (trisomy).

Legator and *Flamm* [171] have reviewed the chemical mutagens inducing point mutations and the nature of their mutagenic effect. They discuss (1) base analogues (bromodeoxyuridine (BUDR), bromouracil, chlorouracil, iodouracil, and two amino purines). These compounds replace pyrimidine or purine bases in the DNA, thus causing a transition. In fact, BUDR can replace all the thymidine in the DNA and form hydrogen bonds with adenine. However, the presence of BUDR in the DNA will increase the incidence of mistakes in pairing, and the analogue may pair with guanine. (2) Acridine, proflavine, hycanthone, and other intercalating agents (substances that stick in between the bases of the DNA duplex without forming covalent bonds) cause unwinding of the helix, which probably results in deletion or addition of base pairs, and consequently leads to a frameshift mutation. (3) Nitrous acid deaminates adenine, guanine, and cytosine, leading to a variety of transitions. Hydroxylamine under the proper condition leads to the transition of cytosine to thymine with resulting mispairing (Table 1). (4) Alkylating agents which react with DNA will be discussed later. It suffices to point out that their mutagenic effect depends upon whether they are mono- or bifunctional. In the first case they may cause base alterations that lead to transition; in the second they cause cross-links which may or may not be associated with depurination (deletion if not repaired) and frameshift mutations.

Table 1.

Nitrous acid	Adenine → Hypoxanthine	AT → GC
	Cytosine → Uracil	GC → AT
	Guanine → Xanthine	no change
Hydroxylamine	Cytosine → Thymine	CG → TA

The first attempts to test the mutagenic properties of known carcinogens were disappointing mainly because bacteria are devoid of the biochemical machinery necessary to convert the inactive to the active carcinogen. It was only after exposure of the natural compound (e.g., acetylaminofluorene) to microsomal enzymes that compounds were formed which were mutagenic for bacteria. In other cases the active product was extracted from urine after administration of the carcinogen [68]. *Ames* used histidine-negative *Salmonella thyphimurium* strains to test the mutagenicity of carcinogens. These are auxotrophs in which molecular changes of DNA in the histidine operon are well-characterized. *Ames* et al. [3] tested 20 carcinogens in their system and found that 20 of the established carcinogens, or ultimate carcinogens, were also mutagenic.

In other experiments *Escherichia coli* mutants lacking polymerase I (pol A) (31-32) have been used to screen some carcinogens; namely, derivatives of nitrofuran. It was found that the nitrofuran derivatives were mutagenic for *E. coli* but not for *S. thyphimurium.* These results indicate that detection of potential carcinogens by bacterial mutation may require that several different bacterial systems be used.

The (pol A) strain isolated by *Cairns* from *E. coli* W 3110 (pol A+) was used to test the mutagenic properties of a greenhouse fungicide: Captan. The fungicide is mutagenic in bacteria, is teratogenic in bird embryos, and is an alkylating agent in mammals. Onė can hardly ask for more evidence leading one to suspect that Captan might be carcinogenic for humans.

The simple observation that carcinogens are mutagens is of considerable significance because it provides further evidence for the mutation theory, and it may help to identify the molecular damage caused by carcinogens. One may, however, wonder whether data collected on bacteria can be extrapolated to man. The uptake, metabolism, and excretion of carcinogens in mammals varies considerably from that in bacteria, and therefore, the correlation can only be qualitative. Moreover, the effect of a carcinogen in vivo may be altered by promotors which may themselves not be mutagenic. There are, in fact, several carcinogens that were inactive as mutagens in bacteria but are carcinogenic in humans, and a few compounds that are mutagenic in bacteria are not carcinogenic in mammals because they are effectively detoxified. Therefore, studies of bacterial mutagenesis can only serve as a preliminary screening technique which must be followed by testing on mammalian cells, and ultimately on animals that behave metabolically as closely as possible to humans.

Because of these limitations of the bacterial system, other systems have been devised; namely, mammalian cells in culture- and host-mediated assay. Cells obtained from individuals with known hereditary defects (galactosemia, etc.), cells which have changed while in culture and have become drug resistant, have acquired new nutritional requirements and have changed in thermal sensitivity, have all been used [171, 303].

Auxotroph mutants are also often used. Most cells in culture can synthesize purines and pyrimidines, but as the line developed, some cells lost their ability of synthesizing bases, amino acids, etc., as a result of a spontaneous mutation. Thanks to the advances made in tissue culture techniques, such as single cell plating, clone isolation, and the development of chemically defined media, it is possible to isolate the auxotrophic mutant and define its dependency. In absence of the needed nutrient the dependent cells will not grow. If BUDR is added to the medium, the wild type continues to grow and picks up the BUDR. All cells whose DNA contain BUDR can readily be killed by exposure to near visible light. By transferring the surviving cells to media containing the defective nutrient, the dependent cell lines can be obtained. Auxotrophic mutants dependent on carbohydrate, purine, pyrimidines, amino acids, etc., have been prepared.

Heidelberger and his associates [129] demonstrated in vitro transformation of rodent cells with K region epoxides derived from benz(a)anthracene, dibenz(a)anthracene, and methylcholanthrene. The mutagenicity of the epoxides in a clone of Chinese hamster cells was established by following the production of clones resistant to 8-azoguanine [133].

When a population of 5×10^6 cells was examined for spontaneous mutation with the BUDR technique, none were observed, but a variety of auxotrophic mutants were found when the cells were treated with ethylmethane-sulfonate, N-methyl N^1-nitro-N-nitrosoguanidine, UV light, acridine mustard, N-nitrosodimethylamine, N-nitrosomethylurea, and N-nitrosomethylurethan. These mutants are stable and have a diploid chromosome population [150].

In the host-mediated assay an animal is injected with the mutagen or carcinogen and with a test organism such as *S. tryphimurium, Neurospora crassa*, or *Sacchoromyces cervisiae.* The test system is, after an appropriate time, removed from the host and mutations are scored. Using such an approach it has been established that certain types of mutation occur with certain types of carcinogens. For example, according to *Malling* and *Chu*, polycyclic hydrocarbons induce predominantly base insertion and deletion, while alkyl carcinogen (dimethylnitrosamine) induces base pair substitutions [191].

In the face of these observations, it seems reasonable to assume that the transformation of a normal into a cancer cell results from either a single or multiple somatic mutation coding for proteins whose alteration will lead to repeated errors in transcription. Two such alterations have been proposed: DNA polymerases with a high incidence of replication mistakes have been found in leukemic cells [186]. Mammalian repressor molecules could be made of two units: one whose amino acid sequence is common to many repressors, another which is unique for each repressor. An alteration in the amino acid sequence of the common sequence would interfere with the expression of many genes [314]. Of course, many of the mutations might be lethal, but in some cases the combination of several mutations may provide the normal cell with the survival advantages of the cancer cell.

IV. Interreaction Between Carcinogens and Macromolecules

1. Reaction with DNA

Figure 5 gives some favorite sites for adduct formation by carcinogens on nitrogenous bases. For review, see [95, 131, 163].

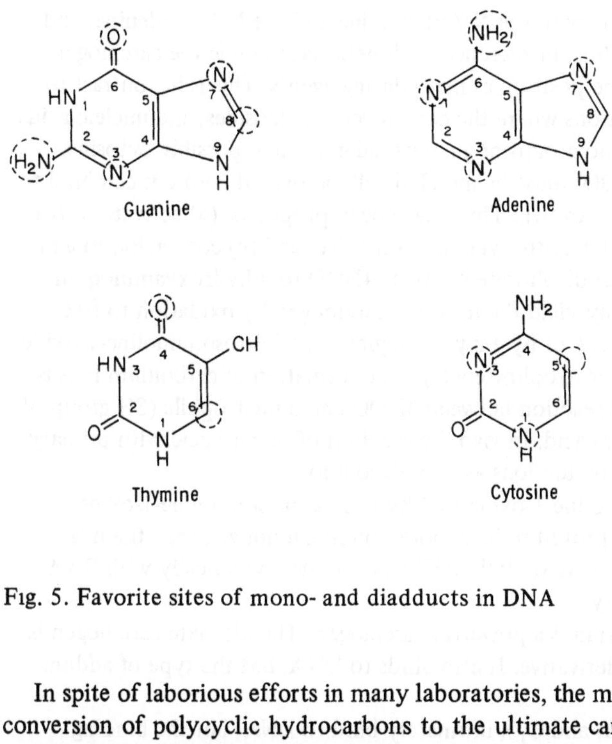

Fig. 5. Favorite sites of mono- and diadducts in DNA

In spite of laborious efforts in many laboratories, the molecular reaction causing the conversion of polycyclic hydrocarbons to the ultimate carcinogen are not known with certainty. Two mechanisms have been proposed: formation of the radical cation, and formation of K region epoxides.

In some cases the binding of carcinogens to DNA has been studied as it occurs in vivo by purifying the DNA, hydrolyzing it, and identifying the type of complex formed by the carcinogen with the nitrogenous bases. In other cases in vitro models were used to study the binding with DNA. In the latter case it is imperative to determine whether the findings made in vitro obtain in vivo.

The work of *Ts'o* and his collaborators [304, 315], although done in vitro, may shed some light on the mode of binding of the carcinogen in vivo.

In these experiments polycyclic hydrocarbons, such as benzo(a)pyrene and 3-methylcholanthrene, react with DNA in two steps: intercalation and covalent binding as a result of a metabolic activation which yields free radical formation, or through radiation activation. Yet, the molecular structure of the adducts formed by benzo(a)pyrene and 3-methylcholanthrene in vivo with DNA remains unknown. However, in vitro a covalent bond with the C-6 of thymine is formed.

Methylated hydrocarbons such as 7,12-dimethylbenz(a)anthracene are believed to be converted to the ultimate carcinogen by the hydroxylation of the hydroxymethyl, followed by esterification of the hydroxymethyl. These reactions lead to the formation of a reactive carbonium ion. However, identification and enzymic formation of such esters has not been demonstrated: 7-bromomethylbenz(a)anthracene and 7-bromomethyl-12-methylbenz(a)-anthracene yield 2 aminoguanine, 3 aminoadenine, and 4 aminocytosine adducts in vitro.

Dipple et al. [84] have shown that in vitro 7-bromomethylbenz(a)anthracene reacts in the presence of dimethylacetamide with the N-7 of guanine and the N-1 of adenine, and the N-3 of cytosine derivatives. Thus, in presence of dimethylacetamide the carcinogen reacts with nucleosides in the same position as methylating agents. This is in contrast to what is observed in aqueous solutions where the carcinogen, nucleosides, and nucleic acids lead to adducts involving the amino groups of guanine, adenine, and possibly cytosine.

4-Nitroquinoline-1-oxide (4NQO) must be metabolically activated before it can bind to DNA. The mode of activation is unknown. Three have been proposed: (a) activation to a carbonium ion, (b) reduction to the hydroxyamino derivative, and (c) conversion to a nitroso derivative. Bacterial and liver diaphorase converts 4NQO to 4-hydroxyaminoquinoline-1-oxide (4OHAQO) which may yield the ultimate carcinogen by oxidation to free radicals by the formation of diesters, or by enzymic acylation. 4-Nitrosoquinoline-1-oxide has been detected by rapid scan spectrophotometry. The formation of carbonium ions is believed to involve a replacement reaction between 4NQO and a nucleophile (SH group of proteins) with liberation of nitrous acid. In vivo the reaction of nitrous acid with primary amine diazonium ions forms carbonium ions as they decompose.

Rat tissues convert 4-nitroquinoline-1-oxide to 4-hydroxyaminoquinoline-1-oxide (4OHAQO). The new metabolite proved to be a more potent carcinogen than the nitro derivative (42). The diacetyl derivative of 4OHAQO reacts nonenzymatically with DNA, RNA, and methionine at neutrality.

The antibacterial agent nitrofuran is a powerful carcinogen. The ultimate carcinogen is believed to be a hydroxyamino derivative. It also binds to DNA, but the type of adduct formed is unknown.

The fungal toxins, aflatoxin (B_1 and C_1), ultimately bind to DNA, but the binding sites and intermediates are unknown.

The hepatocarcinogen extracted from sassafras oil, safrole, binds to DNA after conversion to the hydroxy derivative 1'-hydroxysafrole. All the adducts are not known, but the hydroxysafrole seems to bind to dGMP, and the major product involves a substitution on 0-6 of guanine.

We have already mentioned that 2-acetylaminofluorene and related compounds need to be converted to the N-hydroxy derivatives prior to reacting with cellular macromolecules. A second step is believed to be involved in the activation of the molecules prior to binding to DNA; namely, esterification of the N-hydroxy derivative. A number of esters, the N-hydroxy derivatives, have been found in tissues: glucuronide, sulfate, and O-acetyl derivative diacetyl coenzyme A, carbamyl phosphate, and acetyl phosphate. It is not known which of these esters is in vivo involved in the binding to DNA. What is certain is that the major adduct of the acetylaminofluorene derivatives to DNA occurs at the C-8 position of guanine (80%). A minor adduct occurs in the same position, but instead of involving the amino nitrogen of the fluorene ring, it is believed to involve an interreaction between fluorene rings [162]. By using physicochemical methods such as circular dichroism, proton magnetic resonance spectroscopy, or molecular models of polynucleotides bound to AAF, it has been proposed that the binding of 2AAF to guanine brings the guanine to rotate around the glycosidic bond at the N-9. This results in a distortion for the DNA double helix [221].

Before it can bind to DNA, 2-naphthalamine needs to be converted to an ultimate carcinogen. The proximate carcinogen is believed to be 2-amino-1-naphthol. The binding of β-naphthalamine derivatives to DNA is unknown.

The in vivo activation of 4-dimethylaminoazobenzene (DAB) involves (1) mono-N-demethylation, (2) hydroxylation, and (3) esterification to yield a sulfur ester. When N-methyl-4-aminoazobenzene labeled in the prime ring was injected into rats, *Miller* et al. [214] demonstrated binding to DNA with the formation of an N-(deoxyguanosine-8-yl)-N-methyl-4-aminoazobenzene [257].

Ortho-aminoazotoluene, the first carcinogen shown to induce liver cancer in rats [263] has also been found to bind covalently to DNA, but the nature of the interreaction is unknown.

Dimethyl- and diethylnitrosamine, probably after conversion to an active derivative through oxidative dealkylation, yield the 7-methyl-, or the 7-ethylguanine as the major adducts, among a number of others However, structure of the ultimate carcinogen has not been established; it is suspected to be a diazomethane.

Methylating and ethylating alkylating agents primarily produce 7-alkylguanine adducts in the DNA (over 70%). The exact reasons for the high sensitivity of the N-7 of guanine are not clear. Two different mechanisms have been proposed: (a) Guanine donates 2 hydrogen bonds and accepts only one. This property increases the electron density over the guanine ring and renders the N-7 nucleophilic. (b) It has also been proposed that the reactivity of the N-7 of guanine results from its peripheral position in the wide groove of the Watson-Crick model. However, substitution on 0-6, N-3, and $2NH_2$ of guanine, of N-1, N-3, N-7, and $6NH_2$ of adenine, and N-1, N-3, and C-5 of cytosine, as well as 0-4 of thymine, have been reported. Although there have been claims that alkylation of the RNA takes place with the various alkylating agents, *Lowley* and *Brookes* [168, 169] believe that RNA alkylation is negligible.

The alkylating agent, β-propiolactone, does not require activation for alkylation; 7-ethyl (2-carboxyethyl) guanine is formed in both DNA and RNA.

After a catabolic activation of unknown nature, 1-2 dimethylhydrazine yields 7-methylguanine in DNA.

Methylazoxymethanol (MAM), the aglycone of cycasin, causes the nonenzymic formation of methylguanine.

Urethan is not a direct carcinogen; it is N-hydroxylated to yield N-hydroxyurethan which is further esterified to an ethyoxycarboxylating agent which, after losing carbon dioxide, forms the ethylating species. The administration of radioactive urethan yields various DNA adducts in rats, but the molecular structure of the adducts remains unknown. It should be remarked, however, that *Boyland* and *Williams*, in 1969 [29], demonstrated the formation of ethyl ester of cytosine-5-carboxylate in mouse liver RNA after the administration of 1-^{14}C-ethylcarbamate, or ethyl-(carboxyl-^{14}C) carbamate.

The nitrogen in the sulfur mustards esterifies the primary phosphate group of DNA and produces interstrand and intrastrand crosslinks by forming diguanine-ethylsulfide. The incidence of interstrand crosslinks in DNA duplexes is higher than that of intrastrand crosslinks in denatured DNA. As may be expected the incidence of guanine alkylation will vary with the GC content of the DNA; the greater the GC content the greater the guanyl alkylation.

Although all four bases of the DNA can be the site of adducts, the phosphodiester backbone of DNA is also a target for at least some carcinogens, and there is evidence that both UV light and X-radiation cause damage to the deoxyribose sugar [277].

Chemicals may bind to DNA noncovalently through intercalation between the base pairs of the DNA duplex, or through adlineation. In adlineation the chemical is disposed perpendicular to the base pair and reacts with base sites not involved in base pairing. In intercalation the chemical is disposed in the plane of the base pairs.

Intercalation occurs with acridine dyes, antibiotics, alkaloids, polycyclic hydrocarbons, and actinomycin D. Inasmuch as intercalation has been mainly studied in vitro, it is difficult to evaluate its extent in vivo. In vivo the conformation of the DNA and the interreaction between DNA and other macromolecules may restrict intercalation. *Lerman* has shown that the intercalation of acridine orange to phytohemagglutinin-stimulated lymphocyte DNA is 4 times that of nonstimulated lymphocytes [174]. Hycantone (a schistosomiasis antibiotic) is also intercalated between the strands of the DNA duplex. If, after intercalation, crosslinks are formed, they should reduce the melting point of the DNA (Tm). Although this is the case when the drug is reacted with DNA in vitro, it is not in vivo. Thus, when the drug is administered in vivo, although some of the intercalating chemicals like hycantone and benzo(a)pyrene are carcinogenic, the role of intercalation in carcinogenicity remains poorly understood.

External binding takes place with 4NQO, aflatoxins and polycyclic hydrocarbons. These are weak bindings and their role in carcinogenesis is unknown.

This list does not include all the unknown, or even the known carcinogens present in the environment. Even among those carcinogens listed, the mode of action of many remains unknown. In addition, there are probably many other chemical carcinogens whose mode of action will be difficult to reconcile with present knowledge. For example, in the case of the N-oxides or purines (adenine-1-oxide, guanine-3-oxide, and xanthine-3-oxide), no interreaction with DNA has been established.

Metals, in particular nickel, are known to be carcinogenic. Their mode of action in vivo is unknown. Plastic sheets inserted in the peritoneal cavity of rats cause sarcomas; again through unknown mechanisms. We will not discuss here the mechanism of carcinogenesis of "implanted plastic films," a mechanism which has to date defied all theories on nonviral carcinogenesis [2, 31, 141, 160, 228, 350].

When a carcinogen has affinity for DNA, it is not restricted to nuclear DNA, but it may also involve mitochondrial DNA. Since mitochondrial DNA which represents only a very small percentage of the total cellular DNA, will bind relatively more carcinogens than nuclear DNA, *Wunderlich* has shown that after the injection of MNU and DMN, the binding of the carcinogen is 7 times greater in mitochondrial than nuclear DNA [346, 347].

2. Reaction with RNA and Proteins

Any study of the mechanism of action of chemical carcinogens cannot ignore that they also bind to RNA and protein. To date the significance of the binding to RNA to the cells' economy is difficult to appreciate because only in recent years have the Hn RNA, the polyadenylic acid components of messenger RNA, the tRNA, the mRNA, and the ribosomal RNA been separated. Many of the studies done on the binding of carcinogens to RNA were done on either all nuclear, or all cytoplasmic RNA, or even on total cellular RNA.

Therefore, a comprehensive analysis of the binding to RNA will not be presented here. At least three factors will modulate the binding of the carcinogen to RNA: The reactive sites, the type of carcinogens, and the conformation of the RNA. Binding of carcinogens to RNA has been studied most extensively with alkylating agents, ethionine and AAF.

It is generally agreed that the sensitivity of the various purine and pyrimidine sites to alkylating agents decrease in the following order: N-7 of guanine, N-1 of adenine, N-3 of cytosine, N-7 of adenine, N-3 of adenine, and N-3 of guanine. Thus, the N-7 of guanine is the most sensitive, and the N-3 of guanine the least sensitive reactive site.

DMN methylates liver RNA; MNU methylates DNA and RNA from various tissues, probably because it does not require activation for binding MMS methylates like MNU. However, more binding takes place in DNA than in RNA [292].

The affinity of the nitroso compound for the base sites differs somewhat from that of DMS and MNS. The nitroso compounds attack the N-7 of adenine and the N-3 of guanine, while DMN and MNS are more reactive for N-1 of adenine, N-3 of cytosine, and possibly N-3 of uracil. Thus, for the nitroso compound the decreasing order of affinity would be N-7 of guanine, N-7 of adenine, N-3 of guanine, N-3 of cytosine, and N-3 of adenine.

Ethionine reacts preferentially with liver tRNA and yields a variety of monoethyl, or diethyl derivatives, most of which (35%) involve guanine. The mechanism of transfer of the ethyl group to the base is believed to involve methyl transferases which require S adenosyl methionine for activity [71, 94, 97].

The configuration of the nucleic acid is critical in determining the site and the amount of binding. Using single- and double-stranded RNA viruses, it was shown that much more binding took place on single-stranded than on double-stranded RNA. Perhaps the most enlightening study demonstrating the role of the nucleic acid configuration in the binding of carcinogen to RNA is the critical study that was made using ^{14}C-AAF and purified tRNA f meth. This tRNA has, like other tRNA's, the shape of a cloverleaf. It contains 25 guanine residues, 18 of which are involved in hydrogen bonding, and are, therefore, believed to be unable to offer a binding site for AAF. Among the 7 remaining guanine residues, 5 are buried inside the tertiary structure of the tRNA. Thus, only 2 guanine residues are left free to react with AAF. They are located in the dihydrouridine loop and form the sequence G-G-Dh. AAF, after metabolic conversion, binds to one guanine residue in the tRNA f meth. The N-acetyl group is retained in the binding to tRNA (but not in binding to DNA) [1, 87]. AAF also binds to rRNA [161].

Dawson [81] has established that a single dose of AAF produces a rapid inhibition of DNA and RNA synthesis in the male mouse. Sucrose gradient analysis of the RNA showed that both ribosomal and transfer RNA synthesis were affected. Table 2 summarizes the type of adducts found in DNA after exposure to some carcinogen.

Table 2

Carcinogen	*Adduct*
N-methylnitrosourea (MNU) Dimethylnitrosamine (DMN) Methyl methylsulfonate (MMS) Dimethylsulfate (DMS) Dimethylnitrosamine [a] (DMN) Diethylnitrosamine [b] (DEN) N-methyl-N-nitro-N-nitrosoguanidine (MNNG)	7-Methyl- or ethylguanine (70%) + other adducts
β-Propiolactone	7-(2 carboxyethylguanine)
1-2-Dimethylhydrosine	N-7 Methylguanine
Methylazoxymethanol (MAM)	N-7 Methylguanine
Urethan	DNA adducts of unknown structure Cytosine-5-carboxylate in RNA
Ethionine	7-Ethylguanine
Nitrogen and sulfur mustards	Esterification of primary phosphate groups of DNA Formation of intrastrand / interstrand di (guanine-7-ethyl-sulfide) (25%)
Benzo(a)pyrene (in vivo) Cation radicals or K epoxides 3-Methylcholanthrene (in vivo) Cation radical or K epoxide	DNA binding site unknown)
Benzo(a)pyrene (in vitro)	Covalent bond with C6 of thymine
7-Bromomethyl-12-methylbenz(a) anthracene (in vitro)	Substitution in vitro 8 carbon of guanine
7-Bromomethyl-12-methylbenz(a) anthracene, 7-12-Dimethylbenz(a)-anthracene	2-amino group of guanine 6-amino group of adenine 4-amino group of cytosine
4-Nitroquinoline-1-oxide (4NQO) (in vivo)	Type of DNA adducts unknown Ultimate carcinogen unknown
Nitrofuran	Type of adduct with DNA unknown
Safrole (−1-hydroxysafrole)	Binds to DNA primarily to the O-6 of guanine

[a] Produces the methyl derivative. [b] Produces the ethyl derivative.

Carcinogens bind to DNA, RNA, proteins, and polysaccharides. Binding to one or more types of these macromolecules could be significant to carcinogenesis. Little is known of the binding of carcinogens to polysaccharides. Yet, it is conceivable that binding to the carbohydrate moieties of cell membrane glycopeptides could profoundly change the membrane properties. If the environmental input at the level of the membrane is in some way communicated to the genome, it is not inconceivable that the genome may respond by a permanent reorganization of its mode of expression, which may in some cases provide the survival advantages that are the hallmark of the cancer cell.

Binding of carcinogens to proteins was first discovered with the azo dyes. In general, the *Millers* and their associates [11, 204, 206, 213] found a good correlation between the level of binding of the carcinogens to proteins and carcinogenicity. Similar results were obtained in *Heidelberger*'s laboratory using polycyclic hydrocarbon [128, 132, 133]. Binding of carcinogens to protein was, in fact, at the origin of the deletion theory of cancer, and later after *Jacob* and *Monod* [145] had attributed specific roles to some proteins in gene expression (repressors), *Pitot* and *Heidelberger* [243], proposed a theory of carcinogenesis which more directly involved the proteins.

In the intact cell most of the DNA is closely associated with proteins. The latter have been divided into two major categories: the histone and the nonhistone proteins. The exact function of each type of protein is unclear. The histones are believed to maintain DNA in a permanent repressed state. The acidic proteins are believed to modulate transcription of DNA that is not permanently repressed. If such assumptions prove to be correct, then alteration of the distribution of histones and acidic proteins along the DNA duplex could have profound effects on transcription.

Alkylating agents and other types of carcinogens could alter the proteins in various ways by leading to their denaturation, preventing or accelerating their breakdown by proteolytic enzymes, or establishing crosslinks within the proteins or between the proteins and the DNA.

Total deletion, or even deletion of a polypeptide segment of one or more proteins playing a key role in gene expression, could well result in complex distortion of gene expression, or in other words, in alteration of the status of differentiation. It could be argued that such alteration would not be transferred from one cell generation to another without concurrent changes in the DNA sequence. This a fallacy. Indeed, bone marrow and gastrointestinal cells have the same DNA, and yet the intestinal mucosa yields intestinal cells generation after generation without ever yielding marrow cells, and the bone marrow cells yield marrow cells generation after generation. When a rat is subjected to 12 successive partial hepatectomies [239], it is still liver cells that are found after the last operation. Therefore, the transfer of the genotype in the phenotype is not only dependent upon the DNA sequence, but other factors are involved. These other factors might include the relative distribution of histones and acidic proteins on the DNA, possibly humoral factors, or even chemical mediators still unknown.

Of interest are *Jungmann*'s and *Schweppe*'s studies [146], which investigated the binding of radioactive N-hydroxy-N-2-fluoranylacetamide, ^{14}C, and 7,12-dimethyl-benzanthracene ^{3}H to rat liver histones, acidic proteins, and DNA. After a single injection of the carcinogen, maximum binding of radioactivity to nuclear acidic protein and histones was found within 60 to 90 minutes. The histone fractions most susceptible to binding were the F2A1, the F2A2 for the P-dimethylaminoazobenzene, F1, and the F2B for the 7,12-dimethylbenzanthracene.

Baserga [10] has proposed that it is not the binding of carcinogen to DNA, but to chromatin that is critical to the conversion of a normal into a cancer cell. The binding of N-acetoxy-N-2-acetylaminofluorene is lower in the inactive than the active chromatin.

At present so little is known of the role and structure of the cell membrane and of the function and structures of the proteins that regulate gene expression that the simplest approach to the study of the distortion of gene expression after the administration of nonviral carcinogens is to investigate the alterations of DNA. There are good reasons to believe that DNA might be the primary target of nonviral carcinogens. UV radiation and ionizing radiation, which are effective carcinogens, exert their effects at doses where the primary molecular injuries involve the nucleic acids, in particular DNA. Especially in the case of UV, the wave lengths that are preferentially absorbed by nucleic acid bases are more effective in causing mutagenesis and carcinogenesis than those wave lengths that are more readily absorbed by proteins. Although such experiments are not unequivocal, they suggest that nucleic acids are the primary target for radiation damage [142].

Damage to proteins can, however, influence damage to DNA with UV light. A study of the action spectrum and dose response of unscheduled DNA synthesis in normal human fibroblasts exposed to UV light has suggested that both DNA and proteins are concerned in the absorption of UV, which leads to DNA damage and excision repair.

Because of RNA turnover and because the alteration in gene expression is transferred from one cell generation to another, RNA is not a likely candidate for the primary injury caused by carcinogens.

The fact that the incidence of cancer is increased in patients (xeroderma pigmentosum, Fanconi's syndrome, Higashi-Chédiak disease, etc.) in whom DNA repair is impaired, further support this view (see below).

Studies with chemical carcinogens are not in conflict with the notion that damage to DNA might constitute the initiating event in the conversion of a normal into a cancer cell. Although a good quantitative correlation between DNA binding and carcinogenesis is seldom found, except in the case of β-propiolactone [27], the site of the binding might be more critical than the amount of binding. Therefore, the factors modulating the site and level of binding of carcinogens to DNA may be of considerable significance.

V. Factors Modulating the Binding of Carcinogens

A number of factors modulate the incidence of carcinogenesis,–diet, hormones, immunosuppression (for review, see [316]). They will not be discussed here because they are not likely to be related to the initiating event in carcinogenesis. However, a discussion of the factors that modulate the binding of the carcinogens to the DNA are most relevant. The amount of carcinogen that binds to DNA will be a function of (1) the dose of carcinogen administered, (2) the uptake of the carcinogen by the target tissues, (3) the activity of the enzymes that convert the carcinogen to the ultimate carcinogen, (4) the amount of total DNA present in the target tissue at the time of administration of the carcinogen, (5) the conformation of the DNA at the time of administration of the carcinogen, and (6) the ability of the repair enzymes of the target tissue to repair all, or part, of the damaged DNA. We will ignore the first three of these factors because they belong more appropriately in a discussion of the metabolism of carcinogens. The amount of DNA available in the target organ and the configuration of the DNA may be modified during cell replication, and of course, potential DNA repair is likely to markedly modify the damage caused to DNA by nonviral carcinogens. Therefore, after considering a few miscellaneous factors we will center this discussion on the factors modulating the binding of carcinogens to DNA on two parameters: cell replication and DNA repair.

1. Miscellaneous

AAF produces a higher incidence of tumors in male hooded rats or Sprague-Dawley rats than in female rats of the same strain. Two factors that have been cited to explain this higher incidence are: (1) the higher level of N-hydroxy-AAF sulfotransferase activity which is in males 5 to 10 times what it is in females and (2) the mitotic index which is markedly increased after 3 weeks of administration of the carcinogen [144].

It would obviously be of interest to determine whether the level of binding of the carcinogen to the DNA is different in males from females.

A single dose of N-diethylnitrosamine (DEM) to male and female Sprague-Dawley rats caused an increased number of kidney carcinomas and decreased survival especially in female rats [215].

The arylhydroxylases which contribute in converting some carcinogens to the ultimate carcinogen are inhibited by 7,8-benzoflavone. *Kinoshita* and *Gelboin* [156, 157] have demonstrated that the administration of the inhibitor interferes with tumorigenesis by 7,12-dimethylbenz(a)anthracene, but not by benzo(a)pyrene. Benzoflavone markedly interfered with the binding of 7,12-dimethylbenz(a)anthracene, but not with that of benzo(a)-pyrene to macromolecules including DNA.

Covalent binding of DMBA to proteins, DNA, and RNA is also prohibited by 7,8-benzoflavone and the incidence of tumors in mouse skin is reduced. There is a peak of binding to all three types of macromolecules 12 h after injection, and the binding slowly drops from there on until 72 h. At 72 h after injection there is still significant binding in both nucleic acids and proteins.

When nitrosomethylurea (NMU) and nitrosoethylurea (NEU) are injected into the hepatic vein of rats, tumors develop in several organs, but the liver seems to be preserved. Purification of liver DNA and RNA after administration of labeled NMU and NEU showed no detectable alkylation of liver DNA, but 7-methylguanine adducts were present in the RNA of NMU-treated rats, but not in the NEU-treated rats. NMU and NEU are powerful mutagens and carcinogens. They produce tumors in a variety of species and several organs: the stomach, the kidney, the brain, the spinal cord, and the skin. No tumors have, however, been observed in liver, whether the alkylating agent be administered intragastrically, by injection, or by direct injection in the portal vein [182]. The findings strongly suggest that DNA binding is essential to carcinogenesis.

In contrast to 4-dimethylaminoazobenzene 2-methyl DAB is a weak carcinogen. Yet, after a single administration, its binding to DNA is higher than that of DAB [114]. However, if one pursues the feeding of 2-methyl DAB long enough, the binding to liver DNA is somewhat lower than that of DAB.

2. Cell Replication

A single application of urethan to the skin, followed by repeated applications of croton oil, results in the formation of papillomas and occasionally invasive carcinomas. Chemicals that are noncarcinogenic in adults are frequently found to be carcinogenic when given to embryonic or newborn animals. Urethan which is nonhepatocarcinogenic or yields a low incidence of liver cancers in adult mice or rats becomes a potent hepatocarcinogen in newborn rats and in adult rats subjected to partial hepatectomy. An increase in liver tumors was observed in suckling mice compared to adult mice with the administration of methylcholanthrene [249, 336]. This and other experiments brought *Pound* [249] to conclude that proliferating epidermal cells are more susceptible to the carcinogenic effect of urethan in proliferating than normal cells. To extend this notion, mice were subjected to one-third partial hepatectomy, and then injected with 25 mg urethan or 1 mg of DMBA in 0.25 M peanut oil by gastric instillation at intervals up to 6 days after hepatectomy. The animals were killed 12 months later. Twenty seven percent of hepatomas and 33% of hemangiomas occurred among 120 survivors. The non-hepatectomized animals had only 5.5% hepatomas and 7% hemangiomas. When dimethylbenzanthracene was used, 29% hepatomas and 8.5% hemangiomas were found in the hepatectomized animals against 16% hepatomas and 0.3% hemangiomas in the non-hepatectomized animals. However, *Pound*'s experiments do not exclude that the effect of partial hepatectomy results primarily from a dose response type of effect [249, 252].

Indeed the carcinogen will, through the portal system, first hit the liver which is rich in enzymes that convert a carcinogen to the ultimate carcinogen. The activity of the liver remnant after partial hepatectomy may be sufficient to convert most, if not all the carcinogens administered to the ultimate carcinogens.

Heston et al. [135] demonstrated that urethan increased the incidence of hepatomas in C^3H mice, a finding confirmed by *Liebelt* et al. [177]. Similarly, a one-third partial hepatectomy or a two-third partial hepatectomy reduce the latent period for the hepatocarcinogenic effects of acetylaminofluorene in rats [111].

Hollander and *Bentvelzen* [138] also concluded that partial hepatectomy enhanced the incidence of hepatomas after the administration of urethan. These authors partially hepatectomized mice 2 months old. In the first group they injected urethan 7 days before the operation; in the second group, 4 days after the operation. In the first group 50% of the mice developed hepatomas; in the second group 95% of the mice developed hepatomas. Among the non-hepatectomized animals 33% of the mice developed hepatomas. The increase is significant only when mice are injected after partial hepatectomy. The authors do not, however, take into account the dose response effect, but since the injection was done 4 days after the partial hepatectomy, the liver may have largely recovered. Moreover, it is unlikely that much cell division took place at that stage.

The arguments that a dose effect is responsible for higher incidence of cancer after partial hepatectomy cannot be used in experiments in which the incidence of liver cancer increases after treatment of animals with X-rays or neutrons. Neither X-ray nor neutrons produce hepatomas, or kidney tumors, when given to intact adult mice. Tumors are, however, produced if the animals are irradiated after partial hepatectomy or following unilateral nephrectomy [78, 275]. Similarly combinations of X-radiation and carbon tetrachloride yield a high incidence of hepatomas, even when the carbon tetrachloride is administered 9 months after the administration of X-irradiation. This indicates that a nonproliferating liver cell is the target for the carcinogen action of radiation, as it is for the block of DNA synthesis [78], and that the lesion remains latent and unrepaired in nonreplicating cells for long periods when animals are treated with aminoazo compounds and then partially hepatectomized, again, as is the case for X-radiation [78]. Similar results were obtained by *Cole* and *Nowell* [65] and *Rosen* and *Cole* [260].

The experiment with carbon tetrachloride is, however, difficult to interpret because carbon tetrachloride is itself a carcinogen and could have acted synergetically with X-radiation as shown by *Kiplinger* and *Kensler* [158].

The role of cellular proliferation in the activation of skin carcinogenesis has been derived from other types of experiments in which antibiotics and actinomycin D were found to reduce the level of tumor incidence and the covalent binding of DMBA to DNA. These experiments are difficult to interpret because actinomycin D and antimetabolites exert various toxic effects besides blocking DNA synthesis. For example, actinomycin D interferes with messenger RNA synthesis, and ultimately with protein synthesis.

Although 2-methyl, 4-dimethylaminoazobenzene is nonhepatocarcinogenic in rats, high yields of hepatomas are obtained if the chemical is included in the diet and fed to adult rats immediately following a two-thirds partial hepatectomy [72, 73].

Tomsak and *Cook* [299] studied the subcellular distribution and the binding of the DNA of ^{3}H-7,12-dimethylbenz(a)anthracene in partially hepatectomized and nonhepatectomized rats. Two weeks after injections the intact liver contained only 9% of the total radioactivity present at 4 h after injection, while the regenerating liver contained 60% of the radioactivity present 4 h after injection. This occurred in spite of the fact that the mass of the liver had tripled at that time. Part of the DMBA found in the regenerating liver was hexane extractable and was cochromatographed with ^{3}H-7,12-dimethylbenz(a)-anthracene. Direct binding to DNA was not measured. The authors suggest that DMBA is more rapidly metabolized in regenerating than in normal liver.

In contrast, *Pound* and *Lawson* [250] found that dimethylnitrosamine given 1, 6 and 12 h after partial hepatectomy, increased the yield of tumors in the liver. Microsomal dimethylnitrosamine demethylase was depressed after partial hepatectomy for up to 6 days. The finding of *Pound* and *Lawson* confirmed findings by *Craddock* [72, 73]. The effect of DMN was greater 24 h to 3 days after hepatectomy than during the first 12 h. It was concluded that the proliferating liver cells are more susceptible to the carcinogen. It is interesting that this is the time when the repair enzyme activities are lowest [321], but there is also a decrease in cytochrome P450, and numerous enzymes of the endoplasmic reticulum are involved in the conversion of the carcinogen. The depressed level occurs between the third and the fifth day after partial hepatectomy, and returns to normal only slowly. Similarly, as already mentioned, there is a depression of DMN methylase during that period after partial hepatectomy.

Marquardt et al. [195 - 197] have shown that DMBA increases the susceptibility of rat liver to the induction of hepatomas. But they also established that the hexane extractible DMBA was twice that in intact rat liver. Consequently, the increased incidence in hepatomas might result from the fact that the remnant is smaller than the intact liver.

Marquardt [194] has found that the induction of malignant transformation in the S phase in a clone line of M2 of mouse fibroblast synchronized by the use of double thymidine block was greater than observed in cells treated during G1. Various carcinogens were used, including 7,12-dimethylbenzanthracene, and N-nitrosoguanidine.

3. DNA Repair

a) Introduction

Our environment contains physical and chemical agents that cause damage to DNA, the molecule responsible in both prokaryotes and eukaryotes for the storage of genetic information. If the integrity of the genome is to be preserved, DNA must be repaired. Table 3 gives a brief classification of the lesions observed in DNA.

Table 3. Type of lesions in DNA by nonviral carcinogens

I. Breaks	
Single-strand	X-ray
Double-strand	UV
II. Depurination	Alkylating agents
III. Single base alteration	
N-4 OH cytidylate	Hydroxylamine
Deamination	
Purine	Nitrous acid
Pyrimidine	
Cytosine	Bisulfite
Formation of purine and pyrimidine adducts	Carcinogens, mutagens (see above)
Formation of 4,6-dihydroxy-dihydrothymine	OSO_4, ionizing radiation
IV. Crosslinks	
Intrastrand	
Pyrimidine dimers	UV light
Interstrand	Quinacrine + light Psoralen + light
Interduplex	Some alkylating agents
Intermolecular (e.g., DNA protein)	Alkylating agents Ionizing radiation
V. Intercalation	Acridine orange Some polycyclic hydrocarbons
VI. Adlineation	

Except perhaps for lesions of type V and VI, all other forms of DNA damage, unless repaired, cause permanent injury to DNA, interfere with DNA synthesis, and cause cellular death or mutations through base replacement in the replicating DNA.

Grossman [118] has described the lesions listed under II and IV as mono- and diadducts primarily because the mechanism of enzymic repair is likely to be different in each case.

DNA, a child of the sun, may also well have been its first victim. Thus, from its inception at the beginning of life, replicating DNA, for better or worse, became the target of various agents, for example, UV light, ionizing radiation, and possibly small chemical agents present in the atmosphere. When DNA became in some way associated with other macromolecules to form the replicative and the transcribing unit of the cell, the integrity of the genotype and the phenotype depended upon the immutability of the DNA [315].

Since, at least on the surface of the earth, exposure to UV and ionizing radiation was inescapable, persistance of life required that the living unit accommodate itself either with extensive mutability or create defense mechanisms against DNA damage. In the course of evolution bacteria have developed various mechanisms for repairing DNA; they include photoreactivation, base excision, and postreplicative repair (for review [118, 140, 273, 285]). There is evidence that some of these mechanisms also exist in some mammalian cells [89, 118, 139].

However, most of the work on repair of UV and ionizing radiation has been done in bacteria, and much of our knowledge on DNA repair in mammalian cells is derived from studies on bacteria. Therefore, it is appropriate to briefly review what is known about DNA repair in bacteria.

b) DNA Repair in Bacteria

Photoreactivation repair of UV-irradiated cells, or bacteriophage, by visible light was first observed by *Kelner* [152] and *Dulbecco* [89]. The event was later established to require a specific enzyme (photolyase or photoreactivating enzyme). The enzyme in presence of 320-370 nm light monomerizes the thymine dimers formed by ultraviolet light [269, 270, 329]. This enzyme must have played an important role in the evolution and the maintenance of unicellular organisms and plants and the more primitive forms of life. *Sutherland* [291] claims to have found the enzyme in human lymphocytes. If present in lymphocytes, the role of the enzyme needs to be clarified. Unfortunately, to date no confirmation of the existence of photoreactivating enzymes in mammalian cells is available.

A photoreactivating enzyme specific for pyrimidine dimers formed in RNA has been described in plants by *Gordon* [115].

In conclusion, although photoreactivation plays a critical role in DNA and RNA repair in primitive forms of life, there is no unequivocal evidence that it exists in higher mammals, including humans, and therefore, this discussion will center on other forms of DNA repair.

4. Dark Repair in Bacteria

a) Introduction

Some mutants of *E. coli* are extremely sensitive to ultraviolet light; these mutants are, therefore, called uvr. At least five different genes may be responsible for a mutation that leads to greater UV sensitivity, and thus the UV sensitive *E. coli* mutants have been classified as uvr A, uvr B, uvr C, uvr D, and uvr E. In each case the defective gene has been identified and located in a discrete position on the chromosome (for review, see [118]). In addition to being sensitive to UV, these mutants are also sensitive to mitomycin and have lost their ability to reactivate UV sensitive phage. Uvr A, B, and C have similar sensitivities to UV and γ-radiation. Uvr A and uvr B lack an endonuclease needed for the repair of pyrimidine dimers, but uvr C contains normal amounts of the enzyme. Uvr D and uvr E differ somewhat in their relative sensitivity to UV light and γ-radiation, but there is no point in dwelling on this matter. It suffices to conclude that the existence of uvr mutants of *E. coli* makes it obviously clear that some mutants have lost their ability to repair DNA damage caused by UV light or ionizing radiation, and that such mutants have provided a remarkable tool for the study of DNA repair in bacteria.

Some *E. coli* are not defective in any of the uvr genes, yet they are sensitive to UV and ionizing radiation as well as to mitomycin. They also are defective in genetic recombination (for review [58]). Again several genes have been identified and mapped out: rec A, rec B, and rec C. Cells defective in the rec A gene present a spontaneous degradation of their DNA, and their cell division process is defective. The product of rec A genes has not been identified. The rec B and rec C genes are responsible for the coding of a complex enzyme made of two subunits, exhibiting a quadruple catalytic activity: (1) ATP-dependent single-stranded exonucleolytic activity; (2) ATP-dependent double-stranded exonucleolytic activity; (3) ATP-dependent single-stranded endonucleolytic activity: (4) ATP-dependent DNAase activity [103, 113, 222, 226]. Thus the study of bacterial mutants suggests that in addition to photoreactivation repair in bacteria, there is also a dark excision repair mechanism and a postreplication repair mechanism.

The study of the dark repair mechanism in bacteria has centered on the elimination of thymine dimers (diadducts) from the DNA of UV-irradiated DNA. The elimination involves three separate steps: (1) an endonucleolytic incision close to the diadduct, (2) the excision of the diadduct and several bases, and (3) the reinsertion of new bases complementary to the base sequence of the old strand. Each step of the repair process may involve one or more enzymes (Fig. 6).

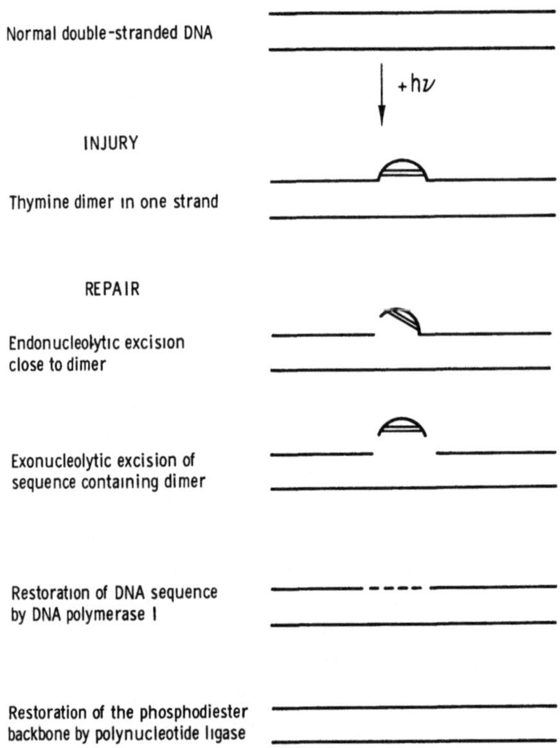

Fig. 6. Excision repair of UV-irradiated DNA

The incision is performed by an endonuclease, or an enzyme that can attack the core of double- or single-stranded DNA, and does, therefore, not require a terminus for activity. Endonucleases are phosphodiesterases and may yield a 3'OH, or 3'PO_4 terminus. Their specificity is directed toward specific sequences, altered bases (pyrimidine dimers), and conformational distortions. Most identified endonucleases act on DNA, but some have been found to act on RNA.

b) Enzymology

Table 4 gives some of the properties of the bacterial endonucleases that participate in the correction of DNA damage. Most are small enzymes whose activity may or may not be stimulated by magnesium. These endonucleases are believed to cause single-strand breaks close to the damaged portion of the DNA. Clearly too few of the bacterial enzymes have been sufficiently purified to appreciate their exact substrate specificity and their physical properties. This discussion on DNA repair does not intend to describe in detail the properties of all the bacterial and bacteriophage endonucleases that have been discovered, but to illustrate the complexities of the problem.

The enzyme involved in the first steps of the repair of DNA irradiated with UV light was investigated in Japan by *Takagi* and his associates [293] and in the United States by *Grossman* and his collaborators [148, 164, 165]. In 1962 *Strauss* showed that crude extracts of *Micrococcus lysodeikticus* selectively inactivated the transforming DNA preexposed to UV radiation. Later, *Strauss* [287] showed by zone sedimentation analysis that the extract induced endonucleolytic breaks in UV-irradiated DNA. *Nakayama* and his associates [220] established in 1967 that the degradation of UV-irradiated DNA requires two fractions: fraction A, which exhibits only endonuclease activity without releasing acid-soluble material (endonuclease?), and fraction B, which has exonuclease activity.

Takagi [293] and his associates purified (360-fold), and endonuclease specific for UV-irradiated DNA from crude extracts of *M. lysodeikticus.* The enzyme acts on double-stranded and single-stranded DNA. Magnesium is not necessary for the reaction.

The endonuclease was further purified by *Grossman* and coworkers [148, 164, 165] who also devised an assay based on the fact that the enzyme causes single-strand breaks by hydrolyzing phosphodiester bonds in UV-irradiated double-stranded ^{32}P DNA. In addition, *Grossman* purified (approximately 100 times) an exonuclease, which also attacks only the UV-irradiated DNA. Thus, the combination of the UV endonuclease incision enzyme and the UV exonuclease excision enzyme appears to release the portion of the polynucleotide chain that contains the dimer. An important feature of the endonuclease is that its specificity is not dictated by the molecular configuration of the dimer, but rather by distortions of the DNA molecule. Therefore, the endonuclease might well be the enzyme responsible for the repair mechanism observed in DNA extracted from bacterial or animal sources treated with nitrogen mustards, X-irradiation, etc. However, direct evidence for such an activity of the bacterial endonuclease is lacking, except in the case of X-irradiated DNA.

Friedberg [101] has reviewed the properties of the UV endonuclease of T4 phage. The enzyme is coded for by a so-called V gene. The molecular weight of the enzyme (18,000 daltons) was determined by both gel filtration in Sephadex and gel electrophoresis.

Table 4. Bacterial correctional endonucleases

	Synonym	*Substrate*	*Mode of action*	*Size*	*Inhibitor*	*Reference*
Bacterial correndonuclease I	***E. coli*** endonuclease II	Alkylated DNA (methyl methane sulfonate)	Specific for apurinic sites	s:3.2 Stimulated by Mg^{2+}	EDTA	103
		X-irradiated DNA	Yields 3'OH	17,000 daltons Stimulated by Mg^{2+}		148, 164, 165
	M. luteus endonuclease	X-irradiated DNA	Yields $3'PO_4$	?		40, 58, 103, 113, 125, 148, 164, 165, 222, 226, 339
		Pyrimidine dimers	Terminus	?		40
	E. coli endonuclease	γ-irradiated poly d(A-T)	5_7b-dihydroxy-dihydrothymine	?		147, 339
		DNA treated with osmium tetroxide	apyrimidinic sites	14,000 daltons No Mg^{2+}	Caffeine	30
					Acriflavine	
	Micrococcus luteus uracil endonuclease	DNA containing uracil, for example after deamination of cytosine	?	17,000 daltons No Mg^{2+}		255
				15,000 daltons		102
	E. coli mismatch endonuclease	Duplex, T4 phage with heteromutations on both strands	?			
Bacterial correndonuclease II	uvr A, uvr B endonuclease	Pyrimidine dimers mitomycin bound DNA	3'OH	See text		
	Micrococcus UV endonuclease	Pyrimidine dimers No effect on X-irradiated DNA	3'OH	See text		
	Bacteriophage T_4 endonuclease	Pyrimidine dimers	3'OH	See text		

Like the *M. luteus* endonuclease and the mammalian enzyme to be described later, the T4 enzyme is rather unstable. The enzyme is inactivated by freezing. Yet, it can be stored in 3% polyethylene glycol at 4°C or in 50% glycerol at -26°C.

The enzyme is free of exonuclease and polymerase activity, but causes single-strand breaks in UV-irradiated double-stranded DNA. It is believed that a 3'OH and a $5'PO_4$ termini develop at the site of the break.

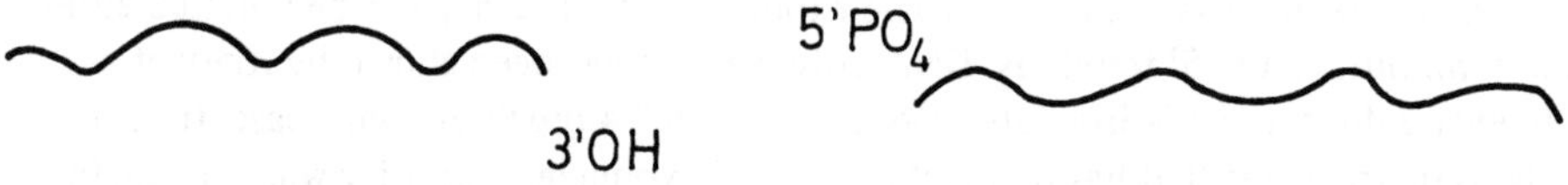

Fig. 7. T4-caused single-strand break

Although the enzyme has no species specificity for UV-irradiated DNA, and will thus cause single-strand breaks in T4 or *E. coli* UV-irradiated DNA, it is believed to specifically distinguish thymine dimers. The arguments supporting this notion are: (1) the enzyme causes one break per lethal radiation hit. Therefore, it is postulated that the catalysis of single-strand breaks is dimer specific. There is, however, no convincing evidence that only thymine dimers, at the exclusion of all other photoproducts, are lethal. (2) Proflavine interferes with the formation of thymine dimers at 254 nm. Therefore, by modulating the dose of proflavine during UV-irradiation of DNA, it is possible to modulate the amount of thymine dimers formed. (3) Treatment of DNA with various dimer content demonstrated a direct correlation between the susceptibility of the substrate and the dimer content. However, these experiments did again ignore the formation of other photoproducts whose concentration might increase, or even exceed that of thymine dimers when DNA is irradiated in presence of proflavine.

More compelling evidence for the specificity of the T4 endonuclease is derived from experiments with photoreactivating enzymes and the use of V gene deficient mutants. The photoreactivating enzyme is assumed to be strictly specific for thymine dimers. After exposure of T4 UV-irradiated DNA to photoreactivating enzymes, all T4 endonuclease susceptible sites disappear.

Tfv_1, an endonuclease defective mutant, is no more sensitive to X-radiation and bifunctional alkylating agents than the wild form (T4).

T4 phage is double-stranded. If one of the strands is either extended, or has suffered a deletion when it is formed, an heteroduplex is believed to form a loop. The loop is believed to be excised by a repair process similar to that responsible for the dark repair process [13].

In conclusion, although the T4 endonuclease undoubtedly causes single-strand breaks in UV-irradiated DNA, and thymine dimers seem to be the principal target of the enzyme, it cannot be excluded that the T4 endonuclease is not also involved in the formation of single-strand breaks for other photoproducts.

The sequence of steps that follow the endonuclease step in vivo are not known. *E. coli* contains an exonuclease (exonuclease VII) with a molecular weight of 88,000. The enzyme does not require Mg for activity. The enzyme is specific for single-stranded DNA, but it can also hydrolyze the loose ends extending from a nicked double-stranded DNA. Such hydrolysis results in the appearance of small nucleotides ranging from dinucleotides to oligonucleotides (25 nucleotides long). It initiates the hydrolysis at either the 3' or 5' end of the single-stranded DNA. When UV-irradiated phage DNA is nicked by the *M. luteus* endonuclease and treated with exonuclease VII, the thymine dimers are excised. Whether the exonuclease VII is the enzyme that functions in vivo has not been established.

Inasmuch as the T4 endonuclease yields a 3'OH ending, the nick is susceptible to binding of DNA polymerase I. The complete molecule, in addition to catalyzing polymerization, also exerts exonucleolytic activity. The exonucleolytic activity can be initiated either at the 3' terminus, or the 5' terminus. DNA polymerase I has been shown by *Kelly* et al. [151] to excise dimers of UV-irradiated DNA nicked by T4 repair endonuclease. In fact, the proteolytic fragment that has retained the 5' → 3' exonuclease activity will perform the same reaction [112]. Therefore, it is not surprising that the incubation of nicked DNA with extracts of *E. coli* mutant defective in DNA polymerase I results in low dimer excision. There is, however, enough excision to suspect that other exonucleases might operate.

An exonuclease coded for by the T4 phage develops after infection. The appearance of the exonuclease is inhibited by chloramphenicol, and in contrast to the exonuclease VII of *E. coli*, it requires magnesium for activity, but like the *E. coli* enzyme, it can act in either 3' → 5' or 5' → 3' direction. The enzyme is, therefore, different from the two other T4 exonucleases: one which is associated with DNA polymerase activity and acts in the 3' → 5' direction only [151], and an exonuclease A which does not excise dimers. Thus, it would appear that in repairing T4 UV-irradiated DNA, DNA polymerase I is next to the endonuclease in the sequence of steps involved in the repair process. In absence of exonuclease VII, or an exonuclease coded for by the phage, DNA polymerase I is responsible for the excision of the dimer and then the polymerase must also pursue the repair process.

In conclusion, endonucleases that cause single-strand breaks near the sites of DNA damage caused by UV light, X-radiation, and alkylating agents have been found in bacteria and bacteriophages. Few of these endonucleases are fully characterized, yet some similarities emerge: they are small enzymes stimulated, but not dependent on, Mg^{2+} for activity, and they are often unstable.

The restoration of the continuity of the DNA chain after an endonucleolytic attack close to a lesion of the DNA molecule requires polynucleotide ligase. The ligase requires a 3'OH and $5'PO_4$ termini for activity.

Studies with bacterial mutants suggest that the endonucleolytic attack may be reversed [112]. This is however, not true in all mutants. Indeed, the data indicate that the uvr C gene product prevents resealing of the endonuclease incision before the excision of the dimers has taken place. The product responsible for this activity is unknown.

The incision process is followed by an exonucleolytic attack. Exonucleases are phosphodiesterases like endonucleases, but they require a free end terminus for hydrolysis. They cannot attack the intact DNA core, but they can hydrolyze at a free end. Moreover, if the core has been nicked by an endonuclease generating new free ends, then the exonucleases can exert their catalytic activity. The specificity of the exonucleases is further limited by their capacity to attack a 3' or a 5' terminus. Two kinds of exonucleases have been found to participate in the excision process: free exonucleases and exonucleases associated with DNA polymerase. The correctional exonucleases initiate hydrolysis from both 5' and 3' termini of single-stranded DNA, or at 5' termini at internal phosphodiester bonds.

The best known bacterial free exonuclease is the *M. luteus*, an UV exonuclease which attacks double-stranded DNA nicked by the UV endonuclease, and excises 5' mononucleotides with no oligonucleotide intermediates.

The multiple catalytic properties of DNA polymerase I are well known [88, 136, 229, 246, 247, 312]. In addition to its polymerizing ability, it exerts two exonucleolytic activities: one in the 3' → 5' direction. This activity requires 3'OH for attachment of the enzyme and yields 5' mononucleotides. The enzyme also exhibits a 5' → 3' exonucleolytic activity which is specific for double-stranded DNA and is stimulated by the neighboring 3'OH provided during polymerization of the nucleotide triphosphates. *Kelly* et al. [151] demonstrated that the 5' → 3' exonuclease excises pyrimidine dimers from UV-irradiated DNA incised with pancreatic DNAase, and later *Hamilton* et al. [123] restored the biological activity of transforming DNA of UV-irradiated *Bacillus subtillis* by sequential treatment with an endonuclease, DNA polymerase I, and polynucleotide ligase.

We have seen that the pyrimidine dimers produced in the DNA of uvr mutants are usually excised. Only a fraction of the dimers escape excision and remain in the DNA until at least after the first replication. The repair of the pyrimidine dimers is error-free and depends upon the presence of a rec type of gene. Yet, the incidence of mutation in the wild type strains which contain the rec A and lex genes suggests that the excision repair gap is repaired by an error prone gene that may be rec A^+ or lex^+ dependent. Thus while the error-free mechanism is probably utilizing DNA polymerase I, which is very faithful, the error prone mechanism has been postulated to be caused by the induction of an error prone DNA polymerase [251, 342, 343]. If this hypothesis can be substantiated, it is not without relevance to the cancer problem. Indeed, *Loeb* et al. [186] have found error prone polymerases in leukemic lymphoblasts.

5. Recombination Repair in Bacteria

Howard-Flanders, who pioneered our understanding in excision repair, contributed a great deal to our knowledge of recombination repair [139].

Whenever DNA is damaged in both strands at the same level (e.g., double-strand breaks produced by X-rays, interstand crosslinks, postreplication gaps, opposite thymine dimers), excision repair does not take place. Two-strand damage, whether it be crosslinks or two-strand breaks, are more injurious to the cell than single-strand damage [140]. DNA which carries such damage is incapable of transcription or replication unless the damage is repaired. In bacteria a postreplicative repair mechanism is known to exist, but its molecular mechanism remains obscure.

The phenomenon called multiplicity reactivation gave one of the first clues as to the existence of recombination repair. Already in 1947 *Luria* showed that when a single host cell is infected with several T2 phages exposed to lethal doses of UV radiation, the condemned phages cooperate in some way to yield viable progeny [187]. Later it was shown that recombination defective mutants of *E. coli* were highly sensitive to UV and ionizing radiation [8].

More direct evidence was obtained when it was established that in UV-irradiated bacteria, the DNA synthesized at early times after irradiation has a molecular weight almost identical to that of DNA chains separated by the distance of 2 dimers. This suggested that the presence of the dimer blocked the expansion of the replicating chain by DNA polymerase. DNA polymerase can, however, start new replicating chains at sites removed from the pyrimidine. As a result, early after exposure to UV radiation the newly synthesized strands are incomplete, and the segments of the strands are separated by what is referred to as replicating gaps. However, if one further incubates the UV-irradiated bacteria, after a period of time the gaps are filled and the molecular weight of the new strand is equal to that of UV-irradiated DNA. The filling of the gaps is referred to as postreplication repair. A mechanism for postreplication repair in bacteria has been proposed by *Howard-Flanders* [139]. According to *Howard-Flanders* the pyrimidines are not excised. The postreplication gap is filled by genetic exchange with an intact portion of the parental strand not involved in filling the gap and is excised, leaving a gap in the homologous parental strand (B). The gap in (B) is repaired with DNA polymerase and polynucleotide ligase. The thymine dimers present in the new duplex are excised by the regular excision process (see Fig. 8).

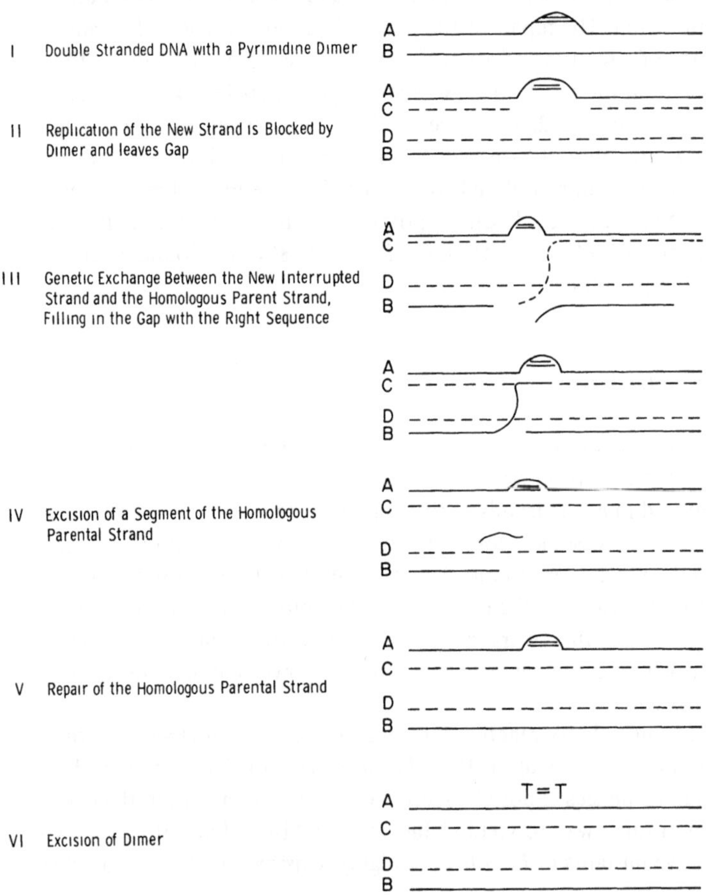

Fig. 8. Recombination repair of DNA-containing pyrimidine dimers. (*Howard-Flanders*, 1968)

In this scheme the dimer remains in the parental strand and is excised through the regular excision process only in step five. However, *Ganesan* has shown that the sites sensitive to the T4 UV endonuclease are not only distributed in parental strands, but are distributed approximately equally between mother and daughter strands, allowing exchange of strands carrying dimers to take place [105].

Psoralen (Fig. 9) intercalates between the two DNA strands. When the DNA containing psoralen is UV-irradiated, the photoproduct crosslinks like a bifunctional alkylating agent (e.g., mitomycin) (Fig. 10). In *E. coli* treated with psoralen and UV-irradiated the molecular weight of the DNA drops in alkaline sucrose gradients. These findings have been interpreted to mean that one of the arms of the crosslink has been detached as shown in Figure 11 (e.g., A as a result of 2 single-strand breaks, each taking place at the opposite site of the crosslink).

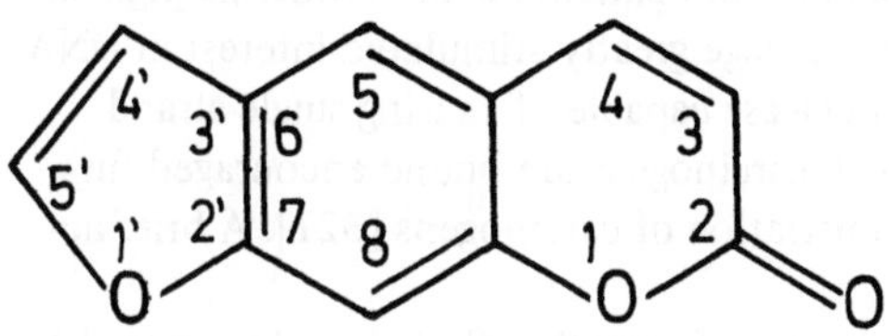

Fig. 9. Psoralen

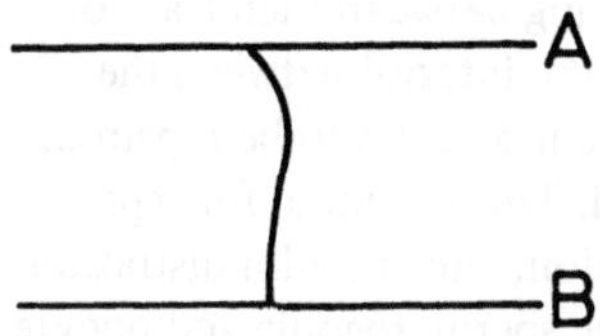

Fig. 10.

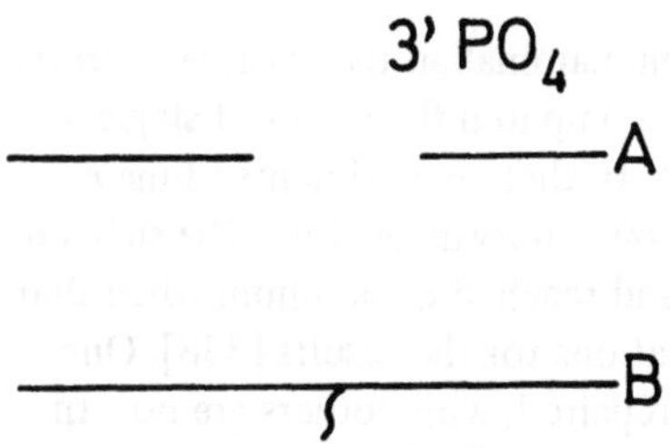

Fig. 11.

This process yields a gap on strand A and a monoadduct on strand B (Fig. 11).

The gap in strand A is believed to be repaired by recombination with sister strands. The adduct on strand B is excised by a core endonuclease in a manner similar to the excision of pyrimidine dimers (Fig. 12).

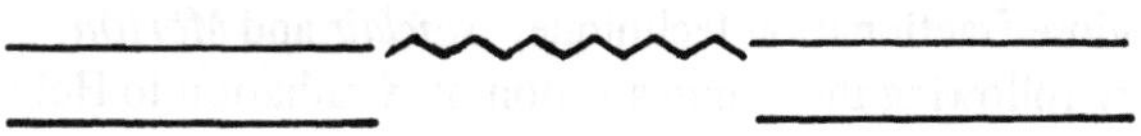

Fig. 12.

6. DNA Repair in Mammalian Cells

a) Introduction

Although parallel in time with studies of DNA repair in bacteria, the investigation of DNA repair in mammalian cells took a somewhat different course. At first they were centered on the response to ionizing radiation, and it is only after the landmarking experiments of *Pettijohn* and *Hanawalt* [240] that they were switched to repair damage caused by UV light. The first experiment which demonstrated that unscheduled DNA synthesis occurred in mammalian cells was reported by *Rasmussen* and *Painter* in 1964 [253]. These authors observed that after UV irradiation of HeLa cells in culture, all cells incorporated ^{3}H thymidine into DNA. This is in contrast to what takes place in control cultures in which only 20 to 30% of the cells pick up the label. These findings were extended by *Painter* and *Cleaver* [232, 233]. Later the discovery that cells obtained from patients with xeroderma pigmentosum were defective in repair of UV radiation damage greatly stimulated interest in DNA repair in mammalian cells [60]. A repair endonuclease capable of causing single-strand breaks in UV-irradiated DNA and DNA to which carcinogens are bound encouraged further investigation of repair of DNA after administration of carcinogens [321]. A brief account of these observations follows.

In 1966 *Kimball* [154] showed that the incidence of mutation (lethal or slow growth) in *Paramecium* differed depending upon the mode of irradiation. Highest incidence was lowest when the same dose was fractionated and the time elapsing between each fraction was greatest. The results were interpreted to mean that during the interval between the fractions of the dose, there was an opportunity for the sensitive molecules to be repaired. Consequently, a mutation caused by irradiation was eliminated. This potential for repair was demonstrated for X-radiation, gamma radiation, UV radiation, and the administration of alpha particles and alkylating agents. Dose rate effects on the spermatogonia and oocyte of mice [261] also led to the conclusion that repair mechanisms must exist in mammalian systems.

The most convincing indication that repair occurred in mammalian tissue came from the remarkable studies of *Elkind* and *Sutton* [91, 92], which compared the effect of single and fractionated doses of irradiation on cells in tissue culture. In these experiments Chinese hamster cells were given two exposures separated by various intervals of time. The survival increased with the time interval between the two doses and reached a maximum when that interval was 1 h. There were at least two possible explanations for the results [338]. One was that all cells are equally sensitive and that some are repaired, while others are not. In that case a molecular mechanism of repair needed to be postulated. The other explanation proposed that the cells at the time of exposure presented different degrees of sensitivity to radiation. The difference in sensitivity could have depended on the stage reached in the cycle for DNA synthesis. Inasmuch as the cell population used in these experiments was not synchronized, it is likely that restoration was the consequence of a combination of molecular repair and differences in sensitivity.

Thus the studies of *Elkind* and *Sutton* did not permit a decision as to whether the changes in survival resulted from repair or selective radiosensitivity of the cell due to its stage in maturation (G-1, S, G-2). Using the two-dose fractionation technique, *Sinclair* and *Morton* [276] were able to distinguish recovery following the administration of X-radiation to HeLa cell from the changing response resulting from varying cell stage. Recovery (with respect to survival) was quite definite for cells in S phase, but whether it occurs when irradiation is administered during other stages of the cell cycle is not certain.

Similarly, *Dewey* and *Humphrey* [83] studied chromosomal restitution of radiation damage in L-fibroblast cells and demonstrated that it occurred over a period of 2 to 3 h when the cells were irradiated in both S and G-1 phases. No clear-cut recovery could be demonstrated while the cells were irradiated in the G-2 phase.

The factors modifying repair after X-irradiation were also investigated in several laboratories.

Phillips and *Tolmach* [241] irradiated synchronous populations of HeLa cells with 300 and 600 rads. Such doses did not cause death but were potentially lethal in that further administration of fluorodeoxyuridine, hydroxyurea, or deoxyadenosine or post-irradiation incubation of these cells at 29°C, decreased the fraction of surviving cells. When administered alone, none of the post-irradiation treatments affected the viability of the cell. The post-irradiation treatments were administered during G-1, and there is no time during G-1 which is more sensitive than any other. When, however, the various treatments described above were administered 5 h after irradiation, the irradiated cells lost their sensitivity to the treatment. A somewhat unexpected observation was that cyclohexamide increased the cell survival when administered post-irradiation. To explain the effects of the DNA inhibitors, it was postulated that repair involved DNA synthesis and that addition of DNA inhibitors interfered with repair. The effect of cyclohexamide was interpreted to result from protection of the repair mechanism as a consequence of selective inhibition of protein synthesis. Thus cyclohexamide might interfere with the biosynthesis of some proteins but not with that of others.

Conflicting results have been obtained by *Djordjevic* and *Kim* [85] who studied the effects of hydroxyurea and puromycin on the modification of the radiation response of synchronized population of HeLa cells.

Dalrymple and his associates studied the effect of 2,4-dinitrophenol on the repair of irradiated L-cells. The radiosensitivity of L-cells cultivated in presence of DNP was lower than that of cells grown in normal culture media. Investigation of the repair component indicated that almost all ATP and macromolecular synthesis had ceased. Such findings led *Dalrymple* to conclude that repair processes occur in spite of a marked block in macromolecular synthesis or interference with the bioenergetic pathway. This led the investigators to suggest that repair mechanisms were in existence before the exposure to irradiation and were automatically released after the injury in the fashion of a corkscrew spring.

Later *Dalrymple* and his associates suggested that X-radiation (1000 r) produced DNA breaks in mouse liver and L cells; the breaks are of the $5'PO_4$ termini type and are reparable in vitro by polynucleotide ligase [79]. These findings, however, conflict with previous studies made by *K. Smith* [149] and *Bopp* et al. [26] in bacteria. Moreover, it seems unlikely that X-radiation would produce only breaks yielding $5'PO_4$ termini, and it cannot be excluded that the breaks detected by *Dalrymple* are not the direct result of radiation, but a step in the repair process [320].

Wolff and *Scott* [345] have studied repair of chromosome breaks in *Vicia faba*, in hamster cells and patients with *xeroderma pigmentosum*. Chromosomal repair took place in absence of unscheduled DNA synthesis which led the authors to conclude that chromosomal repair after X-radiation does not involve the dark repair mechanism.

Although repair of chromosome breaks may not be associated with ^{3}H-thymidine uptake detectable by radioautography, it is not excluded that enzymes involved in dark repair are also responsible for repair of chromosome breaks. If, as proposed by *Kihlman* [153], chromosome breaks are associated with double-strand breaks, the restoration may require excision of the damaged bases, patching by DNA polymerase and polynucleotide ligase. Therefore, more knowledge of the enzymology of DNA repair in mammalian cells may aid in our understanding of the restoration of chromosome breaks.

The number of chemicals that cause chromosome breaks should not be underestimated. These agents have been referred to as chromosomoclasmatogens, or clastogens. Physical, chemical, and biological agents acts as clastogens [4, 39, 64, 77, 274]. Among the physical agents the ionizing radiation plays a central role. More than 200 chemicals are known to act as clastogens. They include: (1) *nucleic acid-related compounds* such as caffeine, theobromine, and theophyllin; (2) *antibiotics* such as streptonigrin, mitomycin, actinomycin D, chloramphenicol, puromycin, streptomycin, tetracycline, metanamine mandalate, and nalidixic acid; (3) *drugs used to modulate* brain activity such as sedatives (e.g., scopolamine and chloralhydrate) and tranquilizers (chlorpromazine and meprobamate, meclizine, dimenhydinate, thalidomide, trimetobenzamide); (4) *food additives*, cyclamate, sodium nitrate which is converted to nitrous acid in the stomach, and possibly monosodium glutamate; (5) *air and water pollutants* (ozone, benzo(a)pyrene, sodium hypochlorite, chloramine T.); (6) *insecticides* (derivatives of melamine or phosphoramide); (7) *captan* (a widely used fungicide); (8) *malic hydrazine* (a plant growth inhibitor).

All these compounds, all with very different chemical properties, cause chromosome breaks in lymphocytes in vitro. However, it has not been possible to demonstrate that these drugs cause chromosome anomalies in mouse or primate embryos. Even in children, who were the victims of congenital anomalies caused by thalidomide, could no increase in chromosome breaks be demonstrated. Among the hallucinogenic drugs, lysergic acid is of key interest. Although there is clear cut evidence that lysergic acid causes chromosome breaks in vitro, there is still no evidence that such chromosome breaks occur in vivo.

The demonstration by *Pettijohn* and *Hanawalt* [240] that UV-irradiation of bacteria was followed by nonsemiconservative replication of segments of the DNA strand greatly enhanced our understanding of repair of UV damage in bacteria. *Painter* and *Cleaver* [232, 233] were the first to utilize a modification of the method of *Pettijohn* and *Hanawalt* to demonstrate nonsemiconservative DNA synthesis in HeLa cells. The dose needed to demonstrate repair replication, however, far exceeded lethal doses. Later *Painter* and *Cleaver* were able to correlate this form of unscheduled DNA synthesis with the incidence of repair from UV- or X-irradiation. There was, however, a notable exception namely that of the skin of patients with xeroderma pigmentosum in which there is no repair and no unscheduled DNA synthesis.

Cleaver [61] further established that compounds which prevent the formation of DNA precursors inhibit semiconservative replication but do not interfere with repair replication. In contrast, compounds that bind to the DNA inhibit both semiconservative and repair replication. The repair process seems to further depend upon metabolic activity since iodoacetate, an inhibitor of glycolysis, and cyclohexamide, an inhibitor of protein synthesis, inhibit both semiconservative and repair replication. Thus, the recovery period after the irradiation damage cannot be explained by interference with the repair mechanism. These results were of considerable significance because they did shed some light on the molecular mechanism of repair in mammalian tissue. The findings suggest: (a) that the nucleotide pool is not rate-limiting, (b) that the enzyme is present and active even when semiconservative DNA synthesis does not take place and (c) that binding enzyme and DNA, and possibly uncoiling of DNA, is needed for repair in mammalian tissue. Moreover, the fact that cyclohexamide inhibits both semiconservative and nonsemiconservative DNA synthesis suggests that new protein synthesis (not likely to be the repair enzyme) is needed for molecular repair. Whether the needed protein is the same in semiconservative and repair DNA synthesis cannot be predicted. This new protein may be an unwinding protein, or a protein similar to that which in bacteria was found to prevent resealing of the endonucleolytic single-strand break.

The discovery that the human xeroderma pigmentosa fibroblasts are like uvr mutant defective in DNA repair (for review [256]) stimulated investigations of DNA repair in mammalian cells exposed to UV light, ionizing radiation, carcinogen, and in various diseases.

Although the absence of DNA repair is in some and probably most cases of xeroderma pigmentosum associated with the development of cancer and might even play a causal role in carcinogenesis, it would be naive to assume that all cancers are associated with defects in DNA repair enzymes. Most humans are not likely to be defective in these enzymes and yet the incidence of cancer of the skin due to UV radiation is high in the general population. Even among patients with xeroderma, some types have been described in which numerous tumors develop with no detectable defect in DNA repair. Most oncologists have seen patients who develop multiple cancers of skin. (I personally recollect one who developed 120 cancers spread all over the body, another 22 cancers of the face). Unfortunately the activity of the repair enzymes were not measured in these patients. Yet, these patients did not present any signs of exceptional sensitivity to exposure to light. Therefore, one must assume that factors other than the activity of the enzymes involved in DNA repair are responsible for initiating carcinogenesis.

b) Excision Repair

Three different methods have been used to determine excision repair in mammals: (1) direct measurement of the removal of thymine dimers, (2) incorporation of ^{3}HTDR into DNA in absence of semiconservative DNA synthesis, and (3) photolysis of incorporated BUDR. The last procedure is based on the following principle.

Cells are incubated with ^{3}HTDR and then UV-irradiated. The irradiated cells are then incubated again with a thymidine analog, bromouridine (BUDR), which during unscheduled DNA synthesis is incorporated into the DNA sequence in positions normally occupied by thymidine. The cells are then exposed to 333 mμ UV radiation which renders the BUDR regions sensitive to single-strand breaks when the DNA is exposed to alkali. This procedure is not only extremely sensitive (it permits the detection of one repair event for 10^8 daltons of DNA), but it also allows one to estimate the size of the oligonucleotide chain excised.

Using the BUDR procedure, *Regan* and *Setlow* [254] distinguished between two types of DNA repair: that involving short sequences and that involving long sequences. The short sequence repair occurs after γ-radiation and probably involves single-strand breaks. The breaks are followed by the exonucleolytic excision of a few nucleotides followed by DNA polymerase restoration of the sequence and sealing of the gap by polynucleotide ligase.

The long sequence repair is associated with excision repair. The prototype is the excision of thymine dimers in UV-irradiated DNA. The agent which injures the DNA does not cause strand breaks, but brings about the endonucleolytic removal of a long polynucleotide chain, DNA polymerase restoration, and polynucleotide ligase sealing of the patch.

A number of chemical agents, such as acetylaminofluorene, ethylmethane sulfonate, and 4-nitroquinolineoxide bind to DNA bases. The damage to the bases can be repaired, at least in part. The process involves long patch DNA repair, and the number of bases excised differs with injury. For example, 100 and 140 nucleotides are excised per average repair region after UV-irradiation and N-acetoxy-AAF respectively.

Studies in adult animals have shown that small fractions of the initial injury caused by irradiation are not repaired. Irreparable damage following exposure of the fetus is considerably greater than that observed in exposed mature animal. The potentialities of many of the fetal cells often remain unexpressed in the phenotype until days or weeks after birth. Although many different mechanisms may work separately or synergetically to bring about differentiation, derepression of repressed DNA is likely to be one of them. If the repressed DNA irradiated in utero is not repaired, harmonious differentiation of all cells is impossible and dysfunction of cells, organs, and entire organisms may take place. It is, therefore, not surprising that the incidence of irreparable injuries would be greater after irradiation in utero [57].

In conclusion, although these early experiments clearly established how damage by X-irradiation is repaired in mammalian tissue, the molecular mechanism of repair remains unknown.

A number of investigators have focused their studies on the repair of single- or double-strand breaks caused by mutagens and carcinogens. *Cox* et al. [70] described a method for measuring DNA damage and repair in the liver in vivo using the incidence of single-strand breaks in alkaline sucrose gradients to determine both the level of damage and degree of repair [80]. They further demonstrated that dimethylnitrosamine, methylazoxymethanol acetate, N-methyl-N-nitrosourea, and N-methyl-N-nitrosourethan caused dose dependent increases in the incidence of single-strand breaks in rat liver DNA. The repair of the strand breaks was complete within a week for MNU, but took more than 2 weeks for DMN and MAM. *Stewart* and *Farber* [284] studied the repair of single- and double-strand breaks after the administration to rats of cyclic nitrosamines, namely nitrosomorpholine, nitrosopiperidine, and dinitroazopiperazone. Surprisingly, the double-strand breaks were more rapidly repaired than the single-strand breaks.

Farber et al. [95] developed a new technique to study the occurrence of single- and double-strand breaks in mammalian cells after layering them on top of neutral or alkaline gradients. If a low molecular weight DNA appears in the neutral sucrose gradient, then double-strand breaks are assumed to develop. In alkaline sucrose gradients the duplex is supposed to dissociate, and the appearance of small molecular weight DNA is assumed to indicate single-strand breaks.

Such methods are not without drawbacks when heavy molecular weight DNA is used, especially as is the case in mammalian chromatin. Indeed, the sedimentation of heavy molecular weight DNA does not follow the expected pattern of distribution observed with low molecular weight DNA. Moreover, the DNA bound to proteins, and RNA are readily removed during cell lysis. However, *Farber* has classified the agents that cause strand breaks in DNA into three groups: (a) those that cause single-strand breaks and are repaired within 4-72 h after exposure, (b) those that cause single-strand breaks and are slowly repaired, and (c) those that cause double-strand breaks. Today the direct relevance of these findings to liver carcinogenesis is still unknown.

Vaught et al. [324] measured DNA breakage in human cells after exposure to 3-methylcholanthrene-11,12-oxide using alkaline sucrose gradient analysis. The study demonstrated the existence of a repair process, but as pointed out by the authors themselves, the data did not provide information on the qualitative aspect of the repair and its fidelity.

Walker and *Sridhar* [330] demonstrated repair of single-strand breaks caused to the DNA of mammalian cells in culture damaged with UV light, methyl methanesulfonate, methyl nitrosourea, and 4-nitroquinoline-1-oxide.

Laishes and *Stich* [166] demonstrated repair synthesis after exposure of human skin fibroblast to dimethylnitrosamine activated by microsomes.

The repair of carcinogen-bound DNA was studied by *Lieberman* and *Dipple* [180] in lymphocytes in cultures using 7-bromomethylbenz(a)anthracene. The unlabeled compound induced excision repair of the DNA in presence of hydroxyurea. The excision repair was completed within 12 h after exposure to the carcinogen. When radioactive carcinogen was used, it was found that only 15-17% of the carcinogen was removed after 12 h. This clearly indicates that excision repair of carcinogen bound to DNA is incomplete. The adenine were more extensively removed than the guanine adducts. Using proximate carcinogens or UV light, *Lieberman* has extended these studies to a number of tissues including skin fibroblasts, and liver and kidney cells [179]. *Norman* et al. have also demonstrated unscheduled DNA synthesis in human tumor cell suspension [225].

Stich et al. [286] and *Takebe* et al. [294] have demonstrated increased sensitivity of xeroderma pigmentosum cells and reduced dark repair when exposed to 4-nitroquinoline-1-oxide (4NQO), 4 hydroxyaminoquinoline 1-oxide (4OHAQO), 2-methyl-4-nitroquinoline-1-oxide (2-Me-4NQO), 3-methyl-4-nitropyridine 1-oxide (3-Me-4NPO).

Similarly, *Cleaver* demonstrated that fibroblasts of patients with xeroderma pigmentosum showed reduced repair after exposure to UV light, 4-nitroquinoline-1-oxide, and 1,3-bis chloroethylnitrosourea [62]. Normal repair is, however, observed when the cells are exposed to the 6-nitroquinoline-1-oxide (noncarcinogenic), or to N-methyl-N-nitrosoguanidine. Similar results were obtained by *Setlow* and *Regan* after exposure of xeroderma pigmentosum cells to N-acetoxy-2-acetylaminofluorene [272]. *Goodman* and *Potter* provided evidence that DNA repair synthesis takes place in rat liver after the administration of 3-methyl-4-dimethylaminoazobenzene [114].

Kitagawa et al. [159] measured unscheduled DNA synthesis radioautographically in primary liver cell cultures obtained from hyperplastic nodules derived from rat livers treated with N-2-fluorenylacetamine. No significant difference in the level of DNA repair was observed between normal liver cells and the hyperplastic nodules. At a first approximation such experiments might lead one to conclude that the transformation of hyperplastic nodules into the malignant state does not result from an alteration in their repair capacity. Radioautographic techniques are rather imprecise for the detection of DNA repair. They will not permit the detection of small quantitative differences that might be of significance to the process of carcinogenesis, and they certainly would not reveal quantitative differences between repair of normal and hyperplastic cells.

Using cell survival assays, *Shipley* et al. have compared the repair of γ-radiation to that of fast neutron radiation in Lewis lung carcinoma cells. The result showed that the repair was lower after the administration of fast neutrons than after that of radiation. This finding is of considerable significance, if high linear energy transfer radiation is to be used in tumor therapy [275].

c) Modulation of Excision Repair

Berenblum and *Shubik* [15, 16] demonstrated that repeated application of croton oil on mouse skin after a single painting with methylcholanthrene was followed by appearance of cancer. Ever since a number of studies established that the administration of tumor promoter after a single application of carcinogen enhanced the incidence and the rapidity of onset of tumors [14], reports have appeared suggesting that promoters interfere with DNA repair synthesis [107, 167, 296]. *Teebor* et al. [296] demonstrated that the carcinogen, phorbol myristate acetate, inhibits the excision of thymine dimers induced by UV-irradiation of HeLa cells. Similar findings were made by *Langenbach* and *Kuszynski* [167]. *Poirier* et al. [244] investigated the effect of some promoters and nonpromoters on DNA repair in diploid human fibroblasts. The cells were damaged with N-acetoxy-2-acetylaminofluorene, croton oil, and 12-O-tetradecanoylphorbol-13-acetate, and anthralin inhibited DNA repair synthesis while the nonpromoting analogues had no such effect. However, the promoters also inhibited DNA replicative synthesis and other types of macromolecular synthesis. Therefore, one cannot be certain that the effect of the cocarcinogens does not result from a broad toxic effect.

Cleaver and *Painter* [63] have excluded the possibility that carcinogens, steroids, and cocarcinogens which bind to DNA selectively inhibited DNA repair.

The synthetic antimalarial agent chloroquine increases the sensitivity of bacteria to ultraviolet light and alkylating agents [348]. Similarly, chloroquine sensitizes tumors to the action of alkylating agents and ionizing radiation [106]. *Michael* and *Williams* [203] tested the effect of chloroquine on DNA repair of rat liver cells damaged with methyl methanesulfonate (MMS), and found that the drug inhibits the repair of single-strand breaks as determined by sedimentation on alkaline sucrose gradient. Although the drug also inhibits protein synthesis, its interference with DNA repair is not likely to result from the interference with protein synthesis since comparable inhibition by cyclohexamide produced a much lesser decrease in DNA repair.

As we have seen, alkylating agents cause alkylation of purines and single-strand breaks in mammalian cells. The damage inflicted by these agents can be repaired either by unscheduled DNA synthesis, or by repair replication (see below). The repair replication for methyl methanesulfonate and N-methyl-N'nitro-N-nitrosoguanidine was the same in normal and xeroderma pigmentosum fibroblasts, indicating that the mechanism of repair of alkylating agents must differ from that of thymine dimers [62].

Unscheduled DNA synthesis in UV-irradiated cells is incomplete, nonrandom, and at least in lymphocytes, occurs at different rates.

Meltz and *Painter* [202] have studied the distribution of repair of thymine dimers using the Cot technique of *Britten* and *Khone*. The results indicate that the repair is random throughout the DNA molecules. Such findings suggest that there are no preferential zones for DNA repair. Thus according to *Meltz* and *Painter*, the damage caused by UV light and the repair of such damage are randomly distributed in spite of the variability in the DNA sequence and in spite of the variable interaction between DNA and proteins. Yet, these findings do not exclude the possibility that small portions of the DNA are inaccessible to repair. Persistent damage is not likely to be detected by the Cot method. Moreover, these findings are in conflict with other studies.

Brunk studied the effects of low doses of UV radiation on the formation of thymine dimers in bacterial DNA and found that most dimers were formed in long pyrimidine tracts and consequently the distribution of the dimers in the UV-irradiated DNA was nonrandom [33].

Only 40-50% of the thymine dimers produced in the DNA after low fluence of UV-irradiation are excised. *Wilkins* and *Hart* [340] irradiated human fibroblasts in vitro and subjected them to the *M. luteus* UV endonuclease to evaluate the repair ability. They found that chromatin treated with 2 M sodium chloride is much more susceptible to the endonuclease than untreated chromatin. The authors concluded that the presence of proteins interferes with the DNA repair.

Using radioautography, *Harris* et al. [126] measured excision repair in DNA directly in nuclei of fibroblasts exposed to UV-irradiation and carcinogens and found that excision repair is nonrandom and occurs in clusters. Repair was more intensive in the inner core of the nucleus, believed to be composed primarily of euchromatin, than in the outer core, which is believed to be composed of heterochromatin.

A working hypothesis which might explain in part the unequal distribution of residual damage in DNA was derived from studies on X-irradiated, partially hepatectomized rats [313], and broadly supported by other experiments to be described.

Irradiation of normal liver interferes with DNA synthesis in the regenerating liver if the animal is partially hepatectomized after irradiation [313]. It is not likely that all the messengers for thymidylic kinase and cytidylic reductase are present in normal livers since these enzymes are never made in extensive amounts, especially not cytidylic reductase [155]. Consequently, a direct effect of X-irradiation on the messenger seemed to be excluded. Among the macromolecules involved in the coding and transcription of proteins involved in the last steps of DNA synthesis in regenerating liver, only DNA appears to be present in liver before hepatectomy. Thus, the inescapable conclusion is that X-irradiation damages the DNA molecules and thereby interferes with their transcribing properties and possibly with their priming abilities for new DNA synthesis. Studies by *Markov* et al. confirmed that X-radiation interferes with the biosynthesis of messenger RNA in regenerating liver and led the authors to conclude that ionizing radiation does not suppress transcription per se, but interferes with gene expression [193].

Direct damage to DNA molecules was demonstrated [302, 318]. Even if DNA molecules are the primary site of radiation injury, it remains to be explained how damage, which is likely to involve all DNA molecules, is ultimately expressed selectively–mainly through interference with transcription of the mRNA of proteins involved in the last steps of DNA synthesis. The solution to this problem can best be understood if one reasons about the difference between the DNA which codes for the proteins involved in the last steps of DNA synthesis and all other DNA molecules in liver; DNA is repressed in normal liver (not transcribed) and becomes derepressed (transcribed) in regenerating liver. Consequently, the repressed DNA must be more sensitive to X-irradiation than derepressed DNA. Some experiments from our laboratory [314] and many previous observations made in others support this conclusion.

Repressed DNA could be more radiosensitive than derepressed DNA for many reasons. For example, the presence of the repressor could radiosensitize the DNA by facilitating crosslink formation. It is also possible that all DNA is damaged but that derepressed DNA is rapidly repaired, whereas the presence of the repressor or other proteins prevents repair [313].

The studies of *Pederson* and *Robbins* [237] may be relevant to the mechanism of restriction of DNA repair. These investigators have measured the binding of [^{3}H]actinomycin to the chromatin of synchronized HeLa cells and found against all expectations that the binding capacity decreases progressively during the S and G2 phases and mitosis. There are stages when one would expect the chromatin to be expanded and DNA to be more accessible to actinomycin than during interphase, when one expects the chromatin to be condensed.

The finding that chlorambucil treatment results in the progressive removal of histones from chromatin, as a result of inhibition of histone synthesis, may explain the results of *Pederson* and *Robbins*. Thus, the effects of both X-irradiation and actinomycin suggest that the DNA of the interphase cell is more susceptible to damage. Whether this is the case because of interference with DNA repair remains to be shown.

It is not known at what stage of the process of DNA synthesis X-irradiation exerts its blocking effect on DNA synthesis. If one accepts modern views that chromosomes are made of multiple replicating units (replicons, see below) and that there are at least three steps in DNA synthesis (initiation, chain growth, and termination), then radioautographic studies suggest that the step sensitive to ionizing radiation is the initiation step rather than chain growth [334].

Results similar to those obtained in liver stimulated to proliferate after partial hepatectomy were obtained when the effect of X-irradiation was studied on organs with normally low or no proliferative activity that were stimulated to proliferate. Such systems include kidneys stimulated to grow after unilateral nephrectomy [297], the uterus stimulated to proliferate after the administration of estrogens [238], and lymphocytes stimulated by phytohemagglutinin [264]. In all cases, irradiation during G0 interferes with DNA synthesis after application of the proliferative stimulus. These findings suggest that in nonproliferative organs of intact mammals, DNA coding for the biosynthesis of proteins involved in DNA synthesis is repressed during interphase. The early observations on the effects of X-irradiation on antibody formation might well be explainable in a similar fashion [12]. Formation of antibodies is blocked if irradiation precedes the administration of antigen, and antibody formation continues if irradiation follows the administration of antigen.

Studies by *Pollard* and *Davis* [245] in which radiation was applied before and after induction of β-galactosidase in special strains of bacteria confirmed the findings made in the mammalian system, except for the fact that much larger doses of irradiation are needed to affect transcription in bacteria than in the mammalian system. Thus, radiation applied before induction interferes with galactosidase synthesis, whereas after induction it has no or little effect on the appearance of the enzyme. *Setlow* and coworkers made similar observations in UV-irradiated bacteria.

Studies by *Stryckmans* and associates [288] have shown that the lymphocytes of leukemic patients submitted to extracorporeal radiation were more radioresistant than normal lymphocytes. The radioresistance has been linked to increased turnover of the lymphocytes. These results illustrate the relevance of our hypothesis to radiotherapy. The survival of lymphocytes could depend on constant replacement of one or more species of RNA. RNA could be coded for on a DNA which is alternately repressed and derepressed. If the periods of derepression are short compared to those of repression, the lymphocytes would obviously be highly radiosensitive. If the mechanism controlling repression were lost, causing the period of repression to be shortened or inhibiting production of repressors, the lymphocytes would be more radioresistant. Interestingly, increased resistance of lymphocytes obtained from patients with chronic lymphocytic leukemia has been observed, and the turnover of the messenger is increased in the leukemic cells. The increased turnover could result from acceleration of transcription, which could be caused by modification in the regulation of repression and derepression in the leukemic lymphocyte [288].

d) Enzymology of Excision Repair

In 1970 *Van Lancker* and *Tomura* purified UV endonuclease from *M. luteus* and demonstrated that it causes additional single-strand breaks in γ-irradiated DNA and increases its template activity for DNA polymerase I [319]. This finding was confirmed by *Hariharan* and *Cerutti* [124], *Paterson* and *Setlow* [235], *Setlow* and *Carrier* [271], and *Noguti* and *Kada* [223, 224]. The observation suggested that even in bacteria the specificity of some correctional endonucleases was not restricted by the lesion inflicted to the DNA. Encouraged by these experiments, we postulated that an enzyme so fundamental to cell survival was likely to be found in all cells including mammalian cells, and that it might have survived through evolution. Consequently, by adjusting the procedure used to purify the enzyme in bacteria, we thought that it would be possible to isolate it from mammalian cells. Moreover, if the enzyme activity was specific for distortion of the DNA helix caused by the diadducts and the monoadducts, then it might also be expected to cause single-strand breaks into DNA to which ultimate carcinogens are bound. Inasmuch as the liver is a major target of ingested carcinogens, or carcinogens injected intraperitoneally, the enzyme was suspected to be present in liver. The mammalian repair endonuclease was purified from rat liver to electrophoretic homogeneity [319]. Unfortunately, the enzyme was found, like many other correctional endonucleases, to be extremely labile, and it took at least 20 purifications to complete a study of its action on single- and double-stranded, apurinic, UV-irradiated, and acetylaminofluorene bound DNA [319].

The elution pattern of the purified enzyme on the 75 Sephadex column suggests a molecular weight between 15,000 and 20,000 daltons. Sedimentation of substrate on neutral and purified enzyme or incubation of the ^{32}P-labeled substrate with purified endonuclease and alkaline phosphatase demonstrates that the enzyme causes single-strand breaks to appear in UV- and acetylaminofluorene-bound double-stranded DNA. The nicked substrate becomes susceptible to an attack by bacterial DNA polymerase I, as shown by the release of material absorbing at 260 nm in the acid-soluble fraction and by increased priming ability of the substrate for DNA polymerase I. The sequential attack of the substrate with the purified endonuclease, alkaline phosphatase, and DNA polymerase I releases in the acid soluble thymine dimers in the case of UV-irradiation and acetylaminofluorene base complexes in the case of acetylaminofluorene-bound DNA.The enzyme was also found to cause nicks to appear in DNA to which 7-bromomethylanthracene is attached [190]. Later it was shown that the mammalian repair endonuclease also caused the appearance of additional single-strand breaks in X-irradiated DNA [301].

The nicks caused by the enzyme are likely to yield 3'PO_4 and OH'5 termini, and therefore, DNA polymerase I binds to the nicked DNA only after treatment of the nicked DNA with alkaline phosphatase. Then DNA polymerase I is free to exert its 5' → 3' exonucleolytic activity, and as a result, mono- and possible oligonucleotides are released in the acid soluble fraction.

Inasmuch as caffeine [176, 327, 341] and cocarcinogens have been claimed to inhibit DNA repair, the effect of these chemicals on enzyme activity were investigated. The enzyme is not inhibited by caffeine, but it is inhibited by two cocarcinogens: anthralin and 12,0-tetradecanoylphorbol-13-acetate. When a phorbol ester, which is not cocarcinogenic was used, it did not inhibit the endonuclease activity. The significance of this effect of the cocarcinogen on the endonuclease is unclear. Whether the finding can help to explain the mechanism of cocarcinogenesis remains to be seen.

In rats the enzyme activity is increased after injection of AAF [323] and markedly increases through de novo synthesis in the first 2 h after partial hepatectomy [322].

It would appear that similar enzymes have been partially purified from human lymphoblasts by *Brent* [32], and from calf thymus by *Bacchetti* and *Berne* [5]. *Feldberg* et al. [98] have purified, from human placenta, a protein that specifically binds to UV-irradiated DNA. The authors excluded binding to pyrimidine dimers and a number of other types of lesions. They believe that the proteins may recognize sugar damage in the UV-irradiated DNA. This finding may be related to the observation of *Payes* made on X-ray-damaged DNA [236]. After irradiation of DNA in solution with kilo rads of X-radiation, a precursor of a malonaldehyde-like substance is formed. The precursor is believed to be derived from damaged deoxyribose moieties, and it remains covalently bound to DNA. The precursor is readily reduced with $NaBH_4$ and consequently when NaB^3H_4 is used, the precursor can be heavily labeled. An enzyme partially purified from rat liver extracts which requires magnesium for activity and is inhibited by Ca^{++}, removes the labeled derivative from DNA.

Lindahl and *Liundquist* [184] isolated an endonuclease specific for apurinic sites from calf thymus. The enzyme, which has a molecular weight of 32,000, is stimulated by magnesium. The enzyme does not normally attack UV- or γ-irradiated DNA. However, when such treated DNA is heated at 70°C for 30 min (thus generating apurinic sites), the enzyme introduces nicks at the 3' site of apurinic acid. A similar enzyme has been purified from liver by *Verly* et al. [325, 326].

The repair endonuclease purified by *Van Lancker* and *Tomura* [319] is one among the mammalian correctional endonucleases to have been identified, and it is not inconceivable that other correctional endonucleases are also present in the mammalian cells. However, the mammalian cells should not be expected to induce endonucleases specific for any type of injury to which they may be exposed. Consequently, correctional endonuclease with specificities broader than those found in bacteria may be expected to be found in mammalian cells. Still several endonucleases differing in their size, their specificity, and their

Table 5. Mammalian DNA polymerase

Source	Intracellular location or types	Properties	Other catalytic activities	References
Human				
Embryo (skin, muscle, kidney, lung)		M.W. 105,000		*Margalith, M.* et al. (1976) [192]
Liver				*Chang, L.M.S.* (1974) [44]
Spleen		M.W. 70,000		*Witkin, S.S.* et al. (1975) [344]
Bone marrow	Nuclear	4S		*Coleman, M.S.*, and
	Cytoplasmic	6-8 S		*Hutton, J.J.* (1973) [67]
Normal lymphocyte	(I)	M.W. 160,000	Exonuclease (+)	*Smith, R.G., Gallo, R.C.* (1972) [278]
	R-	M.W. 30,000	Exonuclease (+)	
	α-	High m.w.		*DeFernandez, M.T.F.* et al. (1975) [99]
	β-	Low m.w.		
				Bertazzoni, U. et al. (1976) [18]
	RPMI-cell			*Smith, R.G.* et al. (1975) [279]
Lymphoblastoid infected with simian sarcoma virus	Cytoplasmic (I)	6-8 S		*Lewis, B.M.* (1974) [175]
	Nuclear and cytoplasmic (II)	3.3 S		
	R- (III)			
Leukemia cell	α-	6-8 S		*Srivastava, B.I.S.* (1974, 1975) [281, 282]
	β	2.5-3.5 S		*Mordoh, J., Friedlander, B.R.* (1975) [217]
	γ			
	Cytoplasmic			*Chang, R.T.* et al. (1976) [50]
Hodgkin's disease				*Piperno, J.R., Kallen, R.G.* (1973) [242]
Burkitt Lymphoma	Cytoplasmic	6-7 S		*Srivastava, B.I.S.* (1974) [281]
Breast Tumor				*Gerwin, R.I., Bassin, R.H.* (1973) [109]
Human milk				*Gerwin, R.I.* et al. (1973) [110]
HeLa cells	α-			*Spadari, S., Weissbach, A.* (1975) [280]
	β-			
	γ-			*Gerard, G.F.* (1975) [108]
	Mitchondrial			*Tibbets, C.J.B., Vindsrad, J.* (1973) [298]

Table 5 (contd.)

Source	Intracellular location or types	Properties	Other catalytic activities	References
	I II	3.5 S: 45,000 10.5 S:150,000		*Chiu, R.W., Baril, E.F.* (1975) [55]
KB cell			Exonuclease (+) Endonuclease (-)	*Greene, R., Korn, D.* (1970) [116]
	Cytoplasmic	M.W. 87,000	Exonuclease (+)	*Sedwick, W.D.* et al. (1972, 1974, 1975) [266-268]
	Nuclear I		Exonuclease (-)	
	Nuclear II		Exonuclease (+)	
	Nuclear	3.4 S		*Wang, T.S.F.* et al. (1975) [332]
Calf	Cytoplasmic	6-8 S		*Bollum, F.J.* (1959, 1962, 1963, 1966, 1968, 1972, 1974, 1975) [19-23, 25, 47, 76]
	Nuclear	3.4 S (3.3 S)	44,000	*Chang, L.M.C., Bollum, F.J.* (1973) [48] *Chang, L.M.C.* (1974) [45] *Yoshida, S.* et al. (1974) [349]
	A	M.W. 200,000		*Momparler, R.L.* et al. (1973) [216]
	B	M.W. 150,000		
	α			*Henner, D., Furth, J.J.* (1975) [134]
Rabbit Bone marrow	Cytoplasmic	8 S and 3.4 S	Exonuclease (-)	*Chang, L.M.S., Bollum, F.J.* (1971, 1972) [46, 47]
	Nuclear	3.4 S	Exonuclease (-)	*Chang, L.M.S.* (1973) [43] *Byrnes, J.* et al. (1973, 1975a, b) [35, 37, 38]
Reticulocyte		8 S (11 S)		*Byrnes, J.* et al. (1975a) [37]
Hamster Kidney	Cytoplasmic (I)	6.9 S (6.8 S)		*Lazarus, L.H., Kitron, N.* (1975) [170]

	Nuclear (II)	3.47 S (3.3 S)		*Craig, R.K., Keir, H.M.* (1975a-c) [74-76]
Rat		M.W. 49,000	Exonuclease (-) Endonuclease (-)	*Berger, H., jr., Huang, R.C.C.* (1971) [17]
	I		Exonuclease (+)	*Tomura, T., Van Lancker, J.L.* (1974) [300]
	II		Exonuclease (+)	
	III		Exonuclease (+)	
		3.2 S		*Lynch, W.E.* et al. (1976) [188]
		7.1 S		
		6-8 S		*Morioka, K., Terayama, H.* (1974) [218]
Regenerating Liver	Cytoplasmic	6-8 S		*Chang, L.M.S., Bollum, F.J.* (1972) [47]
	Nuclear and Cytoplasmic	3-4 S		
	I			*Baril, E.* et al. (1973) [9]
	II			
		7.2 S		*Lynch, W.E.* et al. (1976) [188]
LF Hepatoma				*Cazillis, M.* et al. (1975) [41]
A. Rous virus transformed		M.W. 70,000		*Bandyopadhyay, A.K.* (1975) [6]
Ascites hepatoma	P 1	M.W. 117,000		*Tsuruo, T.* et al. (1972a, b, 1974, 1975a, b) [306-310]
	P 2	M.W. 44,000		*Tsuruo, T., Ukita, T.* (1974) [305]
	C	M.W. 142,000		
Brain	A	9 S		*Chiu, J.F., Sung, S.C.* (1971, 1972, 1973) [52-54]
	B (Nuclear)	3.4 S		
	III			
Cardiac muscle				*Claycomb, W.C.* (1973) [59]
Intestinal mucosa				*Pouson, R., Zbarsky, S.H.* (1973) [248]
Nephroma cell	N 1		Endonuclease (+)	*Salzman, L.A., McKerlie, L.* (1975) [262]
	N 2		Endonuclease (+)	
	N 3		Endonuclease (+)	

Table 5 (contd.)

Source	Intracellular location or types	Properties	Other catalytic activities	References
	N 4		Endonuclease (+)	
	C 1		Endonuclease (+)	
	C 2		Endonuclease (+)	
Walker carcinosarcoma 256				*Wang, T.Y.* (1968) [333] *Waalkes, T.P.* et al. (1974) [328]
Mouse				
Liver	Nuclear	3.3 S		*Livingston, D.M.* et al. (1974) [185]
	Cytoplasmic	6-8 S		*Hecht, N.B.* (1975) [127]
	Mitochondrial	M.W. 150,000-170,000		
Mammary gland				*Mukherjee, A.S., Banerjee, M.R.* (1974) [219]
Fibroblasts (Earle's L cell)	Cytoplasmic	6-8 S		*Chang, L.M.S.* et al. (1973) [49]
	Nuclear and cytoplasmic	3-4 S		
Ehrlich ascites tumor cell				*Hirano, H.* et al. (1975) [137]
JLS-V9 cell	N 1			*Bandyopadhyay, A.K.* (1975a, b) [6, 333]
	N 2			
	C 1			
	C 2			
	C 3	M.W. 110,000 (70,000+35,000)		
Myeloma cell	C I	10.5 S (8-6 S)		*Hachmann, H.J., Lezius, A.G.* (1975) [119]
	C II	8.6 S (10.5 S)		
	C III	5.7 S (5.2 S)		
	II	2.5 S		*Matsukage, A.* et al. (1973, 1974, 1975) [198-200]
	III	243,000, 270,000 or 315,000		
	C I	7.2 S		*Hecht, N.B.* (1975) [185]
	C II	3.5 S		

molecular properties have been described in mammalian cells. Their role in the cell physiology remains to be identified [69, 201, 311, 331].

Little is known of the excision process during DNA repair in mammalian cells. Most investigators have used bacterial DNA polymerase I to secure the excision of the DNA monoadducts. As in bacteria, the excision step could involve either a free exonuclease or a polymerase with exonucleolytic activity. *Lindahl* [183] has purified an exonuclease from rabbit tissue referred to as DNAase IV. The enzyme hydrolyzes DNA in a 5' → 3' direction, releasing oligonucleotides containing 5 or 8 nucleotide residues.

Doniger and *Grossman* have purified a repair exonuclease from human placenta [231]. The enzyme attacks single-stranded DNA and hydrolyzes at both 3' and 5' termini, yielding oligonucleotides approximately 4 nucleotides in length. The enzyme has no effect on intact native DNA. Yet, it can initiate excision at single-strand breaks.

Inasmuch as, to date, almost all mammalian DNA polymerase are devoid of exonuclease activity (see Table 5), it seems quite logical to assume that either the deoxyribonuclease IV or the Grossman correctional exonuclease play the central role in the incision phase of the excision repair process.

However, although the evidence is still not conclusive, with use of a nicked UV irradiated DNA cellulose column, we have reported the partial purification from rat liver of a polymerase with a protein complex which excises thymine dimers and patches the gap [86]. A mammalian DNA polymerase with exonuclease activity has also been described by *Byrnes* et al. [36].

If such a polymerase could be purified from mammalian cells, it should surely facilitate the study of excision and the patching steps in the course of DNA repair.

e) Postreplication Repair

Postreplication repair is believed to exist in mammalian systems. However, the mechanism by which it takes place is unknown. Whatever this mechanism may be, it is likely to be much more complex than in bacteria because of the complexity of the structure of the chromosome [153], the mechanism of DNA replication in mammalian cells, and the intricacies of the interference with DNA synthesis by UV ionizing radiation in mammalian cells. The structure of the mammalian chromosome has been reviewed [289]. *Painter* has reviewed the concept of replicons [231], and it will only be briefly summarized here.

The replication of the eukaryotic chromosome is undoubtedly a complicated phenomenon. It involves replication of DNA where histone and nonhistone proteins with quaternary interaction of the macromolecules permit the genotype to be expressed into the phenotype of a specifically differentiated cell. Even if the DNA forms a continuous chain in the chromatid, its synthesis does not start at one end and terminate at the other. There are, in the mammalian chromosome, various points of replication leading to polynucleotides of various lengths. These segments of newly synthesized DNA are called replicons. They have a specific site, called the origin, at which DNA replication begins and both parental strands are duplicated. The elegant experiments that led to the description and the evaluation of the number of replicons per chromosome or per cell will not be described. The reader is referred to the review by *Painter* [231]. It is estimated that 100 replicons exist per chromosome and 1.5 to $2x10^5$ replicating units exist per cell with an average size of 30 μm. Little is known of the initiation of DNA synthesis in the replicon. It is, however, established that the elongation of the chain is discontinuous and that 4 S fragments (200 nucleotides long) are formed and later united (the Okazaki fragments in *E. coli* are 1000 nucleotides long) [227]. Essentially nothing is known of the termination of DNA synthesis in the replicon.

When mammalian cell lines are UV-irradiated, a low molecular weight DNA appears. This finding was believed to indicate that gaps occurred in the daughter strands as in bacteria [172, 173]. However, *Edenberg* [90] and *Painter* [230] have suggested that similar results would be obtained if either UV-irradiation or X-radiation blocked the synthesis of complete replicons. The situation is further complicated by the fact that moderate doses of ionizing radiation (100-1000 rads) block replicon initiation. This was proposed by *Van Lancker* [313] from experiments on intact animals, inferred from DNA film radioautography by *Weiss* [337] and *Watanabe* [335], and established by *Painter* and *Young* [234] who demonstrated, by irradiating mouse L, HeLa, S3, and CHO cells, that X-radiation of mammalian cells inhibits the initiation of DNA replicons. At present the only reliable facts concerning postreplication repair in mammalian cells are that: (1) after UV-irradiation a low molecular weight DNA appears, (2) after a longer period of time the molecular weight of the DNA will be the same as in the normal cells, and (3) the restoration of the normal molecular weight is inhibited by caffeine.

In spite of these uncertainties about postreplication repair, several investigators have claimed that it takes place either after the administration of alkylating agents (MNG, MNU, MMS) or ultraviolet irradiation. Nothing is known of the enzymology of DNA postreplication repair in mammalian cells. However, on the basis of damage caused by crosslinking agents in cells obtained from patients with Fanconi's anemia, *Fujiwara* and *Tatsumi* [104] have proposed that the enzymes involved in the excision repair pathway are responsible for the initial steps of postreplication repair.

VI. Repair Carcinogenesis and Disease

At a first glance the very existence of DNA repair should argue against the somatic mutation theory because repair would presumably eliminate all damage caused to DNA. Unfortunately, students of disease are aware that a large number, if not the majority, of diseases result from inadequate response of the defense mechanisms to injury. DNA repair is a molecular form of defense mechanism, and we have seen already that it may contribute to the formation of double-strand breaks [26]. Moreover, there is no evidence that DNA repair is complete and faithful. Experiments with intact animals with X-radiation certainly indicate that some of the DNA damage persists for long periods of time, if not indefinitely [318]. However, studies of *Maher* et al. in which the incidence of azoguanine resistant mutants was determined after UV-radiation of cultured cells suggests that excision repair is faithful. The results do not completely exclude the possibility of the persistance of damage to DNA [189].

Moreover, it is generally believed that postreplication repair is error prone. To appreciate the significance of DNA repair, let's consider the options available to a cell with damaged DNA (Fig. 13). The damage can either be repaired or not. If repaired, the repair may be integral, and the life of the cell goes on unaltered, but the repair could also be faulty and bring about such changes as double-strand breaks and base replacement, which may or may not be compatible with replication of the DNA. If the DNA is not repaired, it may or may not replicate. In absence of replication, one may expect blocking of DNA synthesis and ultimately cellular death. If the damaged DNA is replicated, the new DNA may either be repaired through an error-prone replication process, or contain breaks, gaps, and mutations (point mutations, frameshift mutations). Such lesions may or may not be compatible with transcription. The consequences for the cell phenotype depend upon whether the damaged DNA belongs to the repressed or derepressed pool.

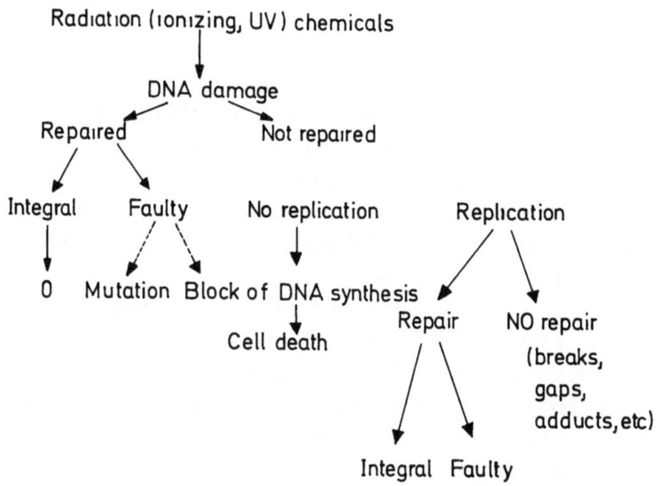

Fig. 13

Damage to DNA which belongs to the repressed pool will not affect the phenotype. Damage to DNA which belongs to the derepressed pool will result in transcription defects. The ultimate consequences will depend upon whether it is the DNA of a germ or a somatic cell that is afflicted (Fig. 14).

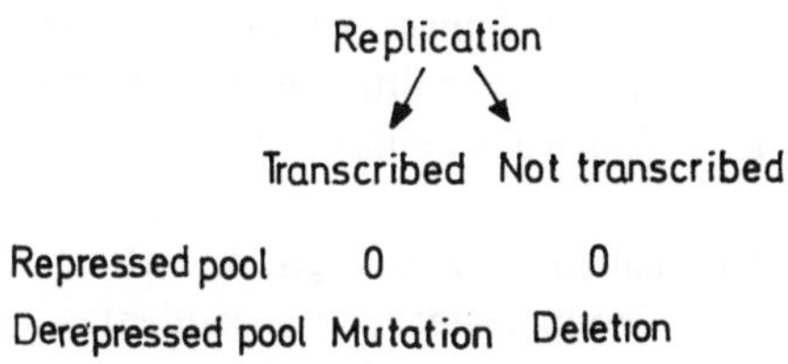

Fig. 14

It is not difficult to imagine how the alterations of DNA described above could in the germ cell lead to sterility, inborn defects, and chromosomal anomalies. In the somatic cell they could lead to cell death or to alterations of gene expression that might be benign—that is, alterations that have few survival advantages, as in the case of benign tumors or some forms of malignant cells that would have a considerable survival advantage over the cells of the host. Moreover, if the cell dies, is it not conceivable that at least portions of the damaged DNA will resist hydrolytic attacks and, as such, could become immunogenic and lead to the formation of antigen antibody complexes? Are some forms of *lupus* diseases in which DNA is not repaired?

Time will tell how much of the speculation is real. What is certain is that absence of repair may lead to cell death or somatic mutation, and the possibility that at least some cancers result from somatic mutation cannot be excluded. Therefore, the study of DNA repair is particularly relevant to mechanisms of carcinogenesis.

VII. Conclusion

Nonviral carcinogens have at least one property in common: they either alter the DNA bases or form mono- or diadducts with the bases. Chemical carcinogens might react with macromolecules directly or after metabolic conversion. Although chemical carcinogens might react with polysaccharides, RNA, and proteins, available evidence points to the need for investigating the role of binding of carcinogens to DNA and the role of base damage caused by UV light and ionizing radiation in carcinogenesis.

Such damage could produce somatic mutations which, under the proper circumstances, would yield one or more cells freed of some of the host control on cell proliferation and differentiation and equipped with remarkable survival advantages over the host (nutritional trap, invasiveness, capacity to metastasize, immunoresistance, etc.). The fact that all these nonviral carcinogens are mutagenic supports this somatic mutation hypothesis.

Although the transfer of the phenotype of the cancer cell from one generation of cells to another is not in conflict with the somatic mutation theory, it is susceptible to other interpretations.

The somatic mutation theory of carcinogenesis is one of the oldest theories. New interest was instilled by the discovery that much of the damage to DNA can be repaired.

In bacteria the discovery of DNA repair followed a rather logical pattern: the discovery of UV resistant mutants followed, at least in some cases, by the systematic isolation of the enzymes involved. Three principal forms of repair were discovered: photoreactivation, dark repair, and postreplication repair.

In mammalian cells the course of the study of repair of DNA damage was more tortuous. However, a multitude of observations on intact animals and cells in culture led to the conclusion that UV, ionizing radiation, and carcinogen DNA damage is repaired by two different processes: excision and postreplication repair.

The study of the enzymology of DNA repair on mammalian cells has begun and at least one correctional endonuclease and two correctional exonucleases have been identified. Time will tell whether a faulty repair mechanism or restrictions of DNA repair contribute to the aging process and the development of disease, cancer among them.

Acknowledgment. The research work done in this laboratory was supported by a U.S. Public Health Service grant, No. 5 RO1 CA 14840-02.

The author is grateful to Dr. *Takanori Tomura* for his constructive suggestions and assistance in searching the literature.

References

1. Agarwal, M.K., Weinstein, I.B.: Modifications of ribonucleic acid by chemical carcinogens. II. In vivo reaction of N-2-acetylaminofluorene with rat liver ribonucleic acid. Biochemistry **9**, 503-507 (1970).
2. Alexander, P., Horning, E.S.: Observations on the Oppenheimer method of inducing tumors by subcutaneous implantation of plastic films. In: Ciba Foundation Symposium on Carcinogenesis, Mechanisms of Action, (Wolstenholme, G., O'Connor, M., eds.) London: Churchill 1959, pp. 12-25.
3. Ames, B.N., Lee, F.D., Durston, W.E.: An improved bacterial test for the detection and classification of mutagens and carcinogens. Proc. nat. Acad. Sci. (Wash.) **70**, 782-786 (1973).
4. Apgar, V.: Drugs in pregnancy. J. Amer. med. Ass. **190**, 840-841 (1964).
5. Bacchetti, S., Benne, R.: Purification and characterization of an endonuclease from calf thymus acting on irradiated DNA. Biochim. biophys. Acta (Amst.) **390**, 285-297 (1975).
6. Bandyopadhyay, A.K.: Partial purification and characterization of the cytoplasmic and nuclear deoxynucleic acid polymerases from baby hamster kidney cells. Biochim. biophys. Acta (Amst.) **407**, 1-13 (1975a).
7. Bandyopadhyay, A.K.: Partial purification and properties of DNA polymerases from JLS-V10 cells. Arch. Biochem. **166**, 83-93 (1975b).
8. Barbour, S.D., Clark, A.J.: Biochemical and genetic studies of recombination proficiency in Escherichia coli. I. Enzymatic activity associated with rec. B^+ and rec. C^+ genes. Proc. nat. Acad . Sci. (Wash.) **67**, 434-441 (1970).
9. Baril, E.F., Jenkins, M.D., Brown, O.E., Laszlo, J., Morris, H.P.: DNA polymerase I and II in regenerating rat liver and Morris hepatomas. Cancer Res. **33**, 1187-1193 (1973).
10. Baserga, R.: Carcinogenesis program. Presented at IV Annual Collaborative Conference 1976.
11. Bauer, H.K.: Mutationstheorie der Geschwulst-Entstehung. Berlin: Springer 1928.
12. Benjamin, E., Sluka, E.: Antikörperbildung nach Röntgenstrahlen. Wien. klin. Wschr. **21(10)**, 311-313 (1908).
13. Benz, W.C., Berger, H.: Selective allele loss in mixed infections with T4 bacteriophage. Genetics **73**, 1-11 (1973).
14. Berenblum, I.: Sequential aspects of chemical carcinogenesis: skin. Cancer (Philad.) **1**, 323-345 (1975).
15. Berenblum, I., Shubik, P.: Role of croton oil applications associated with a single painting of a carcinogen, in tumor induction of mouse's skin. Brit. J. Cancer **1**, 379-382 (1947a).
16. Berenblum, I., Shubik, P.: New, quantitative, approach to the study of the stages of chemical carcinogenesis in the mouse's skin. Brit. J. Cancer **1**, 383-391 (1947b).
17. Berger, H., jr., Huang, R.C.C.: Purification and characterization of a DNA polymerase from rat liver. J. biol.Chem. **246**, 7275-7283 (1971).
18. Bertazzoni, U., Stefanini, M., Noy, G.P., Giulotto, E., Nuzzo, E., Falaschi, A., Spadari, S.: Variation of DNA polymerase-α and -β during prolonged stimulation of human lymphocytes. Proc. nat. Acad. Sci. (Wash.) **73**, 785-789 (1976).
19. Bollum, F.J.: Thermal conversion of nonpriming DNA to primer. J. biol. Chem. **234**, 2733-2734 (1959).
20. Bollum, F.J.: Oligodeoxyribonucleotide-primed reactions catalyzed by calf thymus polymerase. J. biol. Chem. **237**, 1945-1949 (1962).
21. Bollum, F.J.: Studies on the nature of calf thymus DNA-polymerase products. Cold Spr. Harb. Symp. quant. Biol. **28**, 21-26 (1963).
22. Bollum, F.J.: Preparation of deoxynucleotide polymerizing enzymes from calf thymus gland. Progr. Nucleic Acid Res. **1**, 284-295 (1966).
23. Bollum, F.J.: Deoxynucleotide polymerizing enzymes from calf thymus gland. Meth. Enzymol. **12b**, 591-611 (1968).
24. Bollum, F.J.: Mammalian DNA polymerases. Progr. Nucleic Acid Res. Molec. Biol. **15**, 109-144 (1975).
25. Bollum, F.J., Chang, L.M.S., Tsiapalis, C.M., Dorson, J.W.: Nucleotide polymerizing enzymes from calf thymus gland. Meth. Enzymol. **29**, 70-81 (1974).

26. Bopp, A., Cardy, S., Burkhart, B., Hagen, U.: Action of exonuclease III on γ-irradiated DNA. Biochim. biophys. Acta (Amst.) **294**, 47-56 (1973).
27. Boutwell, R.K., Colburn, N.H., Muckerman, C.C.: In vivo reactions of β-propiolactone. Ann. N.Y. Acad. Sci. **163**, 751-764 (1969).
28. Boveri, T.: The Origin of Malignant Tumors. Baltimore: Williams and Wilkins 1929.
29. Boyland, E., Williams, K.: Reaction of urethan with nucleic acids in vivo. Biochem. J. **111**, 121-127 (1969).
30. Braun, A., Grossman, L.: An endonuclease from Escherichia coli that acts preferentially on UV-irradiated DNA and is absent from the uvrA and uvrB mutants. Proc. nat. Acad. Sci. (Wash.) **71**, 1838-1842 (1974).
31. Brend, K.G.: Induction of neoplasms by plastic implants. Bull. Path. **9**, 209 (1968).
32. Brent, T.P.: Partial purification of endonuclease activity from human lymphoblasts. Separation of activities from depurinated DNA and DNA irradiated with ultraviolet light. Biochim. biophys. Acta (Amst.) **407**, 191-199 (1975).
33. Brunk, C.F.: Distribution of dimers in ultraviolet-irradiated DNA. Nature (New Biol.) **241**, 74-76 (1973).
34. Burdette, W.S.: The significance of mutation in relation to the origin of tumors: A review. Cancer Res. **15**, 201-226 (1955).
35. Byrnes, J.J., Downey, K.M., So, A.G.: Bone marrow cytoplasmic deoxyribonucleic acid polymerase variation of pH and ionic environment as a possible control mechanism. Biochemistry **12**, 4378-4384 (1973).
36. Byrnes, J.J., Downey, K.M., So, A.G.: Regulation of the 3' to 5' exonuclease activity associated with mammalian DNA polymerase. Cancer Enzymology VIII. Miami Winter Symp., pp. 100-102, 1975.
37. Byrnes, J.J., Downey, K.M., Esserman, L., So, A.G.: Mechanism of hemin inhibition of erythroid cytoplasmic DNA polymerase. Biochemistry **14**, 796-799 (1975a).
38. Byrnes, J.J., Downey, K.M., So, A.G.: Regulation of the 3' to 5' exonuclease activity associated with mammalian DNA polymerase. Miami Winter Symp. **8**, 100-102 (1975b).
39. Carr, D.H.: Chromosome studies in spontaneous abortions. Obstet. Gynec. **26**, 308-326 (1965).
40. Carrier, W.L., Setlow, R.B.: An endonuclease from *M. luteus* that acts on uracil in UV-irradiated BrUra-containing DNA. Fed. Proc. **33**, 1599 (Abst.) (1974).
41. Cazillis, M., Derecondo, A.M., Frayssinet, C.: High molecular weight deoxyribonucleic acid polymerase of LF hepatoma. Biochim. biophys. Acta (Amst.) **407** 133-146 (1975).
42. Cerutti, P.A.: Photochemistry and photobiology of nucleic acids, Wang, S.Y., Patrick, M. (eds.). New York: Gordon and Breach 1974, Vol. II.
43. Chang, L.M.S.: Calf thymus chromatin DNA polymerase. Fed. Proc. **32**, 415-451 (1973).
44. Chang, L.M.S.: Replication of initiated polyriboadenylic acid by mammalian low molecular weight DNA polymerase. J. biol. Chem. **249**, 7441-7446 (1974a).
45. Chang, L.M.S.: Purification and properties of low molecular weight DNA polymerase from mammalian cells. Meth. Enzymol. **29**, 81-88 (1974b).
46. Chang, L.M.S., Bollum, F.J.: Low molecular weight DNA polymerase in mammalian cells. J. biol. Chem. **246**, 5835-5837 (1971).
47. Chang, L.M.S., Bollum, F.J.: Variation of DNA polymerase activities during rat liver regeneration. J. biol. Chem. **247**, 7948-7950 (1972).
48. Chang, L.M.S., Bollum, F.J.: A comparison of associated enzyme activities in various DNA polymerases. J. biol. Chem. **248**, 3398-3404 (1973).
49 Chang, L.M.S., Brown, M., Bollum, F.J.: Induction of DNA polymerase in mouse L cells. J. molec. Biol. **74**, 1-8 (1973).
50. Chang, R.Y., Laszlo, J., Keller, P.: Effects of nitrosoureas on human DNA polymerase activities from acute and chronic granulocytic leukemia cells. Biochim. biophys. Acta (Amst.) **425**, 463-468 (1976).
51. Charles, D.R., Luce-Clausen, E.M.: Kinetics of papilloma formation in benzpyrene-treated mice. Cancer Res. **2**, 261-263 (1942).
52. Chiu, J.F., Sung, S.C.: Separation and properties of DNA polymerases from developing rat braint. Biochim. biophys. Acta (Amst.) **246**, 44-50 (1971).
53. Chiu, J.F., Sung, S.C.: Particulate form of DNA polymerase in rat brain. Nature (New Biol.) **239**, 176-178 (1972).

54. Chiu, J.F., Sung, S.C.: Separation and characterization of protein factor of DNA polymerase B from developing rat brain. Biochim. biophys. Acta (Amst.) **331**, 54-60 (1973).
55. Chiu, R.W., Baril, E.F.: Nuclear DNA polymerases and the HeLa cell cycle. J. biol. Chem. **250**, 7951-7957 (1975).
56. Chon, W.E., Leonard, N.J., Wang, S.Y.: Appendix abbreviations for pyrimidine photoproducts. In: Photochemistry and Photobiology of Nucleic Acids. New York: Academic Press (in press), 1976, Vol. II, pp. 403-413.
57. Christensen, G.M., Jackson, K.L.: Irreparable damage following X-irradiation of the fetal mouse. Radiat. Res. **30**, 350-354 (1967).
58. Clark, A.J.: Recombination of deficient mutants of E. coli and other bacteria. Ann. Rev. Genet. **7**, 67-86 (1973).
59. Claycomb, W.C.: DNA synthesis and DNA polymerase activity in differentiating cardiac muscle. Biochem. biophys. Res. Commun. **54**, 715-720 (1973).
60. Cleaver, J.E.: Defective repair replication of DNA in Xeroderma Pigmentosum. Nature (Lond.) **218**, 652-656 (1968).
61. Cleaver, J.E.: Repair replication of mammalian cell DNA: Effects of compounds that inhibit DNA synthesis or dark repair. Radiat. Res. **37**, 334-348 (1969).
62. Cleaver, J.E.: Repair of alkylation damage in ultraviolet-sensitive (Xeroderma Pigmentosum) human cells. Mutation Res. **12**, 453-462 (1971).
63. Cleaver, J.E., Painter, R.B.: Absence of specificity in inhibition of DNA repair replication by DNA-binding agents, cocarcinogens, and steroids in human cells. Cancer Res. **35**, 1773-1778 (1975).
64. Cohen, M.M., Shaw, M.W.: Specific effects of viruses and antimetabolites on mammalian chromosomes. In Vitro **1**, 50-66 (1965).
65. Cole, L.J., Nowell, P.C.: Hepatomas in mice: Incidence increased after gamma irradiation at low dose rates. Science **148**, 96-97 (1965).
66. Cole, R.S., Sinden, R.R.: Repair of cross-linked DNA in Escherichia coli. Basic Life Sci. **5**, 487-495 (1975).
67. Coleman, M.S., Hutton, J.J.: DNA polymerases in human bone marrow. Fed. Proc. **32**, 451 (1973).
68. Commoner, B., Vithayathil, A.J., Henry, J.I.: Detection of metabolic carcinogen intermediates in urine of carcinogen-fed rats by means of bacterial mutagenesis. Nature (Lond.) **249**, 850-582 (1974).
69. Cordis, G.A., Goldblatt, P.J., Deutscher, M.P.: Purification and characterization of a major endonuclease from rat liver nuclei. Biochemistry **14**, 2596-2602 (1975).
70. Cox, R., Damjanov, I., Abanobi, S.E., Sarma, D.S.R.: A method for measuring DNA damage and repair in the liver in vivo. Cancer Res. **33**, 2114-2121 (1973).
71. Craddock, V.M.: Methylation of t-RNA and ribosomal RNA in rat liver in the intact animal and the effect of carcinogens. Biochim. biophys. Acta (Amst.) **195**, 351-369 (1969).
72. Craddock, V.M.: Effect of a single treatment with the alkylating carcinogens dimethylnitrosamine, diethylnitrosamine and methyl methanesulphonate, on liver regenerating after partial hepatectomy. Chem. biol. Interact. **10**, 323-332 (1975a).
73. Craddock, V.M.: Effect of a single treatment with the alkylating carcinogens dimethylnitrosamine, diethylnitrosamine and methyl methanesulphonate, on liver regenerating after partial hepatectomy. Chem. Biol. Interact. **10**, 313-321 (1975b).
74. Craig, R.K., Keir, H.M.: Deoxyribonucleic acid polymerases of BHK-21/C13 cells: Partial purification and characterization of the enzymes. Biochem. J. **145**, 215-224 (1975a).
75. Craig, R.K., Keir, H.M.: Deoxyribonucleic acid polymerases of BHK-21/C13 cells: Heterogeneity, molecular asymmetry and subcellular distribution of the enzymes. Biochem. J. **145**, 225-232 (1975b).
76. Craig, R.K., Keir, H.M.: Deoxyribonucleic acid polymerases of BHK-21/C13 cells: Relationship to the physiological state of the cells, and to synchronous induction of synthesis of deoxyribonucleic acid. Biochem. J. **145**, 233-240 (1975c).
77. Crow, J.F.: Chemical risk to future generations. Sci. Critizen **10**, 113-117 (1968).
78. Curtis, H.J., Czernik, C., Tilley, J.: Tumor induction as a measure of genetic damage and repair in somatic cells of mice. Radiat. Res. **34**, 315-319 (1968).
79. Dalrymple, G.V., Sanders, J.L., Moss, A.J., Baker, M.L., Wilkinson, K.P.: Radiation produces breaks in L cell and mouse liver DNA characterized by 5' phosphoryl termini. Biochem. biophys. Res. Commun. **35**, 300-311 (1969).

80. Damjanov, I., Cox, R., Sarma, D.S.R., Farber, E.: Patterns of damage and repair of liver DNA induced by carcinogenic methylating agents in vivo. Cancer Res. **33**, 2122-2128 (1973).
81. Dawson, K.M.: Time course of the effects of AAF on mouse liver nucleic acid synthesis and its modification by inhibitors. Chem. biol. Interact. **5**, 153-165 (1972).
82. DeLucia, P., Cairns, J.: Isolation of an E. coli strain with a mutation affecting DNA polymerase. Nature (Lond.) **224**, 1164-1166 (1969).
83. Dewey, W.C., Humphrey, R.M.: Radiosensitivity and recovery of radiation damage in relation to the cell cycle. In: Cellular Radiation Biology. Baltimore: Williams and Wilkins 1964, pp. 340-375.
84. Dipple, A., Brookes, P., Mackintosh, D.S., Rayman, M.P.: Reaction of 7-bromomethylbenz(a)anthracene with nucleic acids, polynucleotides and nucleosides. Biochemistry **10**, 4323-4330 (1971).
85. Djordjevic, R., Kim, J.H.: Modification of radiation response in synchronized HeLa cells by metabolic inhibitors: Effects of inhibitors of DNA and protein synthesis. Radiat. Res. **37**, 435-450 (1969).
86. Doniger, J., Grossman, L.: Human correxonuclease: Purification and properties of a DNA repair exonuclease from placenta. (Unpublished).
87. Dube, S.K., Marcker, K.A., Clark, B.F.C., Cory, S.: Nucleotide sequence of N-formylmethionyl-transfer RNA. Nature (Lond.) **218**, 232-233 (1968).
88. DuBois, K.P., Cochran, K.W., Doull, J.: Inhibition of citric acid synthesis in vivo by X-irradiation. Proc. Soc. exp. Biol. (N.Y.) **76**, 422-427 (1951).
89. Dulbecco, R.: Experiments on photoreactivation of bacteriophages inactivated with ultraviolet radiation. J. Bact. **59**, 329-347 (1950).
90. Edenberg, H.J.: Inhibition of DNA synthesis by ultraviolet light. In: Molecular Mechanisms for Repair of DNA. 5B, Part B, pp. 631-633. New York-London: Plenum Press 1975.
91. Elkind, M.M., Sutton, H.: X-ray damage and recovery in mammalian cells in culture. Nature (Lond.) **184**, 1293-1295 (1959).
92. Elkind, M.M., Sutton, H., Moses, W.B.: Post-irradiation survival kinetic of mammalian cell grown in culture. J. cell. comp. Physiol. Part 2, **58**, 113-134 (1961).
93. Enomoto, K., Sato, K., et al.: Reactivity of the diacetyl derivative of the carcinogen 4-hydroxyaminoquinoline-1-oxide DNA, RNA, and other nucleophiles. Life Sci. **7/II**, 1025-1032 (1968).
94. Farber, E., Magee, P.N.: The probable alkylation of liver ribonucleic acid by the hepatic carcinogens dimethylnitrosamine and ethionine. Biochem. J. **76**, 58 (1960).
95. Farber, E., Sarma, D.S.R.: DNA damage and repair during carcinogenesis. In: Control Processes in Neoplasia. New York: Academic Press 1974b, pp. 173-185.
97. Farber, E., McConomy, J., Franzen, B., Marroquin, F., Stewart, G.A., Magee, P.N.: Interaction between ethionine and rat liver ribonucleic acid and protein in vivo. Cancer Res. **27**, 1761-1772 (1967).
98. Feldberg, R., Grossman, L.: Enzymatic repair of DNA: Isolation and characterization of a binding protein specific for UV-irradiated DNA. Biochemistry (in press) (1976).
99. Franze de Fernandez, M.T., Mordoh, J., Fridlender, B.R.: New properties of the DNA polymerase-β isolated from non-stimulated normal human lymphocytes. Biochem. biophys. Res. Commun. **65**, 1409-1417 (1975).
100. Frei, J.V.: Some mechanisms operative in carcinogenesis, (a review). Chem. biol. Interact. **13**, 1-25 (1976).
101. Friedberg, E.C.: DNA repair of ultraviolet-irradiated bacteriophage T4. Photochem. Photobiol. **21**, 277-289 (1975).
102. Friedberg, E.C., Hadi, S.M., Goldthwait, D.A.: J. biol. Chem. **244**, 5879-5889 (1969).
103. Friedberg, E.C., King, J.J.: Endonucleolytic cleavage of UV-irradiated DNA controlled by the V^+ gene in phage T4. Biochem. biophys. Res. Commun. **37**, 646-651 (1969).
104. Fujiwara, Y.: Postreplication repair of alkylation damage to DNA of mammalian cells in culture. Cancer Res. **35**, 2780-2789 (1975).
105. Ganesan, A.K.: Persistence of pyrimidine dimers during post-replication repair in Ultraviolet light-irradiated Escherichia coli K12. J. molec. Biol. **87**, 103-119 (1974).
106. Gaudin, D., Gregg, R.S., Yielding, K.L.: Inhibition of DNA repair by cocarcinogens. Biochem. biophys. Res. Commun. **48**, 945-949 (1972).

107. Gaudin, D., Yielding, K.L., Stabler, A., Brown, T.: The effect of DNA repair inhibitors on the response of tumors treated with X-ray and alkylating agents. Proc. Soc. exp. Biol. (N.Y.) **137**, 202-206 (1971).
108. Gerard, G.F.: Poly(2'-0-methylcytidylate)oligodeoxyguanylate, atemplate-primer specific for reverse transcriptase, is not utilized by HeLa cell γ-DNA polymerase. Biochem. biophys. Res. Commun. **63**, 706-711 (1975).
109. Gerwin, B.I., Bassin, R.H.: Detection and isolation of a new DNA polymerase from human breast tumor cell line HBT-3 by (dT)12-18-cellulose chromatography. Proc. nat. Acad. Sci. (Wash.) **70**, 2453-2456 (1973).
110. Gerwin, B.I., Ebert, P.S., Chopra, H.C., Smith, S.G., Kvedar, J.P., Albert, S., Brennan, M.J.: DNA polymerase activities of human milk. Science **180**, 198-201 (1973).
111. Glazer, R.I., Nutter, R.C., Glass, L.E., Menger, F.M.: 2-Acetylaminofluorene and N-hydroxy-2-acetylaminofluorene inhibition of incorporation of orotic acid-5-^{3}H into nuclear ribosomal and heterogeneous RNA in normal and regenerating liver. Cancer Res. **34**, 2451-2458 (1974).
112. Glickman, B.W.: The role of DNA polymerase I in excision-repair. In: Molecular Mechanisms for Repair of DNA, Hanawalt, P.C., Setlow, R.B. (eds.). New York-London: Plenum Press, Basic Life Sci. **5A**, 213-218 (1975).
113. Goldmark, P.J., Linn, S.: Purification and properties of the rec. BC DNAase of Escherichia coli K-12. J. biol. Chem. **247**, 1849-1860 (1972).
114. Goodman, J.I., Potter, V.R.: Evidence for DNA repair synthesis and turnover in rat liver following ingestion of 3'-methyl-4-dimethyl-aminoazobenzene. Cancer Res. **32**, 766-775 (1972).
115. Gordon, M.P.: Photorepair or RNA molecular mechanism of repair of DNA, Hanawalt, P.C., Setlow, R.B. (eds.). Oxford: Plenum 1975, pp. 115-121.
116. Greene, R., Korn, D.: Partial purification and characterization of DNA polymerase from KB cells. J. biol. Chem. **245**, 254-261 (1970).
117. Gross, J., Gross, M.: Genetic analysis of an E. coli strain with a mutation affecting DNA polymerase. Nature (Lond.) **224**, 1166-1168 (1969).
118. Grossman, L., Braun, A., Feldberg, R., et al.: Enzymatic repair of DNA. Ann. Rev. Biochem. **44**, 19-43 (1975).
119. Hachmann, H.J., Lezius, A.G.: High-molecular-weight DNA polymerases from mouse myeloma. Europ. J. Biochem. **50**, 357-366 (1975).
120. Haddow, A.: Molecular repair, would healing, and carcinogenesis: Tumor producation a possible overhealing? In: Advances in Cancer Research, Klein, G., Weinhouse, S., Haddow, A. (eds.). New York-London: Academic Press 1972, Vol. XVI, pp.181-213.
121. Hahn, G.M., Yang, S.H., Parker, V.: Repair of sublethal damage and unscheduled DNA synthesis in mammalian cells treated with monofunctional alkylating agents. Nature (Lond.) **220**, 1142-1144 (1968).
122. Hahn, G.M., Bagshaw, M.A., Evans, R.G., Gordon, L.F.: Repair of potentially lethal lesions in X-irradiated, density-inhibited chinese hamster cells: metabolic effects and hypoxia. Radiat. Res. **55**, 280-290 (1973).
123. Hamilton, L.D., Mahler, I., Grossman, L.: Enzymatic repair of UV-irradiated DNA in vitro. Basic Life Sci. **5A**, 235-243 (1975).
124. Hariharan, P.J., Cerutti, P.A.: Formation and repair of gamma-ray induced thymine damage in micrococcus radiodurans. J. molec. Biol. **66**, 65-81 (1972).
125. Hariharan, P.V., Cerutti, P.A.: Excision of damaged thymine residues from gamma irradiated poly (dA-dT) by crude extracts of Escherichia coli. Proc. nat. Acad. Sci. (Wash.) **71**, 3532-3536 (1974).
126. Harris, C., Connor, R.J., Jackson, F.E., Lieverman, M.W.: Intranuclear distribution of DNA repair synthesis induced by chemical carcinogens or ultraviolet light in human diploid fibroblasts. Cancer Res. **34**, 3461-3468 (1974).
127. Hecht, N.B.: A DNA-directed DNA polymerase from murine liver mitochondria. Biochim. biophys. Acta (Amst.) **414**, 59-70 (1975).
128. Heidelberger, C.: Studies on the mechanism of hydrocarbon carcinogenesis. Acta Un. int. Cancr. **15**, 107-113 (1959).
129. Heidelberger, C.: Current trends in chemical carcinogenesis. Fed. Proc. **32**, 2154-2161 (1973a).
130. Heidelberger, C.: Chemical oncogenesis in culture. Advanc. Cancer Res. **18**, 317-366 (1973b).

131. Heidelberger, C.: Chemical Carcinogenesis. Ann. Rev. Biochem. **14**, 79-121 (1975).
132. Heidelberger, C., Rieke, H.S.: The synthesis of 3,4-benzpyrene-5-C^{14} and of 2-acetylaminofluorene-9-C^{14}. Cancer Res. **11**, 640-643 (1951).
133. Heidelberger, C., Weiss, S.M.: The distribution of radioactivity in mice following administration of 3,4-benzpyrene-5-C^{14} and 1,2,5,6-dibenzanthracene-9,10-C^{14}. Cancer Res. **11**, 885-891 (1951).
134. Henner, D., Furth, J.J.: Denovo Synthesis of a polymer of deoxyadenylate and deoxythymidilate by calf thymus DNA polymerase. Proc. nat. Acad. Sci. (Wash.) **72**, 3944-3946 (1975).
135. Heston, W.E., Vlahakis, G., Deringer, M.K.: High Incidence of spontaneous hepatomas and the increase of this incidence with urethan in C^3H/f, and C^3H/e male mice. J. nat. Cancer Inst. **24**, 425-435 (1960).
136. Hevesy, G., Forssberg, A.: Effect of irradiation by X-rays on the exhalation of carbon dioxide by the mouse. Nature (Lond.) **168**, 692 (1951).
137. Hirano, H., Higashi, K., Sakamoto, Y.: DNA polymerase in nucleoli isolated from Ehrlich ascites tumor cells. Biochem. biophys. Res. Commun. **67**, 518-524 (1975).
138. Hollander, C.F., Bentvelzen, P.: Enhancement of urethan induction of hepatomas in mice by prior partial hepatectomy. J. nat. Cancer Inst. **41**, 1303-1306 (1968).
139. Howard-Flanders, P.: DNA repair. Ann. Rev. Biochem. **37**, 175-200 (1968).
140. Howard-Flanders, P.: Repair by genetic recombination in bacteria: Overview. Basic Life Sci. **5**, 265-274 (1975).
141. Hueper, W.C.: Cancer induction by polyurethane and polysilicone plastics. J. nat. Cancer Inst. **33**, 1005-1027 (1964).
142. Ichihashi, M., Ramsay, C.A.: The action spectrum and dose response studies of unscheduled DNA synthesis in normal human fibroblasts. Photochem. Photobiol. **23**, 103-106 (1976).
143. Iversen, S., Arley, N.: On mechanism of experimental carcinogenesis. Acta path. microbiol. scand. **27**, 773-803 (1950).
144. Jackson, C.D., Irving, C.C.: Sex differences in cell proliferation and N-hydroxy-2-acetylaminofluorene sulfotransferase levels in rat liver during 2-acetylaminofluorene administration. Cancer Res. **32**, 1590-1594 (1972).
145. Jacob, F., Monod, J.: Genetic regulatory mechanisms in the synthesis of proteins. J. molec. Biol. **3**, 318-356 (1961).
146. Jungmann, R.A., Schweppe, J.S.: Binding of chemical carcinogens to nuclear proteins of rat liver. Cancer Res. **32**, 952-959 (1972).
147. Kaiser, A.D., Hogness, D.S.: The transformation of Escherichia coli with deoxyribonucleic acid isolated from bacteriophage lambda-dg. J. molec. Biol. **2**, 392-415 (1960).
148. Kaplan, J.C., Kushner, S.R., Grossman, L.: Enzymatic repair of DNA. 3. Properties of the UV-endonuclease and UV-exonuclease. Biochemistry **10**, 3315-3324 (1971).
149. Kapp, D.S., Smith, K.C.: Lack of in vitro repair of X-ray-induced chain break in DNA by the polynucleotide-joining enzyme. Int. J. Radiat. Biol. **14**, 567-571 (1968).
150. Kao, F.T.: Cell Mutagenesis studies in vitro using auxotrophic markers. In: Chemical Carcinogenesis,Ts'o, P.O.P., DiPaolo, J.A. (eds.). New York: Marcel Dekker 1974, Part B, pp. 565-573.
151. Kelly, R.B., Atkinson, M.R., Huberman, J.A., Kornberg, A.: Excision of thymine dimers and other mismatched sequences by DNA polymerase of Escherichia coli. Nature (Lond.) **214**, 495-501 (1969).
152. Kelner, A.: Effect of visible light on the recovery of streptomyces griseus candida from ultra violet irradiation injury. Proc. nat. Acad. Sci. (Wash.) **35**, 73-79 (1949).
153. Kihlman, B.A.: Molecular mechanisms of chromosome breakage and rejoining. In: Advances in Cell and Molecular Biology. New York-London: Academic Press 1971, Vol. I, pp. 59-107.
154. Kimball, R.F.: Repair of premutational damage. Advanc. Radiat. Biol. **2**, 135-162 (1966).
155. King, C., Van Lancker, J.L.: Molecular Mechanisms of liver regeneration. Arch. Biochem. **129**, 603-608 (1969).
156. Kinoshita, N., Gelboin, H.V.: The role of aryl hydrocarbon hydroxylase 7,12-dimethylbenz(a)anthracene skin tumorigenesis: On the mechanism of 7,8-benzoflavone inhibition of tumorigenesis. Cancer Res. **32**, 1329-1339 (1972a).

157. Kinoshita, N., Gelboin, H.V.: Aryl hydrocarbon hydroxylase and polycyclic hydrocarbon tumorigenesis: Effect of the enzyme inhibitor 7,8-benzoflavone on tumorigenesis and macromolecule binding. Proc. nat. Acad. Sci. (Wash.) **69**, 824-828 (1972b).
158. Kiplinger, G.F., Kensler, C.J.: Failure of phenoxybenzamine to prevent formation of hepatomas after chronic carbon tetrachloride administration. J. nat. Cancer Inst. **30**, 837-843 (1963).
159. Kitagawa, T., Michalopoulos, G., Pitot, H.C.: Unscheduled DNA synthesis in cells from N-2-fluorenylacetamide-induced hyperplastic nodules of rat liver maintained in a primary culture system. Cancer Res. **35**, 3682-3692 (1975).
160. Klein, G., Sjogren, H.O., Klein, E.: Demonstration of host resistance against sarcomas induced by implantation of cellophane films isologous (syngeneic) recipients. Cancer Res. **23**, 84-92 (1963).
161. Kriek, E.: Difference in binding of 2-acetylaminofluorene to rat liver deoxyribonucleic acid and ribosomal ribonucleic acid in vivo. Biochim. biophys. Acta (Amst.) **161**, 273-275 (1968).
162. Kriek, E.: Persistent binding of a new reaction product of the carcinogen N-hydroxy-N-2-acetylaminofluorene with guanine in rat liver DNA in vivo. Cancer Res. **32**, 2042-2048 (1972).
163. Kriek, E.: Carcinogenesis by aromatic amines. Biochim. biophys. Acta (Amst.) **355**, 177-203 (1974).
164. Kushner, S.R.: UV-Endonuclease: Purification, properties and specificity of the enzyme from *Micrococcus luteus.* The photoproduct excision system of *Micrococcus luteus.* Ph.D. Thesis. Brandeis Univ. Waltham, Mass. 1970.
165. Kushner, S.R., Kaplan, J.C., Ono, H., Grossman, L.: Enzymatic repair of deoxyribonucleic acid. IV. Mechanism of photoproduct excision. Biochemistry **10**, 3325-3334 (1971).
166. Laishes, B.A., Stich, H.F.: Repair synthesis and sedimentation analysis of DNA of human cells exposed to dimethylnitrosamine and activated dimethylnitrosamine. Biochem. biophys. Res. Commun. **52**, 827-833 (1973).
167. Langenbach, R., Kuszynski, C.: Brief Communication: Nonspecific inhibition of DNA repair by promoting and nonpromoting phorbol esters. Cancer Res. **33**, 2166-2172 (1973).
168. Lawley, P.D., Brookes, P.: Further studies on alkylation of nucleic acids and their constituent nucleotides. Biochem. J. **89**, 127-135 (1963).
169. Lawley, P.D., Orr, D.J.: Specific excision of methylation products from DNA of Escherichia coli with N-methyl-N'-nitro-N-nitrosoguanidine. Chem. biol. Interact. **2**, 154-157 (1970).
170. Lazarus, L.H., Kitron, N.: Differentiation and characterization of the cytoplasmic and nuclear deoxynucleic acid polymerases from baby hamster kidney cells. Biochim. biophys. Acta (Amst.) **402**, 309-322 (1975).
171. Legator, M.S., Flamm, W.G.: Environmental mutagenesis and repair. Ann. Rev. biochem. **42**, 683-708 (1973).
172. Lehmann, A.R.: Post-replication repair of DNA in ultraviolet-irradiated mammalian cells. (No gaps in DNA-synthesized late after ultraviolet irradiation). Europ. J. Biochem. **31**, 438-445 (1972).
173. Lehmann, A.R., Stevens, S.: Postreplication repair of DNA in chick cells: Studies using photoreactivation. Biochim. biophys. Acta (Amst.) **402**, 179-187 (1975).
174. Lerman, L.S.: Intercalability, the "y" transition, and the state of DNA in nature. In: Progress in Molecular and subcellular Biology. Berlin-Heidelberg-New York: Springer 1971, Vol. II, pp. 382-391.
175. Lewis, B.J., Abrell, J.W., Smith, R.G., Gallo, R.C.: DNA polymerases in human lymphoblastoid cells infected with simian sarcoma virus. Biochim. biophys. Acta (Amst.) **349**, 148-160 (1974).
176. Lieb, M.: Enhancement of ultraviolet-induced mutation in bacteria by caffeine. Z. Vererbungslehre **29**, 416 (1961).
177. Liebelt, R.A., Liebelt, A.G., Lane, M.: Hormonal influences on urethan carcinogenesis in C^3H/f mice. Cancer Res. **24**, 1869-1879 (1964).
178. Lieberman, M.W.: Demonstration of DNA repair in normal and neoplastic tissues after treatment with proximate chemical carcinogens and ultraviolet radiation. Nature (New Biol.) **241**, 199-201 (1973).

179. Lieberman, M.W.: Approaches to the analysis of fidelity of DNA repair in mammalian cells. Ann. Rev. Cytol. **46**, 1-23 (1976).
180. Lieberman, M.W., Dipple, A.: Removal of bound carcinogen during DNA repair in non-dividing human lymphocytes. Cancer Res. **32**, 1855-1860 (1972).
181. Lieberman, M.W., Poirier, M.C.: Deoxyribonucleoside incorporation during DNA repair of carcinogen-induced damage in human diploid fibroblasts. Cancer Res. **33**, 2097-2103 (1973).
182. Lijinsky, W., Garcia, H., Keefer, L., Loo,J., Ross, A.E.: Carcinogenesis and alkylation of rat liver nucleic acid by nitrosomethyl urea and nitrosoethylurea administered by intraporta injection. Cancer Res. **32**, 895-897 (1972).
183. Lindahl, T.: Excision of pyrimidine dimers from ultraviolet-irradiated DNA by exonuclease from mammalian cells. Europ. J. Biochem. **18**, 407-414 (1971).
184. Lindahl, T., Liundquist, S.: Apurinic and Apyrimidinic sites in DNA. In: Molecular Mechanisms for Repair of DNA, Hanawalt, P.C., Setlow, R.B., (eds.). New York-London: Plenum Press 1975, p. 31-38.
185. Livingston, D.M., Serxner, L.E., Howk, D.J., Hudson, J., Todaro, G.J.: Characterization of a new murine cellular DNA polymerase. Proc. nat. Acad. Sci. (Wash.) **71**, 57-62 (1974).
186. Loeb, L.A., Springgate, C.F., Battula, N.: Errors in DNA replication as a basis of malignant changes. Cancer Res. **34**, 2311-2321 (1974).
187. Luria, S.E.: Reactivation of irradiated bacteriophage by transfer of self-reproducing units. Proc. nat. Acad. Sci. (Wash.) **9**, 253-264 (1947).
188. Lynch, W.E., Short, J., Lieberman, I.: The 7.1S nuclear DNA polymerase and DNA replication in intact liver. Cancer Res. **36**, 901-904 (1976).
189. Maher, V.: Effect of DNA repair of the cytotoxicity and mutagenicity of carcinogenic agents in human cells. In: Carcinogenesis Program. Fourth Annual Collaborative Conference (NCI) February 1976, Orlando, Florida.
190. Maher, V.M., Van Lancker, J.L., Tomura, T.: Mutagenicity of reactive derivatives of carcinogenic hydrocarbons: Evidence of DNA repair. Mutation Res. **23**, 113-128 (1974).
191. Malling, H.V., Chu, E.H.Y.: Development of mutational model systems for study of carcinogenesis. In: Chemical Carcinogenesis, Ts'o, P.O.P., DiPaolo, J.A. (eds.). New York: Marcel Dekker, Inc. 1974, Part B, pp. 545-563.
192. Margalith, M., Gerard, G.F., Green, M.: Study on the DNA polymerase activity contained in particles released from human embryo cell monolayers. Biochim. biophys. Acta (Amst.) **425**, 305-315 (1976).
193. Markov, G.G., Dessev, G.N., Russev, G.C., Tsaney, R.G.: Effects of γ-irradiation on biosynthesis of different types of ribonucleic acids in normal and regenerating rat liver. Biochem. J. **146**, 41-51 (1975).
194. Marquardt, H.: Cell cycle dependence of chemically induced malignant transformation in vitro. Cancer Res. **34**, 1612-1615 (1974).
195. Marquardt, H., Philips, F.S.: The effects of 7,12-dimethylbenz(a)anthracene on the synthesis of nucleic acids in rapidly dividing hepatic cells in rats. Cancer Res. **30**, 2000-2006 (1970).
196. Marquardt, H., Sternberg , S.S., Philips, F.S.: 7,12-dimethylbenz(a)anthracene and hepatic neoplasia in regenerating rat liver. Chem. biol. Interact. **2**, 401-403 (1970).
197. Marquardt, H., Philips, F.S., Bendich, A.: DNA binding and inhibition of DNA synthesis after 7,12-dimethylbenz(a)anthracene administered during the early prereplicative phase in regenerating rat liver. Cancer Res. **32**, 1810-1813 (1972).
198. Matsukage, A., Bohn, E.W., Wilson, S.H.: Multiple forms of DNA polymerase in murine myeloma. Fed. Proc. **32**, 451 (1973).
199. Matsukage, A., Bohn, E.W., Wilson, S.H.: Multiple forms of DNA polymerase in mouse myeloma. Proc. nat. Acad. Sci. (Wash.) **71**, 578-582 (1974).
200. Matsukage, A., Bohn, E.W., Wilson, S.H.: Differential sensitivity of low molecular weight DNA polymerase to sulfhydryl-blocking reagents. Biochim. biophys. Acta (Amst.) **383**, 338-343 (1975).
201. Mechali, M., DeRecondo, A.M.: Co-fractionation of an endonuclease activity during the purification of DNA polymerase from regenerating rat liver. Europ. J. Biochem. **58**, 461-466 (1975).
202. Meltz, M.L., Painter, R.B.: Distribution of repair replication in the HeLa cell genome. Int. J. Radiat. Biol. **23**, 637-640 (1973).

203. Michael, R.O., Williams, G.M.: Chloroquine inhibition of repair of DNA damage induced in mammalian cells by methyl methanesulfonate. Mutation Res. **25**, 391-396 (1974).
204. Miller, E.C.: Studies on the formation of proteinbound derivatives of 3,4-benzpyrene in the epidermal fraction of mouse skin. Cancer Res. **11**, 100-108 (1951).
205. Miller, E.C., Miller,J.A.: Symposium on immunogenetics and carcinogenesis. In vivo combinations between carcinogens and tissue constituents and their possible role in carcinogenesis. Cancer Res. **12**, 547-556 (1952).
206. Miller, E.C., Miller, J.A.: Biochemistry of carcinogenesis. Ann. Rev. Biochem. **28**, 291-320 (1959).
207. Miller, E.C., Miller, J.A.: Studies on the mechanism of activation of aromatic amine and amide carcinogens to ultimate carcinogenic electrophilic reactants. Ann. N.Y. Acad. Sci. **163**, 731-750 (1969).
208. Miller, E.C., Miller, J.A.: The mutagenicity of chemical carcinogens: correlations, problems, and interpretations. In: Chemical Mutagens: Principles and Methods for Their Detection. New York: Plenum Press 1971, Vol. I, pp. 83-119.
209. Miller, E.C., Smith, J., Miller, J.A.: Lack of correlation between the formation of the glucuronide of N-hydroxy-2-acetyl aminofluorence (N-GLO-AAF) and susceptibility to hepatic carcinogenesis. Proc. Amer. Ass. Cancer Res. **11**, 56 (1970) (abstract).
210. Miller, J.A.: Carcinogenesis by chemicals: An overview. G.H.A. Clowes Memorial Lecture. Cancer Res. **30**, 559-576 (1970).
211. Miller, J.A., Lin, J.K., Miller, E.C.: N-(Guanosin-8-yl) and N-(deoxyguanosin-8-yl) N'-methyl-4-aminoazobenzene: Degradation products of hepatic RNA and DNA in rats administered N'-methyl-4-aminoazobenzene. Proc. Amer. Ass. Cancer Res. **11**, 56 (1970).
212. Miller, J.A., Miller, E.C.: The metabolic activation of carcinogenic aromatic amines and amides. In: Progress in Experimental Tumor Research, F. Homburger, (ed.). Int. Symp. on Carcinogenesis and Carcinogen Testing. Basel: S. Karger 1969a, Vo. XI, pp. 273-301.
213. Miller, J.A., Miller, E.C.: Metabolic activation of carcinogenic aromation amines and amides via N-hydroxylation and N-hydroxy-esterifications and its relationship to ultimate carcinogens as electrophilic reactants. In: Physico-chemical Mechanisms of Carcinogenesis, The Jerusalem Symposia on Quantum Chemistry and Biochemistry. Jerusalem: Israel Academy of Sciences and Humanities 1969b.
214. Miller, J.A., Miller, E.C., Finger, G.C.: On the enhancement of the carcinogenicity of 4-dimethyl-aminoazobenzene by fluorosubstitution. Cancer Res. **13**, 93-97 (1953).
215. Mohr, U., Hilfrich, J.: Brief communication: Effect of a single dose of N-diethylnitrosamine on the rat kidney. J. nat. Cancer Inst. **49**, 1729-1731 (1972).
216. Momparler, R.L., Rossi, M., Labitan, A.: Partial purification and properties of two forms of DNA polymerase from calf thymus. J. biol. Chem. **248**, 285-293 (1973).
217. Mordoh, J., Fridlender, B.R.: A modified nuclear DNA polymerase in a case of acute lymphoblastic leukemia. Biochem. biophys. Res. Commun. **67**, 888-896 (1975).
218. Morioka, K., Terayama, H.: A conversion factor for cytoplasmic DNA polymerase of rat liver. Biochem. biophys. Res. Commun. **61**, 518-525 (1974).
219. Mukherjee, A.S., Banerjee, M.R.: RNA-dependent DNA polymerase in preneoplastic nodules and tumors of the mammary gland of BALB/c mice. J. nat. Cancer Inst. **53**, 817-824 (1974).
220. Nakayama, H., Okubo, S., Sekiguchi, M., Takagi, Y.: A deoxyribonuclease activity specific for ultraviolet-irradiated DNA: A chromatographic analysis. Biochem. biophys. Res. Commun. **27**, 217-223 (1967).
221. Nelson, J.H., Grunberger, D., Cantor, C.R., Weinstein, I.B.: Modification of ribonucleic acid by chemical carcinogens. IV. Circular dichroism and proton magnetic resonance studies of oligonucleotides modified with N-2-acetylaminofluorene. J. molec. Biol. **62**, 331-346 (1971).
222. Nobrega, F.G., Rola, F.H., Pasetto-Nobrega, M., Oishi, M.: Adenosine triphosphate associated with adenosine triphosphate-dependent deoxyribonuclease. Proc. nat. Acad. Sci. (Wash.) **69**, 15-19 (1972).
223. Noguti, T., Kada, T.: Studies on DNA repair in *Bacillus subtilis*. Biochim. biophys. Acta (Amst.) **395**, 284-293 (1975a).
224. Noguti, T., Kada, T.: Studies on DNA repair in *Bacillus subtilis*. II. Partial purification and mode of action of an enzyme enhancing the priming activity of γ-irradiated DNA. Biochim. biophys. Acta (Amst.) **395**, 294-305 (1975b).

225. Norman, A., Ottoman, R.E., Chan, P., Kilsak, I.: Mutation Res. **15**, 358-360 (1972).
226. Oishi, M.: An ATP-dependent deoxyribonuclease from Escherichia coli with a possible role in genetic recombination. Proc. nat. Acad. Sci. (Wash.) **64**, 1292-1299 (1969).
227. Okazaki, R.: Short-chain intermediates in DNA replication. In: DNA Replication. Wickner, R.B. (ed.). New York: Marcel Dekker 1974, Vol. VII, pp. 1-32.
228. Oppenheimer, E.T., Willhite, M., Stout, A.P., Danishefsky, I., Fishman, M.M.: A comparative study of the effects of imbedding cellophane and polystyrene films in rats. Cancer Res. **24**, 379-387 (1964).
229. Ord, M.G., Stocken, L.A.: The biochemical lesion in vivo and in vitro. In: Mechanism in Radiobiology. New York: Academic Press 1961, pp. 259-331.
230. Painter, R.B.: Repair in mammalian cells: Overview. In: Molecular Mechanisms for Repair of DNA. New York-London: Plenum Press 1975, 5B, Part B, pp. 595-600.
231. Painter, R.B.: Eukaryotic Replicons. To be published in Handbook of Genetics. New York: Plenum Press, Vol. V.
232. Painter, R.B., Cleaver, J.E.: Repair replication in HeLa cells after large doses of X-irradiation. Nature (Lond.) **216**, 369-370 (1967).
233. Painter, R.B., Cleaver, J.E.: Repair replication unscheduled DNA synthesis, and the repair of mammalian DNA. Radiat. Res. **37**, 451-466 (1969).
234. Painter, R.B., Young, B.R.: Formation of nascent DNA molecules during inhibition of replicon initiation in mammalian cells. Biochim. biophys. Acta (Amst.) **418**, 146-153 (1976).
235. Paterson, M.C., Setlow, R.B.: Endonucleolytic activity from Micrococcus luteus that acts on ray-induced damage in plasmid DNA of Escherichia coli minicells. Proc. nat. Acad. Sci. (Wash.) **69**, 2927-2931 (1972).
236. Payes, B.: Enzymatic repair of X-ray-damaged DNA. I. Labeling of the precursor of a malonaldehyde-like material in X-irradiated DNA and the enzymatic excision of labeled lesions. Biochim. biophys. Acta (Amst.) **366**, 251-260 (1974).
237. Pederson, T., Robbins, E.: Chromatin structure and the cell division cycle. J. Cell Biol. **55**, 322-327 (1972).
238. Perrotta, C.A.: Effect of X-irradiation on DNA synthesis in the uterine epithelium. Radiat. Res. **28**, 232-242 (1966).
239. Personal communication, D. Ingle.
240. Pettijohn, D., Hanawalt, P.C.: Evidence for repair replication of ultraviolet-damaged DNA in bacteria. J. molec. Biol. **9**, 395-410 (1964).
241. Phillips, R.A., Tolmach, J.L.: Repair of potentially lethal damage in X-irradiated HeLa cells. Radiat. Res. **29**, 413-432 (1966).
242. Piperno, J.R., Kallen, R.G.: Evidence for multiple forms of DNA polymerase in Hodgkin's disease. Cancer Res. **33**, 838-842 (1973).
243. Pitot, H.C., Heidelberger, C.: Metbolic regulatory circuits and carcinogenesis. Cancer Res. **23**, 1694-1700 (1963).
244. Poirier, M.C., DeCicoo, B.T., Lieberman, M.W.: Nonspecific inhibition of DNA repair synthesis by tumor promoters in human diploid fibroblasts damaged with N-acetoxy-2-acetylaminofluorene. Cancer Res. **35**, 1392-1397 (1975).
245. Pollard, E.C., Davis, S.A.: The action of ionizing radiation on transcription (and translation) in several strains of Escherichia coli. Radiat. Res. **41**, 375-399 (1970).
246. Potter, R.L., Bethell, F.H.: Oxidative phosphorylation in spleen mitochondria. Fed. Proc. **11**, 270 (1952).
247. Potter, V.: Sequential blocking of metabolic pathways in vivo. Proc. Soc. exp. Biol. (N.Y.) **76**, 41-46 (1951).
248. Poulson, R., Zbarsky, S.H.: DNA polymerase from mitochondria of rat intestinal mucosa. Biochim. biophys. Acta (Amst.) **299**, 404-414 (1973).
249. Pound, A.W.: Carcinogenesis and cell proliferation. N.Z. med. J. **67**, 88-99 (1968).
250. Pound, A.W., Lawson, T.A.: Partial hepatectomy and toxicity of dimethylnitrosamine and carbon tetrachloride in relation to the carcinogenic action of dimethylnitrosamine. Brit. J. Cancer **32**, 596-603 (1975).
251. Radman, M.: SOS Repair hypothesis: Phenomenology of an inducible DNA repair which is accompanied by mutagenesis. In: Molecular Mechanisms for Repair of DNA, Hanawalt, P.C., Setlow, R.B. (eds.). New York-London: Plenum Press, Basic Life Sci. **5A**, 355-367 (1975).
252. Rajewsky, M.F.: Changes in DNA synthesis and cell proliferation during hepatocarcinogenesis by diethylnitrosamine. Europ. J. Cancer **3**, 335-342 (1967).

253. Rasmussen, R.E., Painter, R.B.: Evidence for repair of ultra-violet-damaged deoxyribonucleic acid in cultured mammalian cells. Nature (Lond.) **203**, 1360-1362 (1964).
254. Regan, J.D., Setlow, R.B.: Two forms of repair in the DNA of human cells damaged by chemical carcinogens and mutagens. Cancer Res. **34**, 3318-3325 (1974).
255. Riazuddin, S., Grossman, L.: Unpublished data.
256. Robbins, J.H., Kraemer, K.H., Lutzner, M.A., Festoff, B.W., Coon, H.G.: Xeroderma pigmentosum an inherited disease with sun sensitivity, multiple cutaneous neoplasms, and abnormal DNA repair. Ann. intern. Med. **80**, 221-248 (1974).
257. Roberts, J.J., Warwick, G.P.: Reaction of ^{3}H-labeled butter yellow in vivo and N-hydroxy-methylaminoazobenzene in vitro. Biochem. J. **93**, 189-199 (1964).
258. Roberts, J.J., Crathorn, A.R., Brent, T.P.: Repair of alkylated DNA in mammalian cells. Nature (Lond.) **218**, 970-972 (1966).
259. Roberts, J.J., Pascoe, J.M., Smith, B.A., Crathorn, A.R.: Quantitative aspects of the repair of alkylated DNA in cultured mammalian cells. Part II, Chem. biol. Interact. **3**, 49-68 (1971).
260. Rosen, V.J., Cole, L.J.: Accelerated induction of kidney neoplasms in mice after X-radiation (690 rad) and unilateral nephrectomy. J. nat. Cancer Inst. **28**, 1031-1041 (1962).
261. Russel, W.L., Russel, L.B., Kelly, E.M.: Radiation dose rate and mutation frequency. Science **128**, 1546-1550 (1958).
262. Salzman, L.A., McKerlie, L.: Nuclear and cytoplasmic deoxynucleic acid polymerase from rat nephroma cells. J. biol. Chem. **250**, 5589-5595 (1975).
263. Sasaki, T., Yoshida, T.: Production of carcinoma of the liver by feeding O-aminoazotoluene. Arch. path. Anat. **295**, 175-200 (1935).
264. Schrek, R., Stefani, S.: Radioresistance of phytohemagglutinin-treated normal and leukemic lymphocytes. J. nat. Cancer Inst. **32**, 507-521 (1964).
265. Scribner, J.D., Miller, J.A., Miller,E.C.: Nucleophilic substitution on carcinogenic nacetoxy n-arylacetamides. Cancer Res. **30**, 1570-1579 (1970).
266. Sedwick, D., Wang, T.S.F., Korn, D.: Purification and properties of nuclear and cytoplasmic DNA polymerase from human KB cells. J. biol. Chem. **247**, 5026-5033 (1972).
267. Sedwick, D., Wang, T.S.F., Korn, D.: The DNA polymerase of KB cells. Meth. Enzymol. **29**, 89-102 (1974).
268. Sedwick, W.D., Wang, T.S.F., Korn, D.: Cytoplasmic deoxyribonucleic acid polymerase. J. biol. Chem. **250**, 7045-7056 (1975).
269. Setlow, R.B.: The photochemistry, photobiology, and repair of polynucleotides. Progr. Nucleic Acid Res. **8**, 247-295 (1968).
270. Setlow, R.B.: Relationships among repair, cancer and genetic deficiency. Basic Life Sci. **5**, 711-718 (1975).
271. Setlow, R.B., Carrier, W.L.: Endonuclease activity toward DNA irradiated in vitro by gamma rays. Nature (New Biol.) **241**, 170-172 (Letter) (1973).
272. Setlow, R.B., Regan, J.D.: Defective repair of N-acetoxy-2-acetylaminofluorene-induced lesions in the DNA of Xeroderma Pigmentosum cells. Biochem. biophys. Res. Commun. **46**, 1019-1024 (1972).
273. Setlow, R.B., Setlow, J.K.: Effects of radiation on polynucleotides. In: Annual Review of Biophysics and Bioengineering, Morales, M.F., Hagins, W.A. et al. (eds.). Palo Alto: Annual Reviews 1972, Vol. I, pp. 293-346.
274. Shaw, M.W.: Human chromosome damage by chemical agents. Ann. Rev. Med. **21**, 409-432 (1970).
275. Shipley, W.U., Stanley, J.A., Courtenay, V.D., Field, S.B.: Repair of radiation damage in Lewis lung carcinoma cells following in situ treatment with fast neutrons and γ-rays. Cancer Res. **35**, 932-938 (1975).
276. Sinclair, W.K., Morton, R.A.: Survival and recovery in X-irradiated synchronized chinese hamster cells. In: Cellular Radiation Biology. Baltimore: Williams and Wilkins 1964, pp. 418-422.
277. Smith, K.C., Hanawalt, P.C.: Molecular Biology. An International Series of Monographs and Textbooks. New York: Academic Press 1969.
278. Smith R.G., Gallo, R.C.: DNA-dependent DNA polymerases I and II from normal human-blood lymphocytes. Proc. nat. Acad. Sci. (Wash.) **69**, 2879-2884 (1972).
279. Smith, R.G., Abrell, J.W., Lewis, B.J., Gallo, R.C.: Serological analysis of human deoxyribonucleic acid polymerases: Preparation and properties of antiserum to deoxy-

ribonucleic acid polymerase I from human lymphoid cells. J. biol. Chem. **250**, 1702-1709 (1975).
280. Spadari, S., Weissbach, A.: RNA-primed DNA synthesis: Specific catalysis by HeLa cell DNA polymerase-α. Proc. nat. Acad. Sci.(Wash.) **72**, 503-507 (1975).
281. Srivastava, B.I.S.: Deoxynucleotide-polymerizing enzymes in normal and malignant human cells. Cancer Res. **34**, 1015-1026 (1974).
282. Srivastava, B.I.S.: Modified nucleotide polymers as inhibitors of DNA polymerases. Biochim. biophys. Acta (Amst.) **414**, 126-132 (1975).
283. Stein, G.S., Park, W.D., Stein, J.L., Lieberman, M.W.: Synthesis of nuclear proteins during DNA repair synthesis in human diploid fibroblasts damaged with ultraviolet radiation or N-acetoxy-2-acetylaminofluorene. Proc. nat. Acad. Sci. (Wash.) **73**, 1466-1470 (1976).
284. Stewart, B.W., Farber, E.: Strand breakage in rat liver DNA and its repair following administration of cyclic nitrosamines. Cancer Res. **33**, 3209-3215 (1973).
285. Stich, H.F., Laishes, B.A.: DNA repair and chemical carcinogens. Pathobiol. Ann. **3**, 341-376 (1973).
286. Stich, H.F., San, R.H.C., Kawazoe, Y.: Increased sensitivity of Xeroderma Pigmentosum cells to some chemical carcinogens and mutagens. Mutation Res. **17**, 127-137 (1973).
287. Strauss, B.S.: DNA repair mechanisms and their relation to mutation and recombination. Curr. Top. Microbiol. Immunol. **44**, 1-85 (1968).
288. Stryckmans, P.A., Chanana, A.D., Cronkite, E.P., Greenberg, M.L., Schiffer, L.M.: Studies on lymphocytes. X. Influence of extracorporeal irradiation of the blood on lymphocytes in chronic lymphocytic leukemia: Apparent correlation with RNA turnover. Radiat. Res. **37**, 118-130 (1969).
289. Stubblefield, E.: The Kinetics of DNA Replication in Chromosomes. The Cell Nucleus. Busch, H. (ed.). New York: Academic Press 1974, Vol. II, pp. 149-162.
290. Sugimura, T., Okabe, K., Nagao, M.: The metabolism of 4-nitroquinoline-1-oxide, a 4-hydroxyaminoquinoline-1-oxide in rat liver and hepatomas. Cancer Res. **26**, 1717-1721 (1966).
291. Sutherland, B.M.: Photoreactivating enzyme from human leukocytes. Nature (Lond.) **248**, 109-112 (1974).
292. Swan, P.F., Magee, P.N.: Nitrosamine-induced carcinogenesis: the alkylation of nucleic acids of the rat by N-methyl-N-nitrosourea, dimethylnitrosamine, dimethyl sulfate and methylmethanesulphonate. Biochem. J. **110**, 39-47 (1968).
293. Takagi, Y., Sekiguchi, M., Okubo, S., Nakayama, H., Shimada, K., Yashuda, S., Nishimoto, T., Yoshihara, H.: Nucleases specific for ultraviolet light-irradiated DNA and their possible role in dark repair. Cold Spr. Harb. Symp. quant. Biol. **33**, 219-227 (1968).
294. Takebe, H., Furuyama, J., Miki, Y., Kondo, S.: High sensitivity of Xeroderma Pigmentosum cells to the carcinogen 4-nitroquinolinc-1-oxide. Mutation Res. **15**, 98-100 (1972).
295. Taylor, A.L., Trotter, C.D.: Linkage map of Escherichia coli strain K-12. Bact. Rev. **36**, 504-524 (1972).
296. Teebor, G.W., Duker, N.J., Raucan, S.A., Zachary, K.J.: Inhibition of thymine dimer excision by the phorbol ester, phorbol myristate acetate. Biochem. biophys. Res. Commun. **50**, 66-70 (1973).
297. Threlfall, G., Cairnie, A.B., Taylor, D.M., Buck, A.T.: Effect of wholebody X-irradiation on renal compensatory hypertrophy. Radiat. Res. **27**, 559-565 (1966).
298. Tibbetts, C.J.B., Vinograd, J.: Properties and mode of action of a partially purified DNA polymerase from mitochondria of HeLa cells. J. biol. Chem. **248**, 3367-3379 (1973).
299. Tomsak, R.L., Cook, R.T.: The persistence of unmetabolized ^{3}H-7,12-dimethylbenz-(a)anthracene in regenrating liver. Brit. J. Cancer **32**, 440-450 (1975).
300. Tomura, T., Van Lancker, J.L.: Association of exonucleolytic and DNA polymerase activities in mammalian cells. J. Cell Biol. **63**, 350a (1974).
301. Tomura, T., Van Lancker, J.L.: The effect of a mammalian repair endonuclease on X-irradiated DNA. BBA **402**, 343-350 (1975).
302. Tomura, T., Ariyama, K., Van Lancker, J.L.: Alteration of liver DNA after X-irradiation. Reassociation and hybridization with nuclear RNA. Int. J. Radiat. Biol. **28**, 205-214 (1975).

303. Trosko, J.E., Chu, E.H.Y.: The role of DNA repair and somatic mutation in carcinogenesis. In: Advances in Cancer Research, Klein, G., Weinhouse, S. (eds.). New York: Academic Press 1975, Vol. XXI, pp. 391-420.
304. Ts'o, P.O.P., Lesko, S.A., Umans, R.S.: The physical binding and the chemical linkages of benzpyrene to nucleotides, nucleic-acids and nucleohistones. In: Physicochemical Mechanisms of Carcinogenesis. The Jerusalem Symposia on Quantum Chemistry and Biochemistry. Jerusalem: Israel Academy of Sciences and Humanities 1969, Vol. I, pp. 106-135.
305. Tsuruo, T., Ukita, T.: Purification and further characterization of three DNA polymerases of rat ascites hepatoma cells. Biochim. biophys. Acta (Amst.) **353**, 146-159 (1974).
306. Tsuruo, T., Tomita, T., Satoh, H., Ukita, T.: DNA polymerases of ascites hepatoma cells I. Purification and properties of a DNA polymerase from soluble fraction. Biochem. biophys. Res. Commun. **48**, 769-775 (1972a).
307. Tsuruo, T., Tomita, T., Satoh, H., Ukita, T.: DNA polymerases of ascites hepatoma cells II. Purification and properties of DNA polymerases nuclear membranechromatin fraction. Biochem. biophys. Res. Commun. **48**, 776-782 (1972b).
308. Tsuruo, T., Hirayama, K., Kawaguchi, M., Satoh, H., Ukita, T.: Low molecular weight DNA polymerase of rat ascites hepatoma cells. Biochim. biophys. Acta (Amst.) **366**, 270-278 (1974).
309. Tsuruo, T., Hirayama, K., Ukita, T.: Three DNA polymerase of rat ascites hepatoma cell: Properties of the enzymes and effect of RNA synthesis on the reactions. Biochim. biophys. Acta (Amst.) **383**, 274-281 (1975a).
310. Tsuruo, T., Hirayama, K., Satoh, H., Ukita, T.: A new form of high molecular weight DNA polymerase in the nuclei of rat ascites hepatoma cells. J. Biochem. **78**, 401-408 (1975b).
311. Urbanczyk, J., Studzinski, G.P.: Chromatin-associated DNA endonuclease activities in HeLa cells. Biochem. biophys. Res. Commun. **59**, 616-621 (1974).
312. van Bekkum, D.W.: The disturbance of oxidative phosphorylation and the breakdown of ATP in spleen tissue after irradiation. Biochim. biophys. Acta (Amst.) **16**, 437-438 (1955).
313. Van Lancker, J.L.: Control of macromolecular synthesis in regenerating liver and its alteration by X-radiation. Biochemistry of Cell division. Baserga, R. (ed.). Springfield, Ill.: Charles C Thomas 1969, pp. 155-177.
314. Van Lancker, J.L.: Liver DNA replication, transcription and repair after irradiation and administration of carcinogens. 9th Congres. Int. de Gastro-Enterologie, Regeneration Hepatique, Paris 1972, pp. 99-131.
315. Van Lancker, J.L.: Carcinogenesis and DNA repair. In: Chemical Carcinogens , Ts'o, P.O.P., DiPaolo, J. (eds.). World Symposium on Chemical Carcinogens. Baltimore: Marcel Dekker, Inc. 1974, pp. 427-442.
316. Van Lancker, J.L.: Molecular mechanisms of diseases. In: Radiation. Berlin-Heidelberg-New York: Springer 1976, pp. 947-1116.
317. Van Lancker, J.L.: Molecular mechanisms of diseases. In: Radiation. Berlin-Heidelberg-New York: Springer, 1976, pp. 730-731.
318. Van Lancker, J.L., Tomura, T.: Alteration of liver DNA after X-irradiation. Cancer Res. **34**, 699-704 (1974a).
319. Van Lancker, J.L., Tomura, T.: Purification and some properties of a mammalian repair endonuclease. Biochim. biophys. Acta (Amst.) **353**, 99-114 (1974b).
320. Van Lancker, J.L., Tomura, T.: The enzymology of DNA repair and its relation to carcinogenesis. In: Gene Expression and Carcinogenesis in Cultured Liver. New York: Academic Press 1975, pp. 412-424.
321. Van Lancker, J.L., Tomura, T.: Mammalian repair endonuclease in regenerating liver. Fed. Proc. **35**, 329 (1976a) (Abstract).
322. Van Lancker, J.L., Tomura, T.: Mammalian repair endonuclease in regenerating liver. Fed. Proc. **35**, 329 (1976b).
323. Van Lancker, J.L., Tomura, T.: Mammalian repair endonuclease in regenerating liver. In Preparation.
324. Vaught, J.B., Howard, E.F., Bresnick, E.: Breakage of human cell DNA after exposure to 3-methylcholanthrene-11,12-oxide. Cancer Res. **36**, 272-275 (1976).

325. Verly, W.G.: Maintenance of DNA repair of apurinic sites. In: Molecular Mechanisms for Repair of DNA. Hanawalt, P.C., Setlow, R.B. (eds.). New York-London: Plenum Press 1975, pp. 39-46.
326. Verly, W.G., Paquette, Y.: An endonuclease for depurinated DNA in Escherichia coli B. Cancer J. Biochem. **50**, 217-224 (1972).
327. Vizadlova, M., Janovska, E., Zhestjanifo, V.D.: Effect of dark-repair inhibitors on the survival of *Escherichia coli* under different post-irradiation conditions. Int. J. Radiat. Biol. **20**, 49 (1971).
328. Waalkes, T.P., Sanders, K., Smith, R.G., Adamson, R.H.: DNA polymerases of Walker 256 Carcinosarcoma. Cancer Res. **34**, 385-391 (1974).
329. Walff, D.L., Rupert, C.S.: Disappearance of thymine photo-dimer in ultra violet irradiated DNA upon treatment with a photoreactivating enzyme from Baker's yeast. Biochem. biophys. Res. Commun. **7**, 237-240 (1962).
330. Walker, I.G., Sridhar, R.: The formation and repair of single-strand breaks in DNA of cultured mammalian cells treated with UV-light, methylating agents or 4-nitroquinoline-1-oxide. Chem. biol. Interact. **12**, 229-239 (1976).
331. Wang, E.C., Henner, D., Furth, J.J.: An endonuclease from calf thymus which introduces a limited number of single-strand breaks into DNA. Biochem. biophys. Res. Commun. **65**, 1177-1183 (1975).
332. Wang, T.S.F., Sedwick, E.D., Korn, D.: Nuclear deoxynucleic acid polymerase. J. biol. Chem. **250**, 7040-7044 (1975).
333. Wang, T.Y.: Partial purification of chromosomal DNA polymerase from rat Walker tumor. Proc. Soc. exp. Biol. (N.Y.) **129**, 469-472 (1968).
334. Watanabe, I.: Radiation effects on DNA chain growth.in mammalian cells. Radiat. Res. **52**, 541-556 (1974a).
335. Watanabe, I.: Radiation effects on DNA chain growth in mammalian cells. Radiat. Res. **58**, 541-556 (1974b).
336. Warwick, G.P.: Effect of the cell cycle on carcinogenesis. Fed. Proc. **30**, 1760-1765 (1971).
337. Weiss, B.G.: Perturbations in precursor incorporation into DNA of X-irradiated HeLa S 3 cells. Radiat. Res. **48**, 128-145 (1971).
338. Whitmore, G.F., Gulyas, S., Botond, J.: Radiation sensitivity throughout the cell cycle and its relationship to recovery. In: Cellular Radiation Biology. Baltimore: Williams and Wilkins 1964, pp. 428-441.
339. Wildenberg, J., Meselson, M.: Mismatch repair in heteroduplex DNA. Proc. nat. Acad. Sci. (Wash.) **72**, 2702-2706 (1975).
340. Wilkins, R.J., Hart, R.W.: Preferential DNA repair in human cells. Nature (Lond.) **247**, 35-36 (1974).
341. Witkin, E.M.: Post-irradiation metabolism and the timing of ultraviolet-induced mutations in bacteria. In: Proceedings 10th International Congress of Genetics. Toronto: Montreal. Univ. Toronto Press 1959, pp. 280.
342. Witkin, E.M.: Relationships among repair, mutagenesis and survival: Overview. In: Molecular Mechanisms for Repair of DNA, Hanawalt, P.C., Setlow, R.B. (eds.). New York-London: Plenum Press, Basic Life, Sci. **5A**, 347-353 (1975a).
343. Witkin, E.M.: Thermal enhancement of ultraviolet mutability in a dnaB uvrA derivative of Escherichia coli B/r: Evidence for inducible error-prone repair. In: Molecular Mechanisms for Repair of DNA, Hanawalt, P.C., Setlow, R.B. (eds.). New York-London: Plenum Press, Basic Life Sci. **5A**, 369-378 (1975b).
344. Witkin, S.S., Ohno, T., Spiegelman, S.: RNA fraction of RNA-instructed DNA polymerase from human leukemic spleens. Proc. nat. Acad. Sci. (Wash.) **72**, 4133-4136 (1975).
345. Wolff, S., Scott, D.: Repair of radiation-induced damage to chromosomes, (independence of known DNA dark repair mechanisms). Exp. Cell Res. **55**, 9-16 (1969).
346. Wunderlich, V., Schutt, M., Bottger, M., Graffi, A.: Preferential alkylation of mitochondrial deoxyribonucleic acid by N-methyl-N-nitrosurea. Biochem. J. **118**, 99 (1970).
347. Wunderlich, V., Tezlaff, I., Graffi, A.: Studies on nitrosodimethylamine. Preferential methylation of mitochondrial DNA in rats and hamsters. Chem. biol. Interact. **4**, 81 (1971-1972).

348. Yielding, K.L., Yielding, L., Gaudin, D.: Inhibition by chloroquine of UV repair in E. coli B. Proc. Soc. exp. Biol. (N.Y.) **133**, 999-1001 (1970).
349. Yoshida, S., Knodo, T., Ando, T.: Multiple molecular species of cytoplasmic DNA polymerase from calf thymus. Biochim. biophys. Acta (Amst.) **353**, 463-474 (1974).
350. Zajdela, F.: Production de sarcomes sous-cutanes chez le rat au moyen de membranes cellulosiques de porosite connue. Bull. Cancer **53**, 401-408 (1966).

Soft Tissue Tumors in The Rat

Pathogenesis and Histopathology

C. THOMAS, H.J. STEINHARDT, K. KÜCHEMANN, D. MAAS, and U.N. RIEDE

I. Introduction

Soft tissue tumors represent some of the most interesting neoplasms found in man and animals. Problems remain as to pathogenesis, classification, and nomenclature, the solutions of which are necessary to permit predictions of biologic behavior. Soft tissue tumors have generally been classified as to morphology and histogenesis. In undifferentiated malignant soft tissue tumors, purely morphological diagnoses, such as spindle cell sarcoma, round cell sarcoma or pleomorphic cell sarcoma, are easily made; but such terms do not give reliable information as to prognosis and therapy. For instance, the diagnosis "spindle cell sarcoma" may include the locally invasive but rarely metastasizing fibrosarcoma. A histogenetic classification appears more useful in differentiating these tumors, especially with respect to prognosis and therapy.

In the past decades American authors have tried to classify human soft tissue tumors on the basis of certain histologic criteria. They have chiefly emphasized morphologic features, such as production of specific types of intercellular substances or cytoplasmic structures (myofibrils, fat). In practice, routine diagnostic procedures are usually restricted to histologic investigations. Special studies, such as histochemistry, electron microscopy, immunology, or historadioautography, are not easily done on the routinely fixed tissue.

Can experimental studies in animals help solve some of the questions regarding the pathogenesis and pathology of human soft tissue tumors? Are there fundamental differences in the histologic appearance and the biologic behavior of such tumors in experimental animals when compared to similar lesions in man?

Soft tissue tumors in rats may be induced by subcutaneous (s.c.) or intramuscular (i.m.) injection of carcinogenic hydrocarbons. The carcinogenic activity of these substances was first detected by observations in human medicine. In 1775 the English surgeon *Pott* implicated the local effects of soot and tar as being responsible for the occurrence of cancer of the scrotum in chimney sweeps. This was confirmed in animal experiments 140 years later by the Japanese investigators *Yamagiwa* and *Ichikawa*. They succeeded in producing skin papillomas and carcinomas by repeated applications of coal tar to the rabbit ear. Twenty years later *Kennaway* and coworkers were able to identify the essential carcinogenic substance: by repeated distillation and crystallization they isolated 3,4-benzpyrene from the tar.

Today numerous carcinogenic chemical compounds have been isolated which act either locally at the site of application or produce distant tumors following resorption. In the latter case the chemical is applied to the experimental animal in the form of a transport compound which, when metabolized, releases the "active" agent. With resorptive carcinogens, such as aromatic amines, nitroso-compounds, urethan, and others, large numbers of epithelial tumors can be induced selectively in various organs of experimental animals. However, experimentally induced soft tissue tumors are rare and more commonly are produced by local contact with carcinogenic hydrocarbons, metals, certain dyes, and certain macromolecular compounds.

Some of the most potent carcinogenic agents are 3,4-benzpyrene (3,4-BP), 20-methylcholanthrene (20-MC) and 9,10-dimethyl-1,2-benzanthracene (DMBA), which cannot only be produced in the laboratory, but also may occur as contaminants in oil, air, and even in food and drugs. Such compounds are also thought to have a causative relationship to human tumors such as bronchogenic carcinoma.

The ability of compounds to induce cancer is quite variable in different animal species. Similarly, the incidence of spontaneously occurring tumors also varies with the animal species used. Tumorigenic experiments, therefore, are restricted to certain animal species. We (*Druckrey, Schmähl,* and *Thomas*) have chosen rats as our experimental animals. In this paper, therefore, only causal reference will be made to soft tissue tumors occurring in other species. The main emphasis in this paper will be directed to the pathogenesis and morphology of soft tissue tumors in the rat which have been induced by specific chemical carcinogenic agents. Neither viruses nor radiation have been used in the present experiments. In-vitro malignant transformation of fibroblasts and purely biochemical aspects of tumorigenesis have recently been discussed by *Berenblum* (1974) and others.

II. Pathogenesis

1. Causal Pathogenesis

Soft tissue tumors in experimental animals may occur spontaneously or may be induced by various noxious agents, particularly chemical carcinogens and ionizing radiation.

a) Spontaneously occurring soft tissue tumors have only rarely been described in the rat. Table 1 cites some cases from the literature. Unfortunately, these authors did not consistently identify the animals as to species or sex. "Fibromas", which appear in proximity to the milk line, usually represent completely hyalinized mammary fibroadenomas.

Table 1. Spontaneous soft tissue tumors in rats (literature citations)

Year	Author	Study material	Tumor designations
1885	*B. Sutton*	Wild rats	1 Fibroma of neck
1890	*v. Eiselsberg*	Wild rats	1 Fibrosarcoma right shoulder
1898	*Velich*	Wild rats	1 Spindle cell sarcoma (subperiosteal) right dorsal aspect upper hind leg)
1905	*Legene* *Esmonet*	Rats	1 Fibroma
1906	*Gaylord*	Rats	1 Fibrosarcoma
1909	*McCoy*	100,000 wild rats	16 Fibromas 1 Lipoma 5 Sarcomas
1911	*Woolley* *Wherry*	23,000 wild rats	1 Fibroma 1 Sarcoma
1917	*Bullock* *Rohdenburg*	19,300 rats	1 Fibroma (neck)
1919	*Secher*	Diverse laboratory rats	16 Fibromas 1 Sarcoma
1928	*Suguira*	Single observation	1 Spindle cell sarcoma
1930	*Bullock* and *Curtis*	489 male and female rats	93 Subcutaneous tumors (67 malignant)
1932	*Curtis* and *Bullock*	22,220 rats (various inbred strains)	20 Fibromas 13 Fibromas, some with sarcomatous transformation 13 Sarcomas 3 Lipomas or fibrolipomas
1940	*Ratcliffe*	288 Wistar rats	7 Fibromas 1 Lipoma 10 Fibrosarcomas 2 Myxosarcomas
1958	*Crain*	189 Wistar rats	1 Fibroma 1 Fibrosarcoma
1958	*Davis* et al.	74 female Sprague-Dawley rats	1 Sarcoma
1961	*Thompson* et al.	62 Sprague-Dawley rats	1 Fibrosarcoma

b) Induced soft tissue tumors: Soft tissue tumors in rats may be experimentally induced by local action of a chemical carcinogen or by radiation. Viruses play no significant role in soft tissue tumorigenesis in rats. Tables 2 and 3 (pp. 157) indicate various substances which can induce benign and malignant soft tissue tumors in the rat. The tables indicate the chemical composition of the carcinogen, mode of application, dose, time necessary for tumor induction, incidence, and histological type of the respective tumors. We have used the authors' own designations in these tables.

Locally acting carcinogens include primarily carcinogenic hydrocarbons, alkylating agents, certain metals and macromolecular compounds, various dyes and occasional natural compounds. After application of *carcinogenic hydrocarbons* the yield of soft tissue tumors in rats may approach 100% if the dose is large enough and if the carcinogen has been applied long enough. Soft tissue tumors can also be easily induced in hamsters and mice, but other animals, such as rabbits and guinea pigs, require much larger single and total doses and are generally unsuited for this purpose. For this reason, in the past, the guinea pig has been considered as being "refractory" to cancer. However, later experiments have shown that experimental induction of tumors is also possible in this animal species (see *Druckrey* et al., 1962; *Thomas* and *Schmähl*, 1965; *Bücheler* and *Thomas*, 1971).

Direct-acting alkylating agents include ethylenimines, epoxides, lactones, and other substances, which have been studied by *Walpole* et al. (1954), *Druckrey* et al. (1970a, b), *Dickens* and *Jones* (1961, 1965) and by *Van Duuren* et al. (1963). Their mode of action in inducing tumors, especially by alkylization of nucleic acids, has been described by various authors: *Wheeler* (1962), *Ross* (1962), *Datta* and *Datta* (1969) and *Loveless* (1969). *Druckrey* and coworkers (1970) studied the direct carcinogenic activity of 12 different alkylating agents. 1,3-propansultone, methyliodine, benzylchloride, and dimethyl-trimethylenoxide proved to have considerable neoplastic potential. Dibutyl-n-sulfate,p-toluenesulfonic acid methylester and ethylene sulfide are weak carcinogens. Experiments with veratrylchloride and p-toluenesulfonic-acid butylester had negative results. Their animal experiments also proved that there is no direct relationship between the toxic (tissue necrosis) and carcinogenic potential of a local alkylating agent, but that tumor incidence depends upon the amount of the carcinogenic substance applied. *Druckrey* et al. (1970) were able to induce sarcomas following a single s.c. injection of 1,3-propansultone. The largest number of tumors (18 tumors in 18 experimental animals) was obtained after application of 100 mg/kg b.w. of propansultone. After 30 mg/kg the yield was 12/18, and even after application of only 10 mg/kg, 4 out of 15 rats developed s.c. tumors.

On the basis of these induced s.c. tumors *Druckrey* et al. (1970) have compiled an equation which is thought to permit conclusions as to the carcinogenic potential of various locally active alkylating agents:

$$I = \frac{A}{t \times D} \times 10^3$$

A = yield of sarcomas. t = medium time of death in days. D = total dose based on DL_{50} (Examples: propansultone index 470 to 800, dimethylsulfate index 180).

Among the *carcinogenic dyes* trypan blue when locally applied can induce fibromas and fibrosarcomas at the site of application, and also may produce lymphomas after resorption (*Brown*, 1963). Tumors have been observed after s.c. injection of 4-dimethylamino-triphenyl-methane and parafuchsin (*Druckrey*, 1955; *Druckrey* et al., 1956). Sulfonated triphenylmethan dyes (light green SF) induce sarcomas in rats after local injection (*Gross*, 1961). Other phenol type compounds, however, are not effective (e.g., patent blue). Also ineffective in producing soft tissue tumors in animal experiments were Sudan red G, methyl orange, Ponceau GR, Indigotine, Chinoline yellow and Yellow Orange S (*Oettel* et al., 1965; *Druckrey* et al., 1966). Although the explanation for the different carcinogenic activity of various triphenylmethan dyes is still unknown, the tumors induced are not the result of an "unspecific stimulus" (*Oettel* et al., 1965).

Locally carcinogenic inorganic substances include various metals such as iron, chromium, cobalt, zinc, nickel, and cadmium (*Haddow* et al., 1964). The degree of carcinogenic activity of these substances, however, may be quite variable: in metal-oxygen and metal-sulfur compounds, carcinogenesis is much stronger than in the corresponding oxide compounds (*Gilman*, 1961). While dextran or iron alone have little or no carcinogenic properties, iron-dextran complex (inferon) has significant activity (*Haddow* and *Horning*, 1959). Dextran combined with copper, bismuth, or aluminum, however, exhibits weak or no carcinogenic activity.

Early changes produced by various inorganic substances are quite diverse. Cadmium produces local sarcomas which develop from a markedly vascular and cell-rich granulation tissue. The application of cobalt produces only little necrosis. The histologic type of tumor may depend on the chemical composition of the carcinogen: according to *Gilman* (1961) rhabdomyosarcomas prevail in rats after local application of nickel oxide or nickel sulfide.

Certain subcutaneously implanted *macromolecular compounds* may lead to local development of sarcomas after an adequate latent period. These compounds, which represent foreign bodies, include synthetic products such as polyethylen, polyvinylchloride, bakelite, polyamide, polystrole, polyurethane, etc., that may produce tumors at the site of implantation in rats, mice, hamsters, and dogs. Other foreign body substances which occasionally may induce sarcomas are rubber, quartz, silk, window glass, ivory, thick vellum, and precious metals (gold, silver, and platinum). The small number of tumors produced and the long induction time required indicate only weak carcinogenic activity of these latter substances (for detailed literature see *Ott* et al., 1963; *Ott*, 1970).

The following observations were made in numerous studies of animal experiments with macromolecular substances:

1. The larger the implantation plane of a foreign body, the larger the scar capsule and, correspondingly, the increased risk of developing sarcoma (*Nothdurft*, 1955; *Stout* and *Eirich*, 1955; *Hueper*, 1961; *Ott* et al., 1963). This is only true, however, up to a critical implant size (*Ott*, 1970).

2. The larger the pores of an implanted foreign body, the smaller is its fibrous capsule and, correspondingly, the risk of developing sarcoma decreases (*Contzen*, 1963; *Ott* et al., 1963). This observation, however, does not apply if the porous body exerts a direct fibrosing effect, as, for instance, spongiosa of bone (*Ott* and *Jansen*, 1966).

3. Foreign bodies with concave surfaces are associated with increased number of sarcomas (*Nothdurft*, 1961). This is thought to be due to increased scar formation (*Contzen*, 1963; *Contzen* et al., 1967).

4. Hardness or softness of the implanted substances does not influence the amount of scar tissue produced (*Contzen* et al., 1967; *Nothdurft*, 1961).

5. Chemical carcinogens contained in implants generally do not directly induce sarcomas, as was previously assumed by several authors (*Druckrey* and *Schmähl*, 1952; *Fithugh*, 1953). One of few exceptions to this rule is zinc in metal alloys (*Mohr* and *Nothdurft*, 1958; *Ott*, 1970).

6. Tumor induction by foreign bodies usually requires a certain implantation period, which ranges from 1/3 to 1/2 of the medium life span of the experimental animal (*Ott* et al., 1963).

7. Even after removal of the foreign body, the danger of developing sarcoma is not significantly reduced if the implant has been in place for at least one-third of the medium life-span and if the surrounding scar capsule has not been removed simultaneously with the foreign body (*Ott* et al., 1963). A causative relationship between foreign body implants and the development of sarcoma in man can only be presumed on the basis of the few observations

made. Some human tumors described in the literature have occurred after surgical foreign body implantation (see Table 4). However, it is difficult to precisely determine the exact direct or indirect role of the foreign body, trauma, radiation, and local tissue changes in contributing to the development of the tumor. A final answer to this question may not be known for decades, because the latency period for tumorigenesis could range from 30 (*Hueper*, 1964) to 60 (*Druckrey*, 1960) years. The recent use of larger foreign bodies in prosthetic surgery may help to answer these questions in the future.

Table 4. Human sarcomas after surgical foreign body implantation

Year	Author	Localization trauma foreign body	Latency period (years)	Tumor designations
1956	*McDougal*	Humerus fracture; metal plate	30	Ewing's tumor
1958	*Bürkle de la Camp*	Thigh fracture; nail	4	Alveolar sarcoma
1959	*Struppler*	Femoral neck fracture; nail removed after 2 months	6	Chondrosarcoma
1966	*Nolte*	Tibial fracture, tibial bone graft	21	Rhabdomyosarcoma
1970	*Ott*	Tibial head: surgical implantation of spongiosa from os ileum of father for lack of erythropoiesis	10	Fibrosarcoma
1972	*Burns* et al.	Toe: rupture of femoral artery; Teflon-Dacron prosthesis	10	Fibrosarcoma

Among the natural substances capable of inducing neoplasia, tobacco smoke condensates and tobacco extracts have been most thoroughly investigated. The carcinogenic potential of tobacco smoke condensates (TSC) was first described in 1931 by *Roffo*. His findings were later confirmed by *Schmähl* (see review article 1968), *Wynder* and *Hoffmann* (1967), and several other authors. Those experimental studies were chiefly done in mice, rats, and rabbits. After s.c. injection of TSC the rats developed large grayish-white fleshy tumors with extensive necroses and hemorrhages. The tumors exhibited local infiltration with early ulceration of the epidermis and destruction of the panniculus carnosus. Histologically, these lesions were spindle-cell or pleomorphic sarcomas, which only rarely metastasized to the lungs or lymph nodes (1.8% in our material). The tumors contained inclusions of brownish pigment as well as oil cysts, which represented the residual locally applied tobacco smoke condensate and solvent, respectively.

The tobacco smoke condensates were active only when s.c. injected, and did not produce tumors after oral application (*Schmähl* and *Thomas*, 1964). Alcohol-extracted and nonalcoholextracted TSC obtained prior to smoking had the same carcinogenic potential. The results of these investigations by *Schmähl* (1968) are summarized in Table 5.

Table 5. Tabulation of experimental cancer production in rats (strain BR 46) by various tobacco smoke condensates, tobacco extracts, and tobacco smoke condensates of alcohol-extracted tobaccos (*Schmähl*, 1968)

Experiment	No. animals	Medium life expectancy (months)	Local sarcomas	Medium induction time (months)	Other malignant tumors	Benign tumors	Type of nonlocal malignant tumors
Untreated control	100	24	–	–	3	11	2 Mammary sarcomas 1 Pulmonary sarcoma
Treated Control	181	26	0	–	5	16	3 Mammary sarcomas 1 Retothelial sarcoma 1 Ovarian carcinoma
Extracts							
Virginia	100	24	1	–	3	13	2 Mammary sarcomas, 1 Sarcomatosis of abdominal cavity
Orient	100	23	1	–	3	17	1 Mammary carcinoma 1 Fibrosarcoma of omentum 1 Uterine sarcoma
Blend B	100	23	3	–	4	13	2 Leukemias 1 Reticulosis 1 Mammary carcinoma
Burley	100	21	1	–	3	8	1 Mammary carcinoma 1 Mammary sarcoma 1 Pulmonary sarcoma
Schwarz	100	24	0	–	2	19	2 Sarcomas in abdominal cavity, exact site of origin unknown
Bean leaves	74	20	0	–	2	4	1 Ovarian sarcoma 1 Basalioma
Potato leaves	70	19	0	–	1	5	1 Liposarcoma
Condensates							
Virginia	100	23	27	21±3.5	0	14	–
Orient	100	26	26	22±3.7	1	11	1 Mammary sarcoma
Blend B	100	23	32	22±3.0	1	16	1 Hemangiosarcoma of kidney
Burley	100	23	9	23±6.3	1	7	1 Leukemia
Schwarz	100	22	20	21±4.2	2	10	1 Lymphosarcomatosis 1 Leukemia

According to *Wynder* and *Hoffmann* (1959, 1967) the carcinogenic qualities of TSC were due to the presence of hydrocarbons. The carcinogenic hydrocarbons, especially 3,4-benzpyrene, develop by burning of organic compounds at 900°C in the ember zone of the cigarette. Dectable amounts of this compound vary between 0.8 and 1.8 μ/g TSC, the total dose applied to a rat was about 3 μg/animal (*Schmähl*, 1968). In order to produce a s.c. sarcoma in 20% of the experimental rats, however, 100 times larger doses of 3,4-benzpyrene were required (D_{20} = 300 μg, 3,4-BP/rat). It is assumed, therefore, that the 3,4-benzpyrene present in the TSC cannot be solely responsible for its carcinogenic potential (*Schmähl*, 1969; *Harrold*, 1963; *Lazar* et al., 1966; *Roe*, 1963). In fact, other carcinogens have been found to be present in the remaining fractions of tobacco smoke condensate as well as in recombinations of TSC such as the alpha-radiant ^{210} Pollonium (*Bretthauer* and *Black*, 1967; *Kolb* et al., 1966; *Radfort* and *Hunt*, 1964) and nickel carbonyle (*Sundermann* et al., 1961, 1963, 1965). *Druckrey* and *Preussmann* (1962) and *Boyland* et al. (1964) have pointed out that during cigarette smoking cancerogenic nitroso compounds may develop by the action of nitroso gases on the amines present.

Nitrate-rich tobaccos were proven to contain only very small amounts of nitrosoamines (*Neurath*, 1967: 0.004 μg in the smoke of one cigarette). Arsenic (1 μg arsenic/gm of old tobacco) as a potential carcinogen in tobacco is less significant today because arsenic-containing insecticides are no longer utilized during the growing of the tobacco (*Holland* and *Acevedo*, 1966).

Lung *cancer in man* is considered to be a direct consequence of tobacco smoking, and it is principally found in heavy cigarette smokers (85%), and only rarely in nonsmokers (4.6%) (*Cooper* et al., 1968; *Boll* and *Hill*, 1952; *Hammond* and *Horn*, 1958). The statistical risk of developing lung cancer is dependent upon the number of cigarettes smoked daily combined with the total duration of the smoking habit (*Boll*, 1963). On the other hand, it has recently been suggested that nonsmokers who passively inhale tobacco fumes because of environmental exposure (e.g., waiters) also may have a higher incidence of lung cancer.

In man carcinogens contained in tobacco may also exhibit a primarily local effect. In India, for instance, carcinoma of the oral cavity is only observed in those population groups chewing betel nuts wrapped in tobacco (*Dietz*, 1951; *Eisen*, 1946; *Kenn* et al., 1955; *Kanolkar,* 1950, 1959; *Muir* and *Kirk*, 1960; *Sanghvi* et al., 1955; *Shanta* and *Krishnamurthi,* 1963). The question as to why tobacco smoke does not induce cancer of the oral cavity in cigarette smokers has been answered by fluorescent microscopic investigations. These investigations showed that carcinogenic hydrocarbons only penetrate dry oral mucosa. Normally, the resorption of such hydrocarbons is inhibited by the rinsing and diluting effects of the saliva within the mouth (*Wallenius*, 1966). The location of carcinomas in cigarette smokers (lung), and pipe smokers (lip) may be explained by the different smoking habits: In cigarette smokers that inhale, approximately 90% of the fluorescent material (hydrocarbon-containing compounds) are retained in the lungs (*Schmähl* et al., 1954; *Schmähl,* 1955; see Table 6). Pipe smokers rarely inhale the hydrocarbon-containing smoke into the lungs.

Soft tissue tumors may be produced by *ionizing radiation.* The first experimental cancer produced in rats was induced by local radiation (see *Marie* et al., cited by *Berenblum*, 1974). The effect of the alpha-radiating thorium dioxide after s.c. implantation has been most thoroughly studied (*Roussy* et al., 1934,1936; *Selbie*, 1936). This compound, which previously had been used in human medicine as an x-ray contrast medium, produces sarcomas at the site of local application in experimental animals after a latency period of 1-2 years.

Table 6. Comparison of fluorescence intensity of cigarette smoke after oral smoking and after inhalation

Brand of cigarette	Fluorescence intensity of cigarette smoke in 100 ml benzol, expressed in % of 0.1m% chinine sulfate standard		Retention in lung (%)
	1. Oral smoking	2. Inhalation	
A	233	43	83.5
B	375	5[a]	98
C	403	10[a]	97.5
D	403	9[a]	97.7
E	420	31	92.5

From *Schmähl* (1954, 1955).
[a] Smoked by the same test person.

2. Formal Pathogenesis

Utilizing a histogenetic classification, experimentally induced soft tissue tumors may be divided into fibro-, leiomyo-, rhabdo-, lipo-, or angiosarcoma. In many cases, however, such an exact histogenetic subdivision is not possible. The question arises whether a specific malignant tumor, e.g., rhabdomyosarcoma, can be induced by a specific carcinogen or whether its appearance is only species-dependent. Fibrosarcomas have been described by *Brown* (1963) after application of trypan blue and by *Hagensen* and *Krehbiel* (1936) after application of dibenzanthracene. However, leiomyosarcomas have also been induced using the same compounds. These latter tumors apparently arose from the arrector muscles of hair follicles. The development of rhabdomyosarcoma was repeatedly observed after injection of cadmium (*Heath* and *Daniel*, 1964) or nickel oxide (*Gilman*, 1969). *Zollinger* (1962) described malignant histiocytomas after injection of iron preparations.

The effect of a carcinogen is primarily dependent upon its chemical composition. Alteration in the basic structure of the hydrocarbon is of great importance for its carcinogenic quality; for instance, 9,10-dimethyl-benzanthracene, 3,4-benzpyrene, 1,2:3,4-dibenzanthracene and 20-methylcholanthrene are potent carcinogens, whereas anthracene or 1,2-benzpyrene do not have such tumorigenic properties. The exchange of functional groups may have either a promoting or an inhibiting effect on carcinogenesis. Soft tissue sarcomas are almost exclusively produced by local application of a carcinogen and only rarely by resorption. The locally effective compounds may be administered by either s.c. or i.m. injection.

Both the *single* and the *total dose* of carcinogen are of significance in tumorigenesis. Local sarcomas may be produced with carcinogenic hydrocarbons after a single s.c. or i.m. injection. The required single or total dose varies with the compound (between 1 and 10 mg)

and also varies with the animal species. In mice, hamsters, and rats, for instance, 5-10 mg 3,4-benzpyrene–independent of animal size and body weight–are sufficient for the pro-

Table 7. Muscle tumorigenesis in relation to age, sex, and endocrine status (*Jasmin*, 1965)

Status	Age (days)	Sex	Number of rats	Total number of rats with tumors	Average weight of tumors (g ± S.E.)	Time of appearance of least palpable tumors (days)	Overall average time of appearance (days)	Incidence at necropsy		
								5th month	6th month	7th month
Intact	30	male	14	8	13.4 ± 6.7	139	175	1/2	5/9	2/3
Intact	30	female	10	4	7.4 ± 1.8	158	210	0/5	2/3	2/5
Intact	60	male	10	7	27.0 ±12.6	165	179		1/2	6/8
Intact	60	female	18	15	47.3 ±15.9	126	144	4/6	5/5	6/7
Intact	90	male	8	3	3.8 ± 2.6	172	185		0/2	3/6
Intact	90	female	10	4	20.4 ± 9.8	156	172		0/2	4/8
Castrated	30	male	12	4	2.1 ± 0.6	159	169		2/4	2/8
Castrated	60	female	12	8	13.2 ± 3.5	152	160	0/2	3/4	5/6
Hypophysectomized	55	female	12	3	9.9 ± 7.5	176	185		0/2	3/10

duction of soft tissue sarcoma. In guinea pigs, however, significantly higher doses of the carcinogen are required, varying between 20 and 40 mg per animal. In rabbits the yield for s.c. tumors is usually low, even with very high doses.

Friedrich-Freska (1940) in his studies on dose-effect relationships found that the application of two chemically different carcinogenic hydrocarbons (3,4-benzpyrene and 20-methylcholanthrene) had a summation effect (syncarcinogenesis). The studies of *Nakahara* and *Fukuoka* (1960) also demonstrated that doses of 20-methylcholanthrene and 4-nitroquinoline-N-oxide, each at levels below tumorigenesis threshold, could produce tumors in approximately 40% of the experimental animals if applied in combination.

Factors playing a role in the genesis of soft tissue tumors: In contrast to tumors of the breast and thyroid, age and sex are not important factors in the formation and development of soft tissue tumors in experimental animals. Only *Jasmin* (1965) studied the importance of age and sex in the incidence and induction time of tumors produced by local application of nickel. The results of these studies are shown in the original table (Table 7) of the author. The highest incidence of neoplasms and the shortest latency period were obtained in female rats that were only 60 days old at the beginning of the experiment. However, the number of animals was small, and in our opinion the incidence of tumors in males and females does not differ significantly. A tumor incidence in nickel-treated rats possibly dependent upon strain is discussed by *Gilman* (1969). *Berenblum* (1954) did similar studies in mice. He treated different strains of mice with 3,4-BP and evaluated the number of induced tumors. Tumor incidence was largest in C3H-mice, fairly large in A-mice, and lowest in white-label-C-, IF-, and C_{57}-black-mice.

Carcinogenic hydrocarbons are not water-soluble and, therefore, can only be applied after dissolving in acetone, dimethylformamide, dimethylsulfoxide, sesame, olive-, sunflower-, or poppy oils. The incidence of tumors induced by the solvent itself is uncertain.

The use of s.c. injections of chemical compounds to test carcinogenic potential has been refuted by various authors as representing an "unspecific reaction" (*Oettel* et al., 1965). Local sarcomas have in fact been observed in mice and rats after injection of plant oils or sugar solutions (*Nishiyama*, 1938; *Takizawa*, 1940). *Druckrey* and coworkers (1959, 1966) found that plant oils could be obtained by both mechanical crushing techniques and by chemical extracting with benzine. Upon examination of these oils with ultraviolet (uv) light, a strong natural fluorescence suggested a possible contamination with carcinogenic hydrocarbons. Animal experiments with such contaminated oils showed a high incidence of local sarcomas, while injections of nonfluorescent oils failed to produce tumors even after a total dose of 34 ml.

Similar observations were also made in animal experiments utilizing sugar solutions. Carcinogenic hydrocarbons derived from insufficiently purified charcoal apparently had contaminated the sugar. These solutions displayed marked fluorescence and also a high carcinogenic activity. *Meier* (1942), *Hueper* (1965) and *Druckrey* et al. (1966) could not produce soft tissue tumors in rats with pure nonfluorescent sugar solutions.

III. Pathology

Soft tissue tumors develop in the vicinity of s.c.- or i.m.-deposited carcinogen. Initially the injection evokes tissue necroses, followed by the production of a cellular granulation tissue which proceeds to formation of a scar plate. After application of a chemical carcinogen, sarcomas usually arise within granulation tissue, while after s.c. implantation of a macromolecular foreign body, these tumors originate in poorly cellular scar tissue. The noxious agents responsible for the development of sarcoma can frequently still be demonstrated in the tumor. Thus, carcinogenic hydrocarbons such as 3,4-BP exhibit a blue-yellowish hue under UV light, other agents, such as tobacco extracts or plastics, may be recognized histologically.

The neoplasms display a rapid locally destructive growth pattern. They ulcerate through the surface epidermis and invade the deeper tissue layers.

Larger tumors may actually infiltrate into the thoracic or abdominal cavity. The cut surface of these tumors shows a fish-flesh appearance with bluish-red hemorrhages and yellowish areas of necrosis. The animals usually die from the local effects of the tumor, that is, from hemorrhages or inflammation. Distant metastases are rare.

1. Classification of Soft Tissue Tumors in the Rat

We have collected descriptions of almost 3,000 spontaneous and experimentally induced soft tissue tumors in rats from the literature. In most cases only general diagnoses were given: "subcutaneous tumor, malignant tumor, sarcoma, spindle cell or pleomorphic sarcoma". Histogenetic diagnoses such as "fibro-, leiomyo-, angio- or rhabdomyosarcoma" were only occasionally used. Diagnoses of "liposarcoma" or "histiocytic sarcoma" were rare. *Zackheim* (1973) called undifferentiated sarcomas "mesenchymomas". *Heath* and *Daniel* (1964) in their animal experiments used terms including anaplastic sarcoma, fibrosarcoma, and rhabdomyosarcoma.

In histologically diagnosing our own experimentally induced soft tissue tumors we have favored a histogenetic classification using certain morphologic criteria. In order to confirm our diagnoses, the neoplasms were also studied with electron microscopic, immunohistologic, and histoautoradiographic methods.

For *electron microscopy* the tissues were fixed in glutaraldehyde and postfixed in osmium tetroxide with subsequent epon embedding. For *immunohistology* the tissues were frozen in liquid nitrogen and stored at less than -75° C. Frozen cryostat sections were prepared from this material and were treated with the following antibody-containing sera:

1. Antibodies against cross-striated muscle from patients with myasthenia gravis.
2. Antibodies against smooth musculature from patients with thyroiditis and myositis.
3. Antinuclear antibodies from patients with systemic lupus erythematosus.
4. Antimitochondrial antibodies from patients with pseudo-LE syndrome.
5. Control serum from a normal person.

FITC-labeled antihuman-immunoglobulin serum from a horse (Roboz/USA) was used as an antiserum. Autoantibodies occurring in various immunopathologic diseases are not specific for humans and, therefore, may also be used as an antigen identifying substrate in animal tissue. The results of these studies and a comparison with histologic findings in 14 rats with and without soft tissue tumors are listed in Table 9.

Table 8. Classification of soft tissue tumors in the rat

1. Benign mesenchymal tumors
   1.1. Fibroma
   1.2. Rare tumors: myoma, myxoma, lipoma
2. Malignant mesenchymal tumors (sarcoma)
   2.1. Spindle cell sarcoma
      2.1.1. undifferentiated spindle cell sarcoma
      2.1.2. well-differentiated spindle cell sarcoma
         2.1.2.1. Fibrosarcoma
         2.1.2.2. Leiomyosarcoma
   2.2. Pleomorphic cell sarcoma
      2.2.1. undifferentiated pleomorphic cell sarcoma
      2.2.2. well-differentiated pleomorphic cell sarcoma
         2.2.2.1. anaplastic fibro- or leiomyosarcoma
         2.2.2.2. rhabdomyosarcoma
         2.2.2.3. other tumors: angiosarcoma, liposarcoma, histiocytic sarcoma
   2.3. Round cell sarcoma, lymphocytic or reticular type
   2.4. Mixed sarcomas (of 2.1. and 2.2.)

Table 9. Tabulation of histologic and immunohistologic findings

Animal	SK-A	SM-A	AM-A	AN-A	Control	Histology
1	0	0	(+)	+	0	No tumor
2	(+)	0	0	+	0	No tumor
3	0	0	(+)	+/0	0	No tumor
4	+	0	+	+	0	Sp-Sa and Rh
5	0	0	+/0	+/0	0	E-Sa
6	0	0	(+)	0	0	E-Sa
7	+	0	+	0	0	Sp-Sa and Rh
8	0	0	+	+/0	0	Sp-Sa and F-Sa
9	0	0	+/0	+/0	0	Sp-Sa
10	0	0	+/0	+/0	0	Sp-Sa and F-Sa
11	0	0	+	+/0	0	E-Sa
12	0	0	+	+/0	0	E-Sa and Rh
13	+	0	+	+/0	0	E-Sa and Rh
14	++	0	++	+/0	0	Rh

Sp-Sa: spindle cell sarcoma; Rh: rhabdomyosarcoma; E-Sa: undifferentiated sarcoma: F-Sa: fibrosarcoma; SK-A: skeletal muscle antibody; SM-A: smooth muscle antibody; AM-A: antimitochondrial antibody; AN-A: antinuclear antibody; +/0: occasional cells in the same section are positive, others are negative

2. Pathologic-Anatomical Studies on Experimentally Induced Soft Tissue Tumors of the Rat

Benign mesenchymal tumors have only rarely been observed or described in animal experiments. Such tumors are characterized by a slow, expansile growth forming a peripheral pseudocapsule. Histologically there are no criteria of malignancy. The diagnosis of fibroma (1.1) is made most frequently. When such tumors occur near the milk line in females, they must be differentiated from hyalinized fibroadenomas. Fibromas are firm, variously sized encapsulated nodules. Histologically they consist of sclerotic connective tissue including only occasional nuclei. *Lipomas, myxomas, and myomas* (1.2) are more rare. In the so-called myxomas, stellate cells are found in a loose myxoid stroma. In most cases they probably represent edematous fibromas, because between the tumor cells a delicate network of collagen fibers can often be recognized. Lipomas contain well-differentiated fat cells in a lobular arrangement which may be difficult to differentiate from fat cell hyperplasia. True lipomas display delicate septa and a peripheral capsule. Occasionally, they contain larger oil cysts, which include altered fat that has been derived from disintegrating fat cells.

Malignant mesenchymal neoplasms or *sarcomas* are characterized by rapidly invasive and destructive growth. They show histologic criteria of malignancy including cellular anaplasia and increased numbers of mitotic figures, some atypical. These tumors may have a predominantly spindle cell-, round cell-, or polymorphous cell component.

Spindle cell sarcomas (2.1) are usually quite cellular and consist of elongated cells with ovoid chromatin-rich nuclei. The tumor cells are commonly arranged in bundles showing palisading of nuclei or they may exhibit a whorled appearance similar to that seen in fibrosarcoma protuberans (Fig. 1). Although mitotic figures are frequent, the cellular pattern may be otherwise quite uniform. Spindle cell sarcomas producing a significant amount of collagen fibers are designated *fibrosarcomas* (2.1.2.1.) (Fig. 2). The van Gieson stain shows a delicate intercellular network of collagen fibers surrounding individual cells. In these areas a heavy incorporation of radioactive labeled sulfur can be demonstrated with histoautoradiographic methods. With electron microscopy, two predominant cell types are found:

a) "Mature fibrosarcoma cells" with pleomorphic, sometimes lobulated nuclei. There is almost no chromatin condensation but nucleoli are very prominent. The cytoplasm contains a well-developed, frequently dilated ergastoplasm and a variable number of small, round mitochondria. Occasionally striated immature collagen fibrils are seen in the ergastoplasmic tubules. In addition, scattered rudimentary microfibrils are present. The cells are surrounded by an amorphous ground substance with many collagen fibrils (Figs. 3, 4).

b) "Undifferentiated sarcoma cells" are markedly irregular in shape. There is no chromatin condensation and nucleoli vary in size. The cytoplasm contains only a few, though abnormally large and dilated ergastoplasmic cisternae, which may reach the size of the nucleus. These cisternae include a finely granular material with a ribbon-like or spongy electron-dense mass which can also be observed in the extracellular space (Figs. 5, 6). The deformed mitochondria display irregularly arranged cristal membranes. Some mitochondria contain myelin-like stacks of membranes or amorphous osmiophilic inclusion bodies. In some sarcoma cells the mitochondria may be conspicuous by their large size (Fig. 7). They contain abundant cristal membranes with longitudinal lamellar or concentric patterns. Paracristalline inclusions can be demonstrated in the intercristal space. Cells containing megamitochondria can also be detected with fluorescent microscopy using antimitochondrial serum.

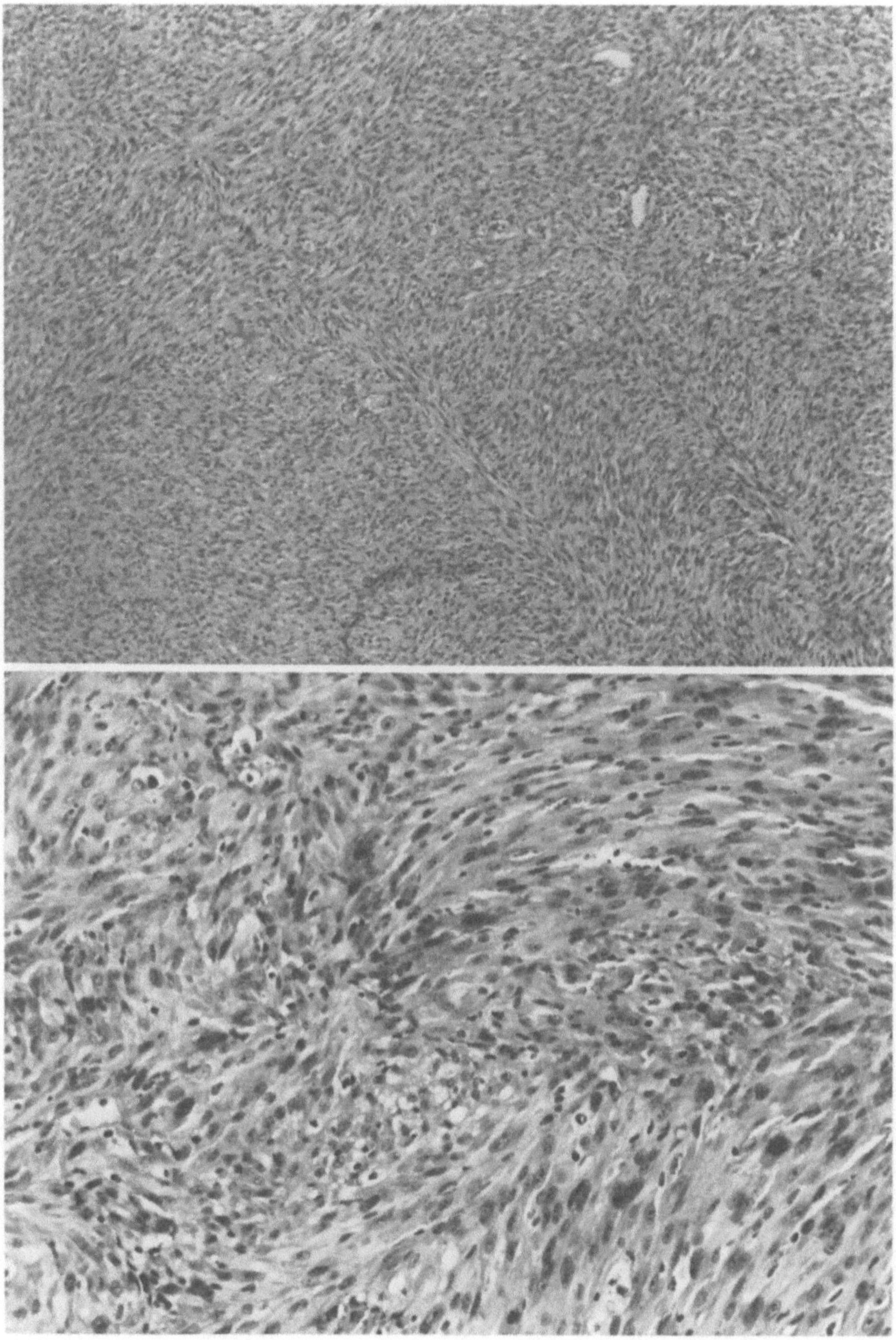

Fig. 1. (a) Fibrosarcoma with storiform pattern resembling dermatofibrosarcoma protuberans. (b) Higher magnification of whorls of tumor cells reveals hyperchromatic and pleomorphic nuclei. H & E x 80

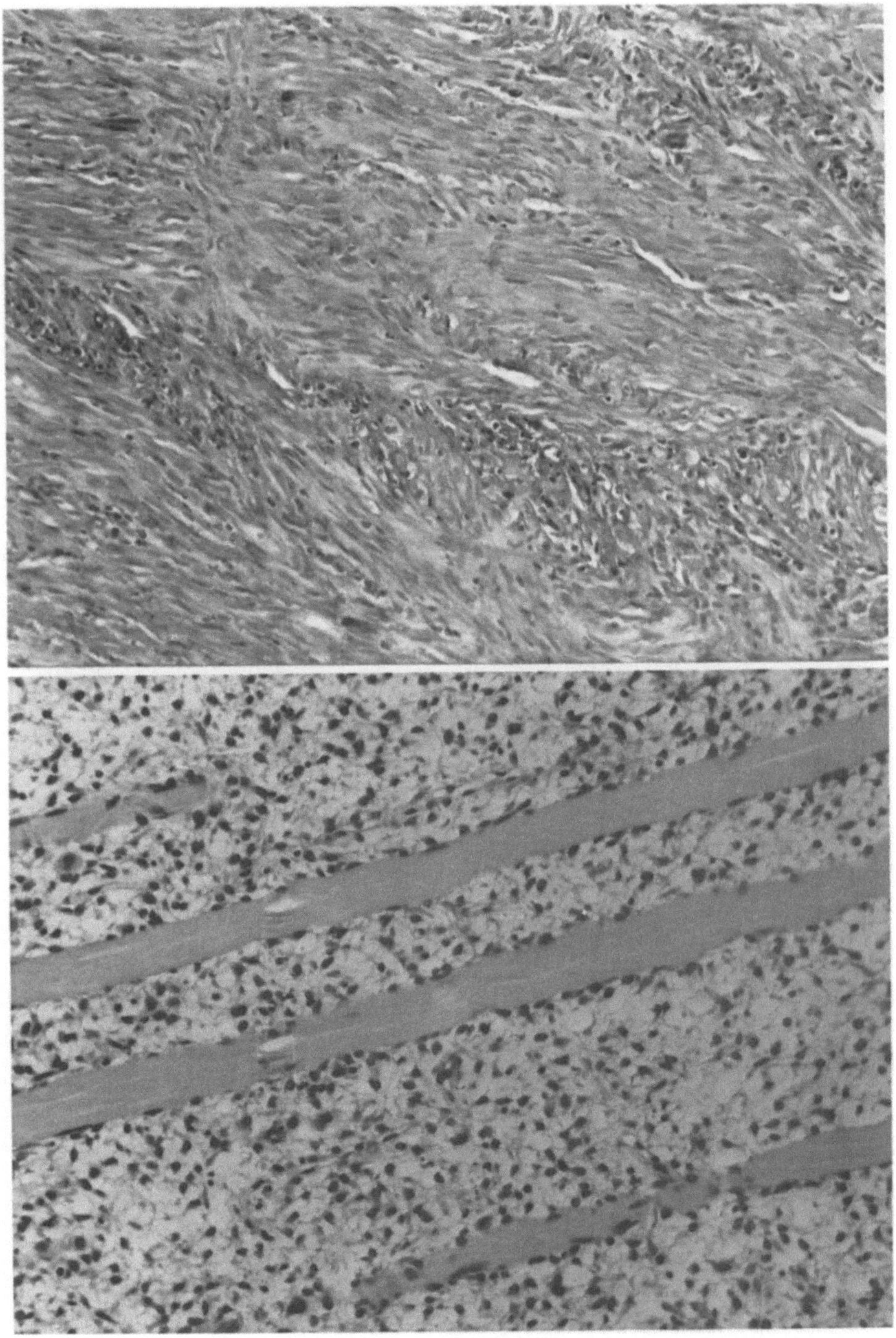

Fig. 2. (a) Fibrosarcoma with abundant collagen fibers surrounding spindle-shaped tumor cells. (b) Myxoid variant of fibrosarcoma with delicate network of collagen fibers interspersed with cross-striated muscle bundles. H & E x 120

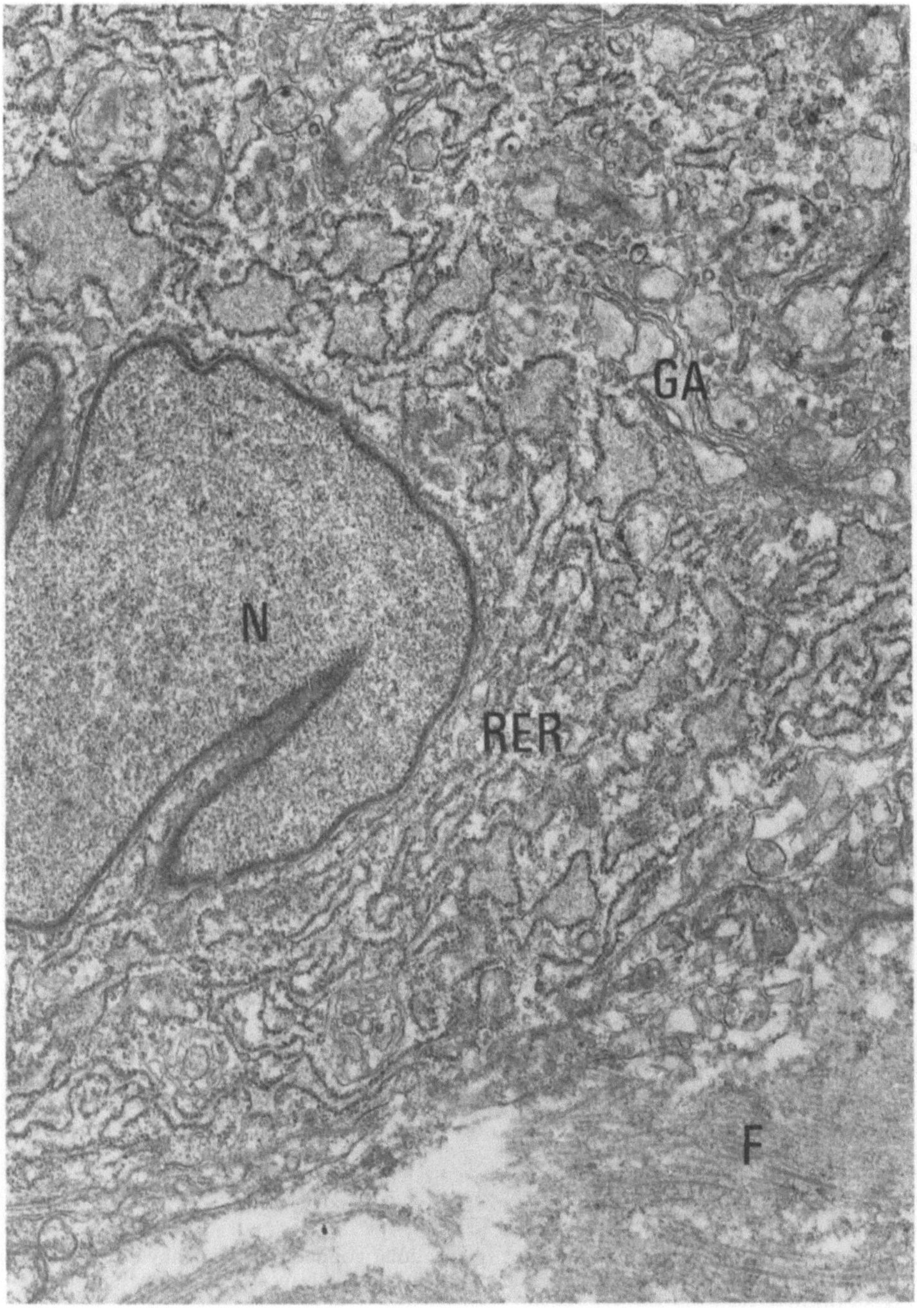

Fig. 3. Tumor cell with abundant ergastoplasm *(RER)*. Note prominent Golgi apparatus *(GA)* and lobulated nucleus *(N)*. Immature collagen fibrils *(F)* are seen in the extracellular space. x 15,000

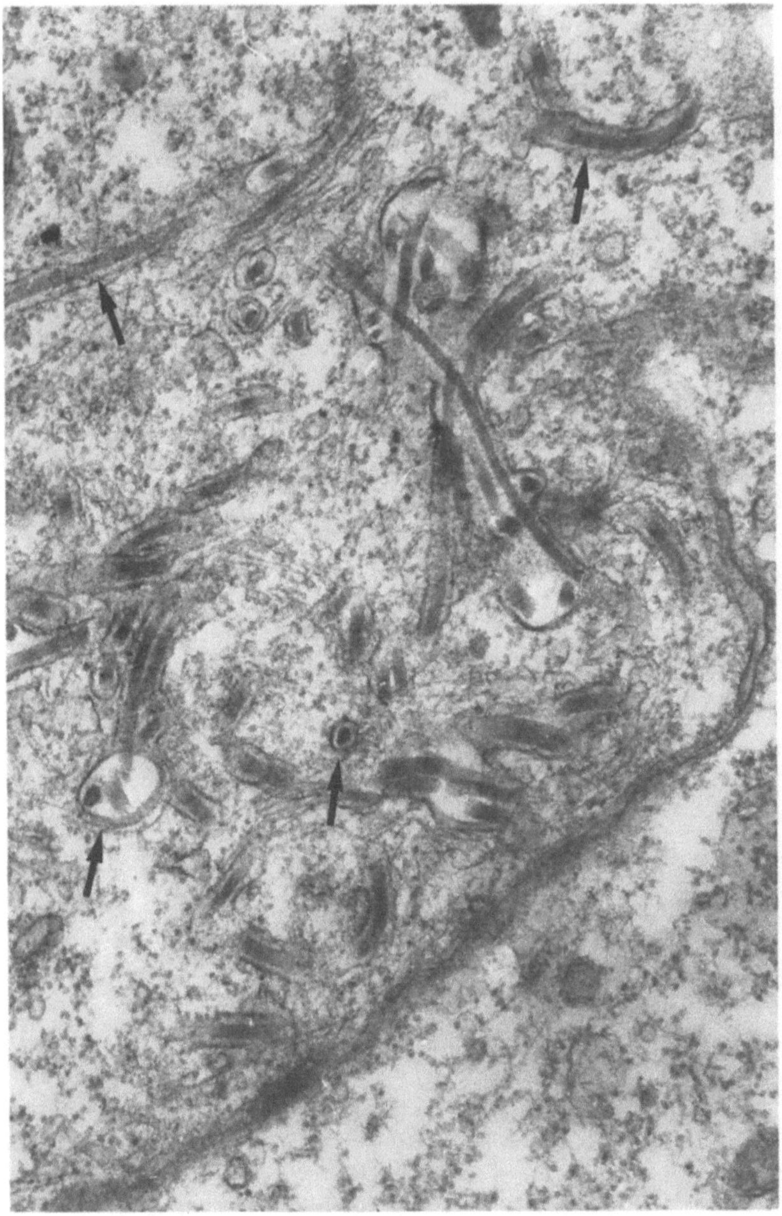

Fig. 4. Fibroblast-like tumor cell containing intracellular collagen fibrils. They are located in channels of the rough endoplasmic reticulum (arrow) and show a typical axial periodicity. x 35,000

The differentiated spindle cell sarcomas include *leiomyosarcoma* (2.1.2.2.), which has been diagnosed more commonly by some authors than by others. We have also occasionally diagnosed a leiomyosarcoma light microscopically, but have always failed to substan-

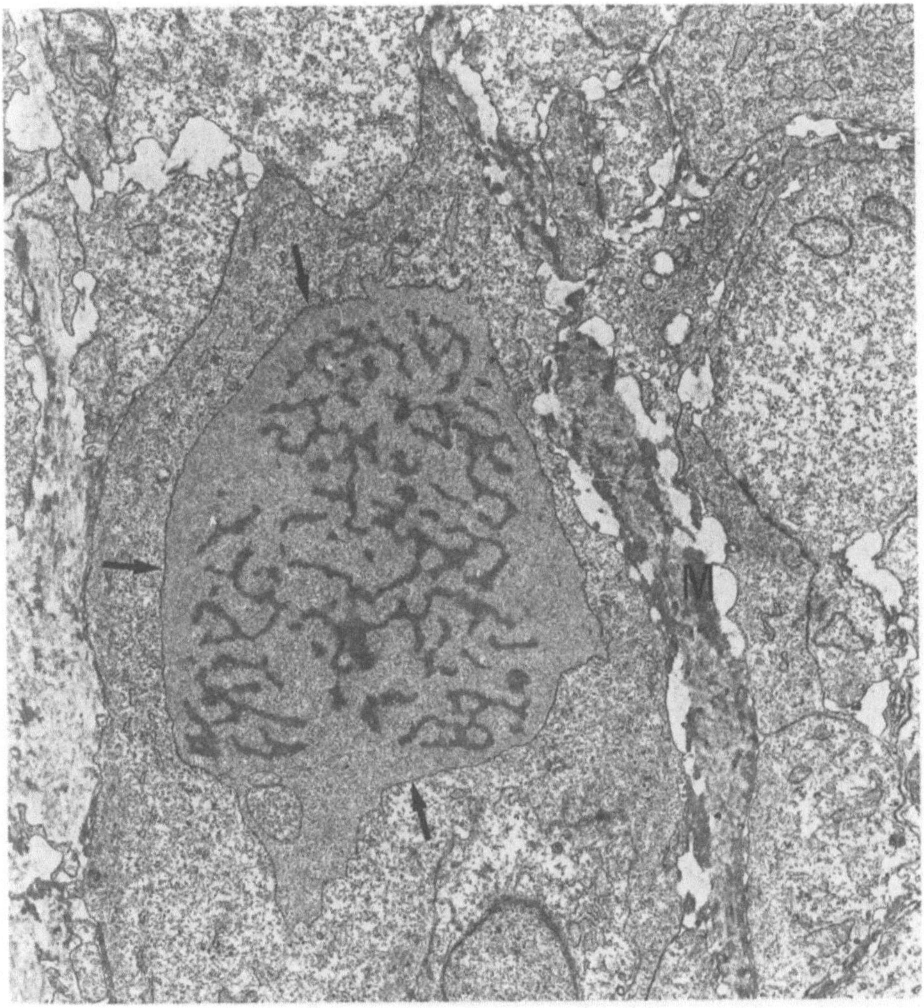

Fig. 5. Fibrosarcoma showing tumor cells with giant ergastoplasmic cisterna (arrow) which contains amorphous finely granular material with a denser spongy component *(M)*. x 65,000

tiate this diagnosis by electron microscopy or by immunohistology with antibodies directed against smooth musculature.

The *pleomorphic sarcoma* (2.2.) is characterized by marked cellular and nuclear pleomorphism. The cells vary significantly in size and may be either elongated or rounded. Usually several nuclei are irregularly distributed in the cytoplasm. The number of mitotic figures is increased and atypical mitotic figures are observed. Cross-striations in the cytoplasm of the tumor giant cells, as occasionally seen in rhabdomyosarcomas, are not demonstrable. The elongated multinucleated giant cells, however, are quite characteristic and are helpful in making a correct diagnosis (Fig. 8). We were able to confirm our histological diagnosis in most cases by electron microscopy and/or immunohistology.

With the electron microscope two cell types can be distinguished in rhabdomyosarcomas:

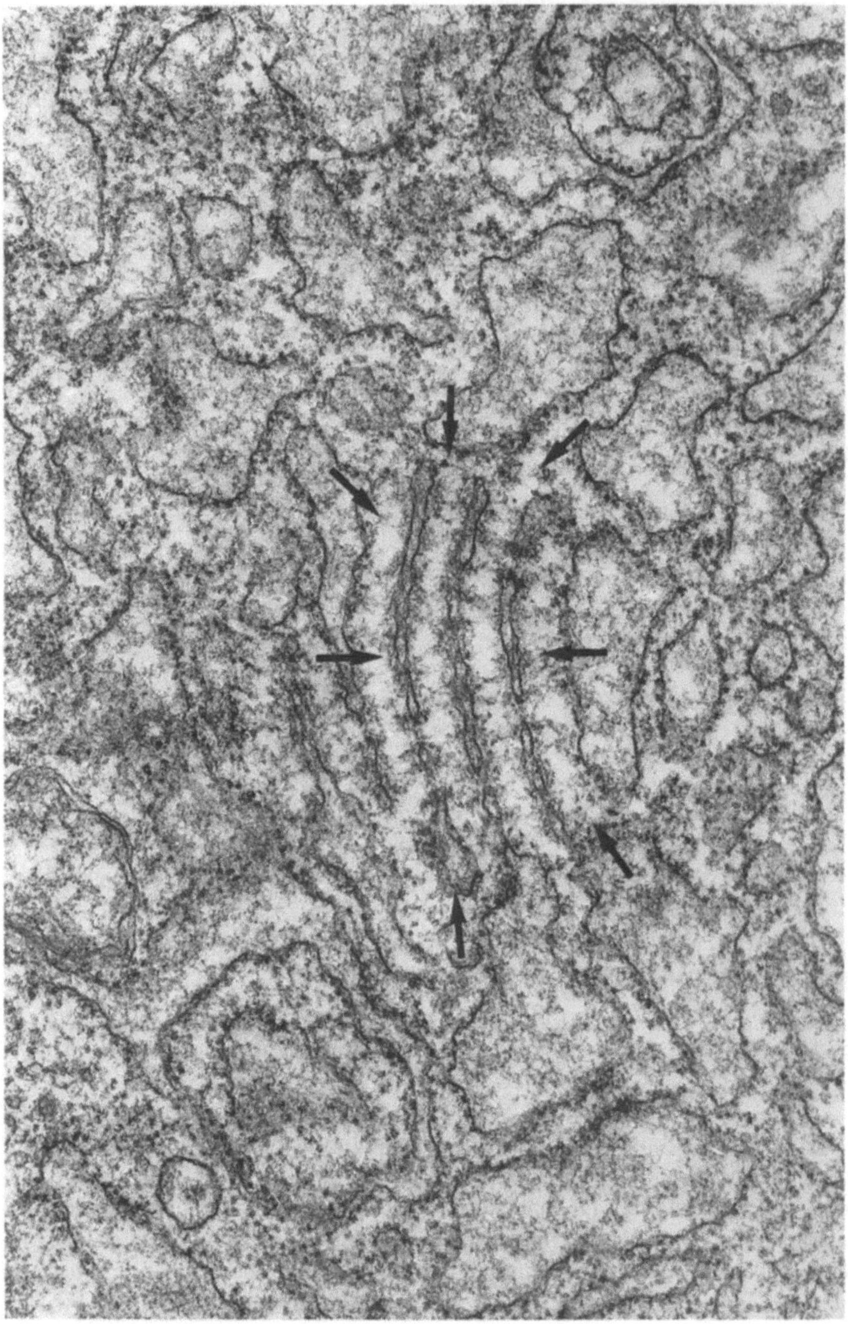

Fig. 6. Cytoplasm of tumor cell displaying annulate lamellae (arrow) of endoplasmic reticulum. x 35,000

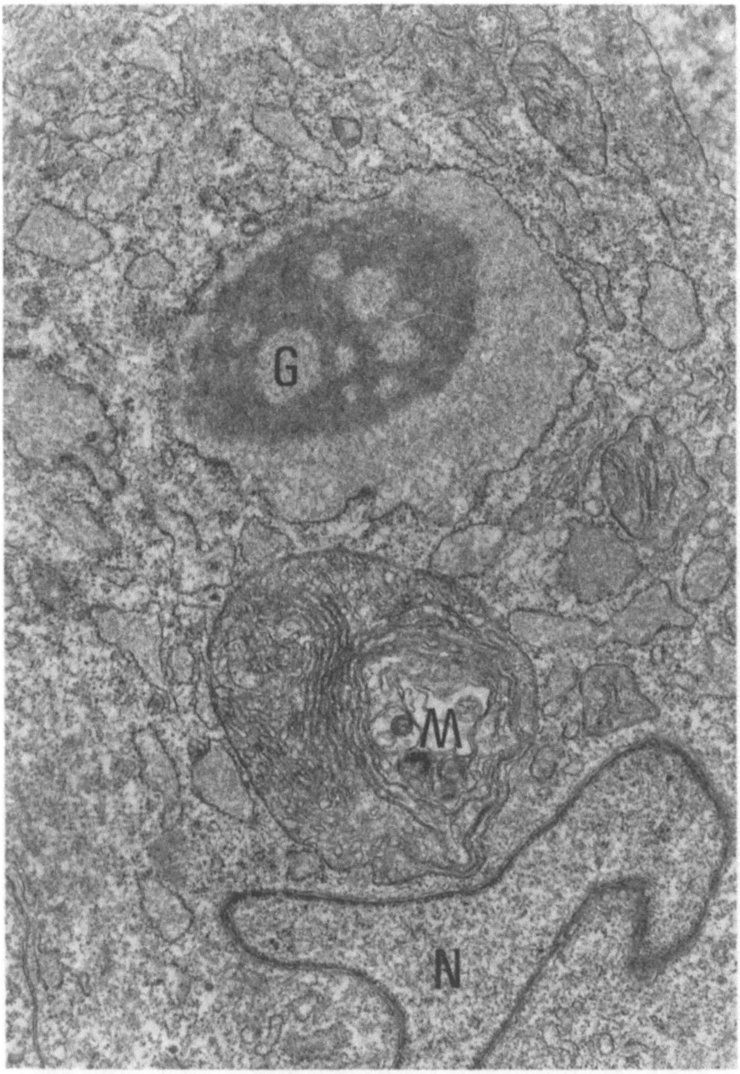

Fig. 7. Tumor cell containing giant ergastoplasmic cisterna *(G)* enclosing electron-dense material. The enlarged mitochondria *(M)* contain concentric stacks of cristal membranes and may reach the size and the nucleus *(N)*. x 15,000

a) "Differentiated rhabdomyosarcoma cells" show little chromatin condensation within the nuclei. The nuclear membrane exhibits small villous projections. The nucleoli are unusually large showing marked segregation of their structural components with prominence of nucleonemata. The cellular membrane appears corrugated and wrinkled and is surrounded by numerous pinocytotic vesicles. The cytoplasm contains moderate numbers of bizarre mitochondria, free ribosomes and little ergastoplasm. The mature rhabdomyosarcoma cell typically includes cytoplasmic fibrils, which are arranged in bundles like sarcomeres interrupted by electron-dense structures similar to Z-bands (Figs. 9, 10).

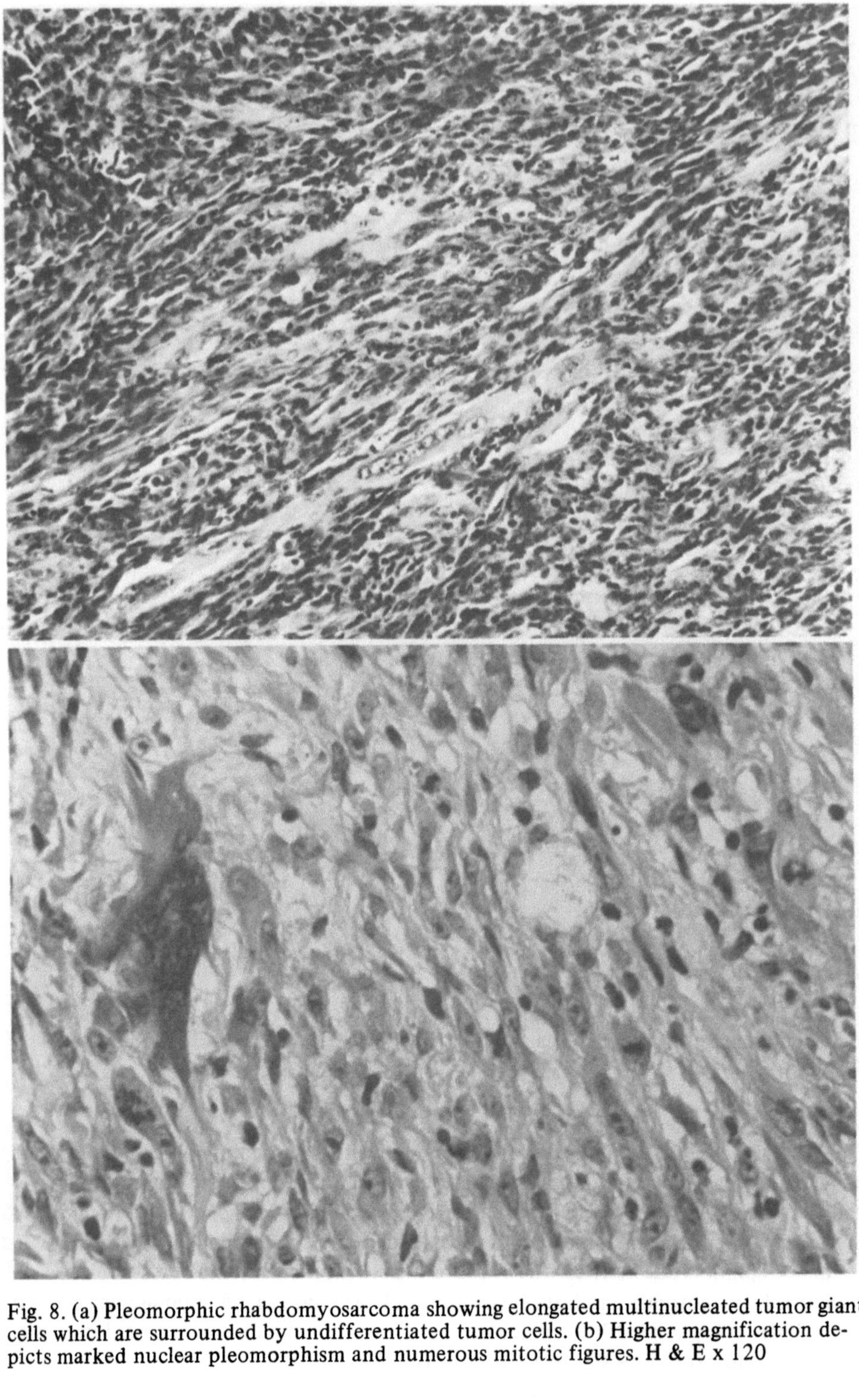

Fig. 8. (a) Pleomorphic rhabdomyosarcoma showing elongated multinucleated tumor giant cells which are surrounded by undifferentiated tumor cells. (b) Higher magnification depicts marked nuclear pleomorphism and numerous mitotic figures. H & E x 120

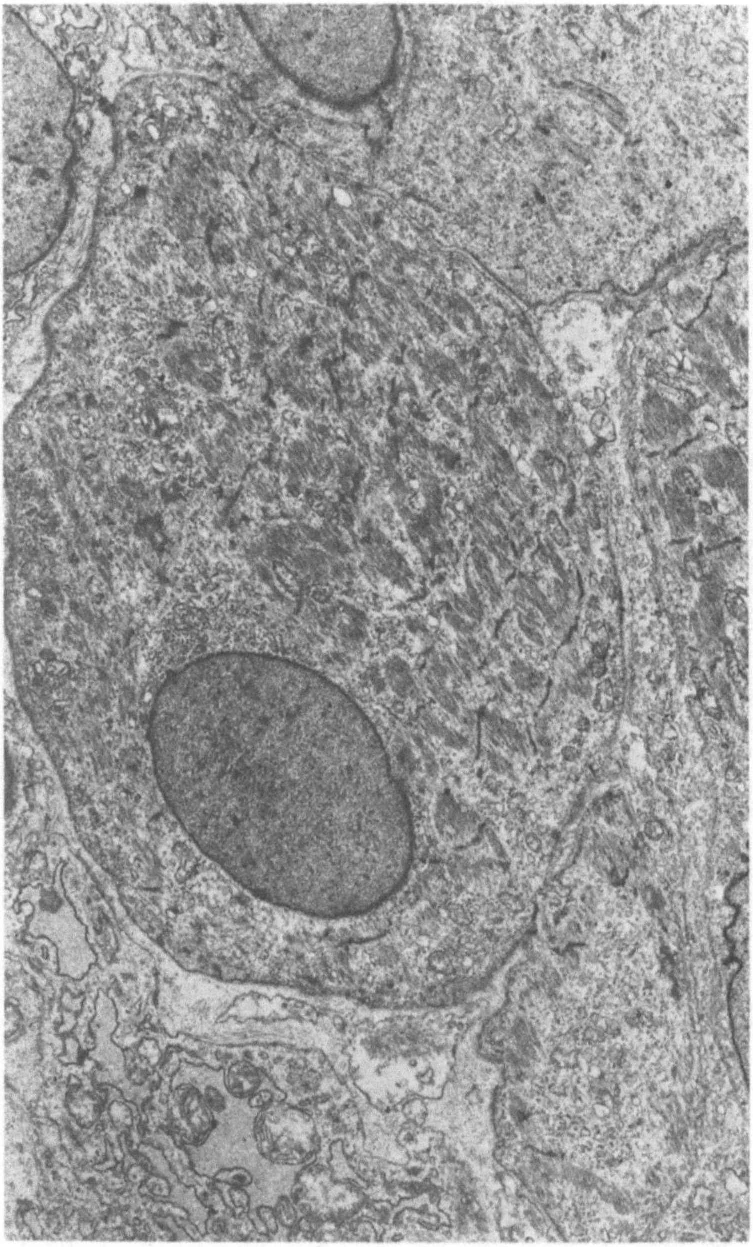

Fig. 9. Tumor cell with scanty organelles is filled with myofibril-like structures. x 5,000

b) The "undifferentiated rhabdomyosarcoma cells" are more spindle-shaped and are tightly packed like epithelial cells. The nuclei are markedly lobulated containing large nucleoli. These cells also contain cytoplasmic microfibrils, however, without sarcomere-

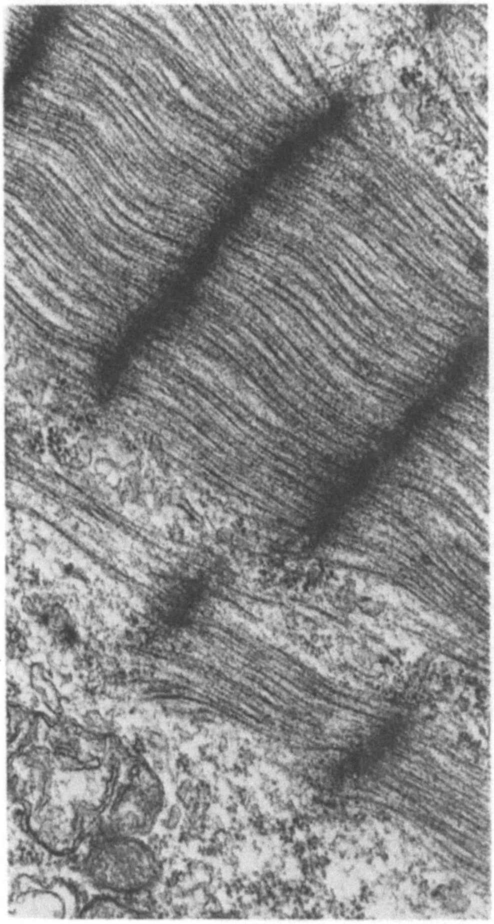

Fig. 10. Portion of tumor cell rich in myofibrils with sarcomere-like arrangement. x 35,000

like structures. The microfibrils are located primarily at the cellular periphery or in the perinuclear area. The ergastoplasm and the Golgi apparatus are well developed. Cisternae of the endoplasmic reticulum contain finely granular material. The Golgi apparatus is characterized by hypertrophy of its lamellar and vesicular components. The extracellular ground substance contains only few, primarily longitudinal collagen fibrils in parallel arrangement (Fig. 11).

Immunohistologically, highly differentiated as well as less well differentiated rhabdomyosarcoma cells show a distinctly positive reaction with antibodies directed against cross-striated musculature. By this method, cells of striated muscular origin are easily identified and can be recognized more reliably than with routine histologic and electron-microscopic methods (Fig. 12). These studies also demonstrated that pleomorphic areas in an otherwise spindle cell tumor probably represent a rhabdomyosarcomatous component and should not be considered as an expression of undifferentiation of a fibrosarcoma. In these neoplasms the nuclei may be quite pleomorphic. Multinucleated giant cells, however, are rare. It is also noteworthy that nuclear changes are not infrequently found in rhabdomyosarcoma cells, preventing these cells from reacting with antinuclear antibodies.

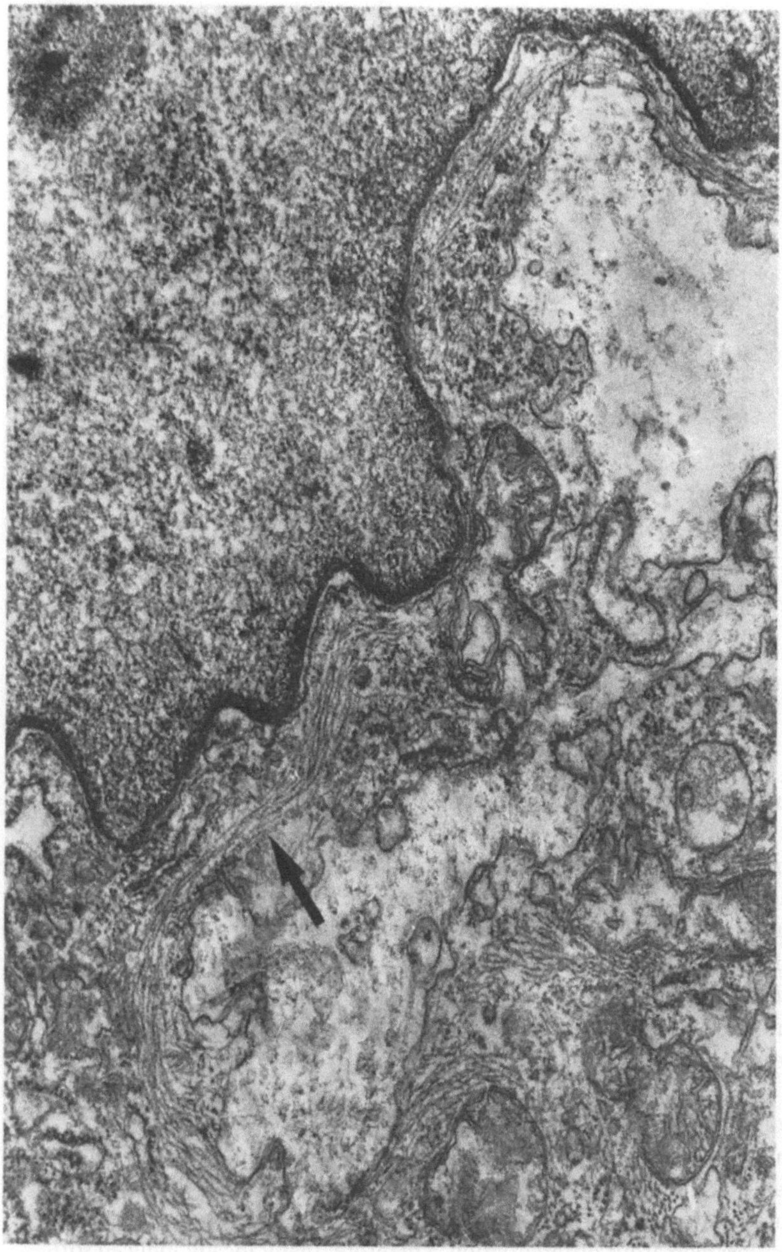

Fig. 11. Cytoplasm of tumor cell shows some mitochondria with disfigured cristae and a few marginal myofibrils (arrow). x 30,000

The term *angiosarcoma* (2.2.2.2.) denotes neoplasms with prominent solid or cavernous vessels showing evidence of malignancy. Pleomorphic *liposarcomas* display variously sized, usually rounded cells with one or several nuclei. They typically contain small or large, oc-

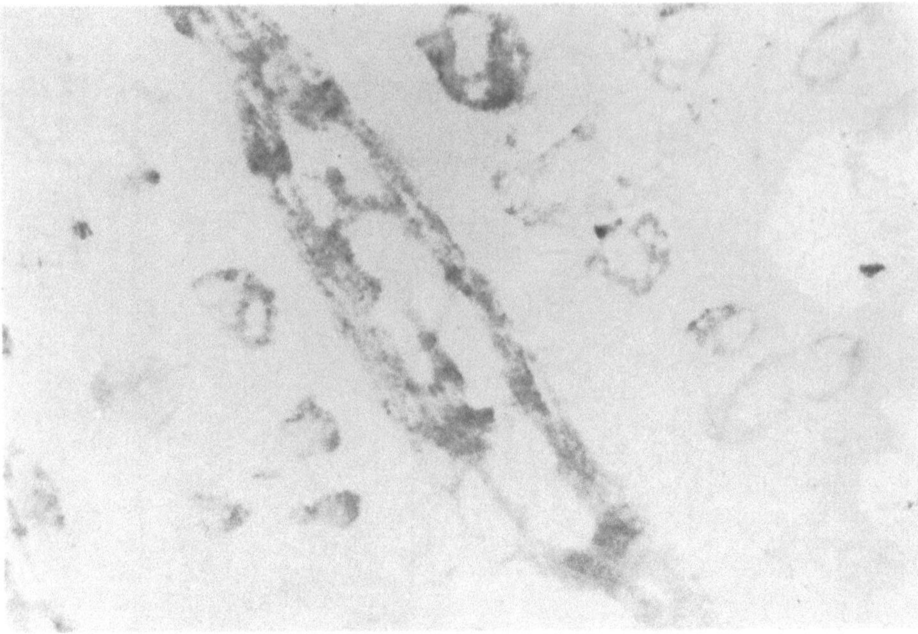

Fig. 12. Rhabdomyosarcoma cell after treatment with antimyosin serum (immunofluorescence). Copied from slide, therefore positive fluorescence black, background gray. x 350

casionally confluent cytoplasmic vacuoles which stain as orange-red droplets with a Sudan III fat stain. The *histiocytic sarcoma* has been induced by some authors after application of trypan-blue or after local injection of iron. This type of sarcoma is also pleomorphic with large rounded or oval cells with one or several nuclei. The cytoplasm often contains accumulated substances, such as the applied carcinogen, indicating phagocytic activity of the tumor cells.

The *round cell sarcoma* (2.3.) which occurs only rarely among experimentally induced soft tissue tumors shows some similarity to lymphosarcoma and reticulosarcoma. Such tumors contain rounded cells in an edematous stroma, especially when there are extensive necroses.

Most of the soft tissue sarcomas that we investigated displayed a *mixed histologic pattern* (2.4.), that is, undifferentiated, spindle cell, fibrosarcomatous, and rhabdomyosarcomatous areas seen simultaneously. Quantitatively, spindle cell areas predominated in most of the tumors. However, careful examination frequently disclosed areas with fibro- or rhabdomyosarcomatous differentiation. Therefore, many locally induced soft tissue tumors in the rat may have to be diagnosed as *fibrorhabdomyosarcomas.*

3. Presarcomatous and Nonsarcomatous Tissue Changes

In contrast to epithelial tumors, such as neoplasms of the liver or of the mucosa of the gastrointestinal tract, where gradual changes may be observed between normal cells and

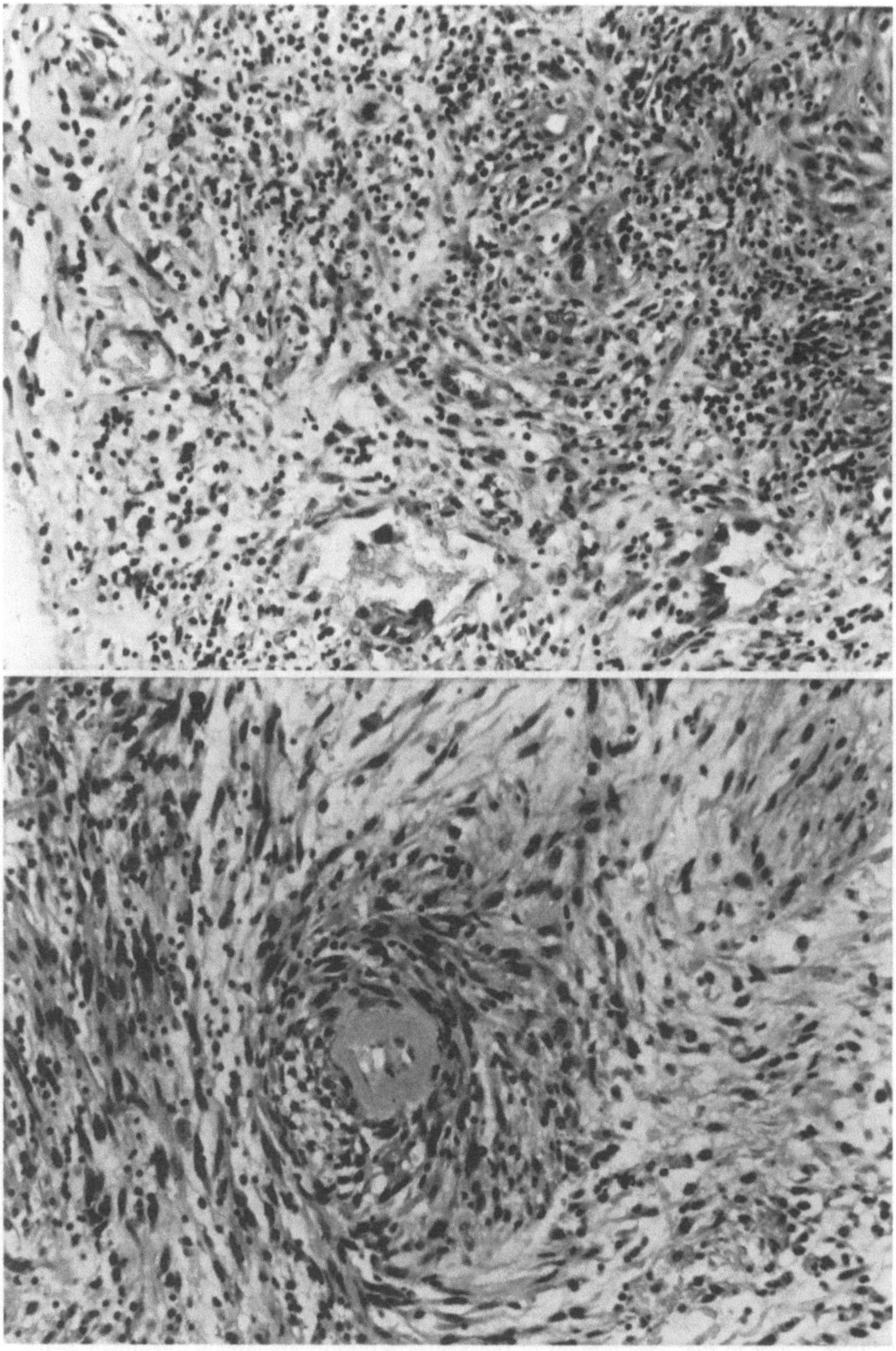

Fig. 13. Granulation tissue at the site of application of DMBA (90th experimental day). (a) Hyperchromatic and pleomorphic vascular and stromal cells are interspersed with round cells of the inflammatory infiltrate. (b) In areas concentric arrangement of stromal cells is seen around fibrinoid-necrotic arterioles. H & E, x 100

cancer cells, only little is known about the presarcomatous stage and the early changes in the genesis of soft tissue tumors. If Wistar rats weighing an average of 100 gm are treated with a single s.c. or i.m. dose of 10 mg 9,10-dimethyl-1,2-benzanthracene and killed at regular intervals, the following tissue changes may be observed:

1. After *30 days* extensive necroses of s.c. fat tissue and adjacent musculature are found in the area of injection. The lumens of blood vessels exhibit thrombotic occlusions. The surrounding connective tissue displays inflammatory infiltration. In addition, oil cyst surrounded by multinucleated giant cells and remnants of the applied carcinogen may bee seen.

2. After *60 days* a fully developed granulation tissue is present. There are numerous newly formed capillaries, surrounded by stellate histiocytes or spindled fibroblasts. Occasional muscle fibers contain several nuclei and prominent cytoplasmic cross-striations: they are interpreted as non-neoplastic muscle giant cells induced by injury.

3. After *90 days* the granulation tissue is still present and scar collagenization is already seen. The cells surrounding capillaries and occasional endothelial cells are irregular in shape and size. There is no evidence of malignancy at this time, although a certain cellular pleomorphism is already present (Fig. 13).

4. After *105 days* small areas are observed which consist primarily of spindle cells showing occasional concentric distribution. The nuclei vary in size and chromatin content. Mitoses are relatively frequent. These foci already exhibit early invasive growth with infiltration of striated muscle. Nearby degenerating muscle fibers show a vacuolar eosinophilic cytoplasm.

5. Between *106 and 150 days* we find larger, fully developed sarcomas consisting primarily of elongated tumor cells. Scattered isolated pleomorphic cells with several nuclei and a partly rounded, partly fibrillar cytoplasm are also identified. Tumor cells exhibiting cross-striations may occasionally be demonstrable. They are, however, difficult to distinguish from non-neoplastic local muscle fibers included in the tumor tissue. Immunohistologically and electron microscopically, myosin and myofibrils may be demonstrated even in undifferentiated tumor cells, proving at least some rhabdomyosarcomatous component. In addition there are some cells ultrastructurally similar to fibroblasts.

6. After *150 days* large, undifferentiated sarcomas are observed with a prominence of spindle cells and focal collections of intercellular collagen fibers. Cellular structures typical for rhabdomyosarcoma are rare at this stage of tumor development. Necroses and edema may simulate an appearance of myxosarcoma. Some neoplastic areas show a prominent vascular component, others contain large tumor cells with vacuolar cytoplasm suggestive of angio- or liposarcoma.

IV. Comparative Pathology and Pathogenesis of Soft Tissue Tumors in Man and Rats

Little is known about the direct cause of soft tissue tumors in man. Classical examples such as occupational cancers or geographic tumors do not offer useful explanations. Also, the preliminary stages in the development of these human tumors are largely unstudied. Malignant soft tissue tumors in man rarely arise from a pre-existing benign tumor such as lipoma or fibroma. These malignant neoplasms usually develop de novo with the exception of fibrosarcomas developing after previous irradiation for inflammatory lesions or for benign or malignant tumors.

Foreign-body- or prosthesis-induced sarcomas, at the present time, are still extremely rare.

The results of animal experiments do not significantly contribute to the problem of causal pathogenesis of similar human tumors. Carcinogenic hydrocarbons or locally alkylizing substances can hardly be considered an important cause of human sarcomas. On the other hand, the attempt to regularly and selectively produce soft tissue sarcomas in animals by resorption of carcinogenic compounds has not been successful. The only corresponding causal pathogenesis of soft tissue tumors in both man and animals at the present time is found in the rare radiation-induced tumors. The possible role of viruses or genetic factors is still obscure. Morphologic features of soft tissue sarcomas in man and animals also show marked differences. For instance, the direction of the differentiation of fibrosarcomas and other types of sarcomas so common in human pathology is not repeated in rats. In rats mesenchymal neoplasms are usually of a mixed type containing a fibrosarcomatous and a rhabdomyosarcomatous component.

Today it can be stated that although we are capable of reliably producing soft tissue sarcomas in animal experiments, significant differences are apparent in the pathogenesis of such neoplasms in man and animals. Crucial questions as to the genesis, diagnosis, prognosis, and therapy are still unanswered.

Acknowledgment. The experimental studies were supported by grants from the Deutsche Forschungsgemeinschaft (Bonn-Bad Godesberg): Th 91/11, R: 271/3.

Table 2. Soft tissue tumors induced by chemical cancerogenic substances (own material)

Substance and dosage	Appl.	t	tumors
Butyl-nitroso-urea			
d = 50 mg/kg/week	s.c.	209±37	5 Undifferentiated spindle cell sarcomas
D = 1150 mg/kg			7 Fibrosarcomas
			2 Undiff. fibrosarcomas
			2 Histiocytic sarcomas
			2 Undiff. spindle cell and polymorph. cell sarcomas
d = D = 400 mg/kg	s.c.	387	1 Fibrosarcoma
Hydrazo-dicarbonic acid-binitroso-methylamide			
d = 15 mg/kg/week	s.c.	296±53	4 Undiff. spindle cell sarcomas
D = 420 mg/kg			4 Fibrosarcomas
			1 Undiff. polymorph. cell sarcoma
			2 Undiff. spindle cell and polymorph. cell sarcomas

Explanations: Dose: d = single dose; D = total dose; D_{50} = median total dose. Appl.= application; s.c. = subcutaneous; i.m. = intramuscular; t = Induction time (days); tumors = classification according to Part III; polymorph. = polymorphic

Table 2 (continued)

Substance and dosage	Appl.	t	tumors
Hydrazo-bis N,N^1-methylnitroso-carbonamide d = 30 mg/kg/week D = 840 mg/kg	s.c.	225±24	1 Undiff. spindle cell sarcoma 4 Fibrosarcomas 1 Round cell sarcoma 3 Undiff. spindle cell sarcomas 2 Spindle cell and polymorph. cell sarcomas (fibrosarcomas)
Nitroso-imidazolidone d = 30 mg/kg/week D = 630 mg/kg	s.c.	321±25	3 Undiff. spindle cell sarcomas 7 Fibrosarcomas 1 Undiff. polymorph.cell sarcoma 2 Liposarcomas
Nitroso-imidazolidone d = 30 mg/kg/week D = 630 mg/kg	s.c.	321±25	3 Undiff. spindle cell sarcomas 7 Fibrosarcomas 1 Undiff. polymorph.cell sarcoma 2 Liposarcomas 1 Histiocytic sarcoma 5 Undiff. spindle and polymorph. cell sarcomas 1 Spindle and polymorph.cell sarcoma (fibrosarcoma)
Nitroso-CH_3-nitroso-guanidine d = 45 mg/kg/week D = 225 mg/kg	s.c.	293±30	1 Undiff. spindle cell sarcoma 3 Fibrosarcomas 1 Histiocytic sarcoma 1 Round cell sarcoma
d = 70 mg/kg/week D = 520mg/kg	s.c.	180	1 Undiff. polymorph. cell sarcoma
Nitroso-CH_3-nitroso-guanidine d = 90 mg/kg/week D = 450 mg/kg	s.c.	307±43	1 Undiff. spindle cell sarcoma 1 Undiff. polymorph.cell sarcoma 2 Rhabdomyosarcomas 1 Round cell sarcoma
O-AE-N-nitroso-morpholine d = 40 mg/kg/week D = 1966 mg/kg	s.c.	350±13	4 Undiff. spindle cell sarcoma 3 Undiff. spindle and polymorph. cell sarcomas
d = D = 100 mg/kg	s.c.	217	1 Spindle and polymorph. cell sarcoma (rhabdomyosarcoma)
d = D = 400 mg/kg	s.c.	384	1 Undiff. spindle cell sarcoma
d = D = 430 mg/kg	s.c.	346	1 Undiff. spindle cell sarcoma
Phenyl-nitroso-urea d = D = 150 mg/kg	s.c. s.c.	456 479	1 Histocytic sarcoma 1 Undiff. spindle cell and poly-morph. cell sarcoma
d = D = 200 mg/kg	s.c.	654	1 Fibroma

Table 2 (continued)

Substance and dosage	Appl.	t	Tumors
Azoethane			
d = D = 500 mg/kg	s.c.	441	1 Undiff. fibroma
		432	1 Fibroma
Azoxyethane			
d = D = 30 mg/kg	s.c.	182	1 Spindle cell sarcoma
		198	1 Undiff. spindle cell sarcoma
d = D = 180 mg/kg	s.c.	259	1 Fibroma
Azoxybutane			
d = 120 mg/kg/week	s.c.	523	1 Undiff. spindle cell sarcoma
D_{50} = 5680 mg/kg		502	1 Undiff. polymorph.cell sarcoma
		325	1 Fibrosarcoma
Bis-(methyl-aniline)methane			
d = D = 850 mg/kg	s.c.	495	1 Fibrosarcoma
Bis-(morpholino)methane			
d = 50 mg/kg/week	s.c.	479±79	2 Fibromas 1 Undiff. spindle cell sarcoma 4 Fibrosarcomas 2 Undiff. spindle cell and polymorph.cell sarcomas 1 Spindle cell and polymorph. cell sarcoma (fibrosarcoma)
Phenylazobutane			
d = 150 mg/kg/week D = 9750 mg/kg	s.c.	681	1 Fibroma
Ethylene sulfide			
2 g% in H_2O d = 8 mg/kg/week D = 320 mg/kg	s.c.	479	1 Fibroma
d = 16 mg/kg/week	s.c.	387	1 Undiff. polymorph. cell sarcoma
D_{50} = 696 mg/kg		420	1 Undiff. spindle cell sarcoma
		474	1 Undiff. spindle cell and polymorph. cell sarcoma
		652	1 Fibrosarcoma
Benzyl chloride			
8 g% in oil d = 40 mg/kg/week D = 2100 mg/kg	s.c.	513 and 605	2 Fibrosarcomas
d = 80 mg/kg/week D = 3681 mg/kg	s.c.	466±105	1 Undiff. spindle cell sarcoma 3 Fibromas 1 Anaplastic fibrosarcoma 1 Round cell sarcoma 1 Undiff. spindle cell and polymorph. cell sarcoma
Diethyl sulfate			
1.25% in oil d = 25 mg/kg/week D = 800 mg/kg	s.c.	415±107	1 Undiff. spindle cell sarcoma 4 Fibrosarcomas 1 Undiff. spindle cell and polymorph. cell sarcoma
2.5% in oil d = 50 mg/kg/week D_{50} = 1545 mg/kg	s.c.	350±126	2 Undiff. spindle cell sarcomas 3 Fibrosarcomas

Table 2 (continued)

Substances and dosage	Appl.	t	Tumors
			1 Undiff. polymorph.cell sarcoma 1 Rhabdomyosarcoma 1 Histiocytic sarcoma 1 Undiff. spindle cell and poly-morph.cell sarcoma 2 Spindle cell and polymorph. cell sarcomas (fibrosarcomas)
Dibutyl sulfate 50g% in oil d = 500 mg/kg/week D = 9500 mg/kg	s.c.	643 and 708	1 Undiff. spindle cell sarcoma 1 Histiocytic sarcoma
Dimethyl sulfate 0.8g% in H_2O d = D = 50 mg/kg	s.c.	480±142	1 Fibroma 1 Undiff. spindle cell sarcoma 1 Round cell sarcoma 1 Spindle cell and polymorph. cell sarcoma (fibrosarcoma)
Dimethyl sulfate 0.8g% in H_2O d = 16.8 mg/kg/week D_{50} = 455 mg/kg	s.c.	498±104	2 Fibrosarcomas 2 Undiff. polymorph. cell sarcomas 1 Undiff. spindle cell and polymorph. cell sarcoma 1 Spindle cell and round cell sarcoma
d = 21.16 mg/kg/week D_{50} = 783 mg/kg	s.c.	335±90	2 Undiff. spindle cell sarcomas 1 Anaplastic fibrosarcoma 1 Rhabdomyosarcoma
Methyl chloride 2g% in oil d = 10 mg/kg/week D_{50} = 491 mg/kg	s.c.	580±124	1 Undiff. spindle cell sarcoma 6 Fibrosarcomas 1 Round cell sarcoma
4g% in oil d = 20 mg/kg/week D_{50} = 885 mg/kg	s.c.	620±135	4 Fibrosarcomas 1 Anaplastic fibrosarcoma 1 Histiocytic sarcoma 1 Undiff. spindle cell and poly-morph. cell sarcoma
Methyl-methane sulfonate 1g% in oil d = 4 mg/kg/week D = 184 mg/kg	s.c.	572	1 Undiff. spindle cell and poly-morph. cell sarcoma
d = 8 mg/kg/week D = 368 mg/kg	s.c.	438 and 555	1 Fibrosarcoma 1 Undiff. spindle cell and poly-morph. cell sarcoma
Propanesultone d = 15 mg/kg/week D = 225 mg/kg	s.c.	280±76	7 Undiff. spindle cell sarcomas 3 Fibrosarcomas 1 Histiocytic sarcoma

Table 2 (continued)

Substances and dosage	Appl.	t	Tumors
			4 Undiff. spindle cell and polymorph. cell sarcomas
d = 15 mg/kg/week D = 210 mg/kg	s.c.	301±39	3 Fibrosarcomas 1 Undiff. polymorph. cell sarcoma 1 Anaplastic fibrosarcoma 1 Rhabdomyosarcoma 1 Spindle cell and polymorph. cell sarcoma (fibrosarcoma)
d = 30 mg/kg/week D = 390 mg/kg	s.c.	269±38	1 Undiff. spindle cell sarcoma 3 Fibrosarcomas 3 Rhabdomyosarcomas 2 Undiff. spindle cell and polymorph. cell sarcomas 1 Spindle cell and polymorph. cell sarcoma (fibromyosarcoma)
Propanesultone			
d = D = 100 mg/kg	s.c.	285±56	5 Undiff. spindle cell sarcomas 1 Fibrosarcoma 5 Undiff. polymorph. cell sarcomas 3 Histiocytic sarcomas
d = 15 mg/kg/week D = 18 mg/kg	s.c.	295	1 Spindle cell and polymorph. cell sarcoma (myosarcoma)
d = 30 mg/kg/week D = 360 mg/kg/week	s.c.	345	1 Spindle cell and polymorph. cell sarcoma (fibrosarcoma)
d = D = 225 mg/kg	s.c.	228	1 Undiff. spindle cell and polymorph. cell sarcoma
d = D = 260 mg/kg	s.c.	260 302	1 Undiff. spindle cell sarcoma 1 Undiff. spindle cell and polymorph. cell sarcoma
d = D = 270 mg/kg	s.c.	258	1 Spindle cell and polymorph. cell sarcoma (myosarcoma)
d = D = 29 mg/kg	s.c.	266	1 Undiff. spindle cell sarcoma
d = D = 300 mg/kg	s.c.	300	1 Spindle cell and polymorph. cell sarcoma (myosarcoma)
Propylene oxyde			
d = 40 mg/kg/week D_{50} = 2088 mg/kg	s.c.	471±127	4 Undiff. spindle cell sarcomas 3 Fibrosarcomas
d = 80 mg/kg/week D_{50} = 4350 mg/kg	s.c.	560±33	3 Fibrosarcomas 1 Anaplastic fibrosarcoma 1 Rhabdomyosarcoma
P-toluene sulfonic acid methylester			
d = 15 mg/kg/week D_{50} = 752 mg/kg	s.c.	472±155	2 Undiff. spindle cell sarcomas 3 Fibrosarcomas 1 Anaplastic fibrosarcoma 1 Undiff. spindle cell and polymorph. cell sarcoma
d = 50 mg/kg/week D = 2350 mg/kg	s.c.	675 673	1 Fibrosarcoma 1 Undiff. polymorph. cell sarcoma

Table 2 (continued)

Substances and dosage	Appl.	t	Tumors
P-toluene-sulfonic acid ethylester d = 50 mg/kg/week D = 2750 mg/kg	s.c.	439	1 Spindle cell and round cell sarcoma
Trichloro-methylethanesultone d = D = 1500 mg/kg	s.c.	222	1 Undiff. polymorph.cell sarcoma
Acetone hydrazone d = 10 mg/kg/week D = 980 mg/kg	s.c.	731	1 Fibroma
Benzaldehyde-methyl hydrazone d = 25 mg/kg/week D_{50} = 1846 mg/kg	s.c.	663±157	1 Fibrosarcoma 2 Undiff. polymorph.cell sarcomas 2 Anaplastic fibrosarcomas
Butyraldehyde-butylhydrazone d = 14 mg/kg/week D = 1180 mg/kg	s.c.	724	1 Anaplastic fibrosarcoma
Nicotinicaldehyde-methylhydrazone d = 25 mg/kg/week D = 1625 and 1925 mg/kg	s.c.	531 567	1 Undiff. spindle cell sarcoma 1 Fibroma
Phenyl-methyl hydrazine d = 100 mg/kg/week D = 6300 mg/kg	s.c.	608	1 Undiff. spindle cell sarcoma
Trichloro-methyl hydrazine d = 25 mg/kg/week D = 1750 mg/kg	s.c.	479	1 Fibrosarcoma
d = D = 170 mg/kg	s.c.	531	1 Undiff. spindle cell and polymorph.cell sarcoma
Phenyl-butylhydrazine d = 60 mg/kg/week D = 318 mg/kg	s.c.	797	1 Fibrosarcoma
Phenyl-methyl-ethanol-triacene d = 25 mg/kg/week D_{50} = 1375 mg/kg	s.c.	457±76	2 Undiff. spindle cell sarcomas 1 Histiocytic sarcoma
d = 50 mg/kg/week D_{50} = 3016 mg/kg	s.c.	503±74	2 Fibrosarcomas 1 Undiff. spindle cell and polymorph.cell sarcoma
Phenyl-dimethyl-triacene d = 25 mg/kg/week D_{50} = 1225 mg/kg	s.c.	465±20	1 Undiff. spindle cell sarcoma 1 Malignant neuroma 1 Undiff. spindle cell and polymorph.cell sarcoma
d = D = 350 mg/kg	s.c.	173	1 Undiff. spindle cell and polymorph.cell sarcoma
Phenyl-methyl-triacene d = 30 mg/kg/week D_{50} = 582 mg/kg	s.c.	198±32	3 Undiff. spindle cell sarcomas 1 Undiff. polymorph.cell sarcoma

Table 2 (continued)

Substances and dosage	Appl.	t	Tumors
			1 Undiff. spindle cell and polymorph.cell sarcoma 2 Spindle cell and polymorph.cell sarcomas (myosarcomas)
d = D = 300 mg/kg	s.c.	235	1 Histiocytic sarcoma
d = D = 340 mg/kg	s.c.	192	1 Liposarcoma
3,4-Benzpyrene d = D = 5 mg/kg	i.m.	172±15	2 Fibrosarcomas 1 Leiomyosarcoma 2 Undiff. spindle cell and polymorph.cell sarcomas
d = D = 10 mg/kg	i.m.	223±108	1 Undiff. spindle cell sarcoma 3 Fibrosarcomas 1 Undiff. polymorph.cell sarcoma 3 Rhabdomyosarcomas 3 Undiff. spindle cell and polymorph.cell sarcomas 2 Spindle cell and polymorph.cell sarcomas (fibrosarcomas)
1,2-Dimethyl-9,10-benzanthracene d = D = 5 mg/kg	i.m.	260±85	3 Fibrosarcomas 1 Undiff. spindle cell and polymorph.cell sarcoma 1 Spindle cell and polymorph. cell sarcoma (fibrosarcoma) 1 Spindle cell and polymorph.cell sarcoma (myosarcoma)
Tobacco condensate d = 33 mg/rat/week D_{50} = 3188 mg/rat	s.c.	689±54	5 Fibrosarcomas 1 Spindle cell and polymorph.cell sarcoma (histocytic)
d = 66 mg/rat/week D_{50} = 4096 mg/rat	s.c.	577±202	1 Fibroma 2 Undiff. spindle cell sarcomas 1 Fibrosarcoma 1 Spindle cell and polymorph. cell sarcoma (myosarcoma)
d = 20 mg/rat/week D = 1200 mg/rat	s.c.	593	1 Anaplastic fibrosarcoma
Nickel sulfide d = D = 10 mg/rat	i.m.	34	1 Undiff. spindle cell sarcoma
Trypan blue d = 100 mg/kg/every 2 weeks D = 1300 mg/kg	s.c.	305 286	1 Undiff. polymorph.cell sarcoma 1 Anaplastic fibrosarcoma

Table 3. Soft tissue tumors induced by chemical cancerogenic substances (literature)

Year	Author	Substance and dosage	Appl.	t	n	Tumors
1934	*Barry* and *Cook*	1,2,5,6 Dibenzanthracene d = D = 2 mg, later 6 mg	s.c.	217	7/10	7 spindle cell-Sa
		1,2,5,6 Dibenzanthracene d = D = 2 mg, later 6 mg	s.c.	182	1/10	1 spindle cell-Sa
		5,6-Cyclopenteno-1,2 Bencanthracene d = D = 2 mg, later 6 mg	s.c.	630	8/10	8 spindle cell-Sa
		Chrysene d = D = 2 mg, later 6 mg	s.c.	626	4/10	2 spindle cell-Sa 2 polymorph. cell
		control with lard	s.c.	462	2/10	2 benign
1934	*Berenblum* and *Kendal*	1,2,5,6 dibenzanthracene d=D= 2.5 mg and 5 mg in aqueous suspension water	s.c.	210	3/12	3 tumors
1934	*Roussy* et al.	Thorotrast d=D= 5 x 1 ml	s.c.	540	4/5	4 sarcomas
1935	*Barry* et al.	20-methylcholanthrene d = 2-3 mg/week	s.c.	193	9/15	9 tumors
1936	*Hagensen* and *Krehbiel*	1,2,5,6 dibenzanthracene d = 1 g/100 cm^3 paraffin/ 2-12 injections/animal	s.c.	128-500	36/198	36 tumors: 9 fibrosarcomas 3 leiomyosarcom 9 rhabdomyosar-comas 15 unspecified
1936	*Dansi*	2,4-benzpyrene d = 2 mg/week dissolved in lards	s.c.	180-235	22/30	22 s.c. tumors
1936	*Domagk*	3,4-benzpyrene 0.1-0.3 ml/week of 0.1% in olive oil	s.c.	231	5/10	5 s.c. tumors
1936	*Dunning* et al.	3,4-benzpyrene d = 0.2-0.4 mg in paraffin/ many injections	s.c.	60	2018/2278	1369 tumors
1936	*Dunning* et al.	3,4-benzpyrene d = 1% paraffin solution	s.c.	201	348/486	348 s.c. tumors
1936	*Oberling*	3,4-benzpyrene				
		d = 0.05 mg in olive oil	s.c.	300	1/10	1 s.c.
		d = 0.1 mg in olive oil	s.c.	450	4/35	4 s.c.
		d = 0.5 mg in olive oil	s.c.	570	9/20	9 s.c.
		d = 1.0 mg in olive oil	s.c.	375	64/75	64 s.c.
1936	*Roussy* et al.	Thorotrast d = 0.5 ml	s.c.	420-720	6/20	6 sarcomas
1936	*Roussy* et al.	Thorotrast d = 0.5 ml D = 0.5-2.5 ml	s.c.	360-510	5/30	5 sarcomas

Table 3 (continued)

Year	Author	Substance and dosage	Appl.	t	n	Tumors
1936	*Selbie*	Thorotrast d = 0.3 ml D = 0.6 ml	s.c.	360	14/60	14 s.c.
1937	*Boyland* and *Brues*	3,4,5,6-dibenzcarbazole 0.05% colloidal suspension 2 x 2 ml/week	s.c.	196	8/10	8 sarcomas
1937	*Brunshwig* and *Tschetter*	20-methylcholanthrene d = 2 mg dissolved in lard	s.c.	220	17/25	17 s.c.
1937	*Thomas*	20-methylcholanthrene d = 1.8-2 mg in olive oil	s.c.	195-330	40/45	40 s.c.
1938	*Bogliolo*	Thorotrast d = 0.2 ml D = 30 injections 2 times/day	s.c. s.c.	390	16/29	16 spindle cell sarcomas
1940	*Dunning* et al.	20-methylcholanthrene	s.c.	180-330	387/724	387 s.c.
1941	*Davenport*	9,10-dimethyl-1,2-benzanthracene 8 mg in rat fat 8 mg in linseed oil 8 mg in lard 8 mg in lanoline	 s.c. s.c. s.c. s.c.	 264 152 310 164	 3/7 5/5 7/10 6/6	 3 s.c. 5 s.c. 7 s.c. 6 s.c.
1941	*Magat* et al.	20-methylcholanthrene d = 2 mg in paraffin and sunflower oil D = every 1 to 3 months	i.m.	230-300	–	156 s.c.
1942	*Badger* et al.	3,4-benzphenanthrene	s.c.	533	1/6	1 sarcoma
1942	*Esmarch*	20-methylcholanthrene	s.c. i.m.	450 420	7/9 7/10	7 s.c. 7 s.c.
1942	*Schairer* and *Rechenberger*	3,4-benzpyrene 1.4 mg rat; dissolved in olive oil	i.m.	132	67/80	67 sarcomas
1943	*Hill* et al.	20-methylcholanthrene 1 or 3 mg in maize oil	i.m.	181-260	43/86	43 fibro-sarcomas
1943	*Schairer*	3,4-benzpyrene d = 1 ml of 1.5% solution olive oil	i.m.	40-60	112/180	112 tumors
1947	*Hartmann*	3,4-benzpyrene d = 0.2 ml of 0.5% solution in bencene/week	i.m.	–	30/30	30 tumors
1947	*Jaffé*	20-methylcholanthrene d = 2 mg	s.c.	450	11/53	11 sarcomas

Table 3 (continued)

Year	Author	Substance and dosage	Appl.	t	n	Tumors
1948	*Haerem*	Sulfapyridine d = 0.015 g, 2 injections in different places	s.c.	630	3/40	3 spindle cell sarcomas
1948	*Haerem*	Sulfanilamide 0.015 g, 2 injections in different places	s.c.	480	2/10	2 spindle cell sarcomas
1948	*Kuppermann* and *Greenblatt*	3,4-benzpyrene d = 1 mg every second day	s.c.	185	4/4	4 sarcomas
1948	*Lacassagne* et al.	2-aminofluorene d = 50 mg/month D = 6 x 50 mg	s.c.	687	2/6	2 fibrous bone tumors (one sarcoma)
		d = 50 mg/month D = 5 x 50 mg	s.c.	687	4	4 fibrosarcomas
1949	*Dyad'Kova*	9-10-dimethyl-1,2-benz-anthracene d = 0.5-3 mg	s.c.	225	47	sarcomas
1950	*v.d. Schueren*	20-methylcholanthrene 0.5 or 0.2 ml of 0.2% solution in oil D = 2 or 10 injections/week	s.c.	120-150	14/40	14 sarcomas
1951	*Hendry* et al.	Stearoyl-ethylenimine d = 10 mg/100 g b.w. 2 x/week D = 5 x 10 mg/100 g for 5 weeks	s.c.	330	5/10	5 sarcomas
1951	*Hendry* et al.	Trimethylolmelamine d = D = 200 mg/100 g b.w.	s.c.	660	1/10	1 sarcoma
1951	*Klinke*	3,4-benzpyrene D = 10-30 mg	i.m.	191-250	152/245	152 tumors
1952	*Bonser* et al.	2-amino-1-naphthol-hydrochloride d = 5 mg/100 g b.w. each 2nd week	s.c.	560	5/14	5 sarcomas
1952	*Cook* and *Schoental*	8-methoxy-3,4-benzpyrene d = 2 mg D = 4 mg/13 weeks	s.c.	210	11/11	11 sarcomas
1952	*Eldregde* and *Luck*	20-methylcholanthrene D = 2 x 5 mg	s.c.	–	2/2	sarcomas
1952	*Ezeyza*	Trypaflavine d = 1 mg, D = 40 mg	s.c.	600	1/24	1 sarcoma
1952	*Hansen* and *Bichel*	Sulfathiazole d = D = 0.1 g	s.c.	540	4/10	4 tumors (1 spindle cell sarcoma, 1 sarcoma in the control)

Table 3 (continued)

Year	Author	Substance and dosage	Appl.	t	n	Tumors
1952	*Oppenheimer* et al.	Polyethylene 0.002-mm-thick films	s.c.	–	14/98	8 fibrosarcomas 1 rhabdomyo-sarcoma 1 osteogenic sarcoma
1952	*Oppenheimer* et al.	Vinyl-chloride polymer film	s.c.	660	14/45	12 fibrosarcomas 1 liposarcoma 1 unclassified sarcoma
1952	*Umeda*	Rhodamine B d = 1 cm^3 of 0.05-0.15% solution of water 2-3 x/week	s.c.	630	5/20	5 fibrosarcomas
1952	*Walpole* et al.	4-aminodiphenyl d = 3.6-5.8 mg/kg b.w. 5 injections in the week	s.c.	537	4/23	4 sarcomas
1953	*Ghiringheli* et al.	20-methylcholanthrene 0.25% in oil d = 1 ml each 2nd week D = 4 ml	s.c.	140	8/10	8 sarcomas
1954	*Lindner* and *Müller*	10-methyl-1,2-benz-anthracene D = 3.2 mg in 0.1 ml Tricaprylin	s.c.	720	3/10	3 sarcomas
1954	*Lindner* and *Müller*	10-bromo-1,2-benzanthra-cene d = D = 4.1 mg	s.c.	720	1/10	1 sarcoma
1956	*Nothdurft*	Implantation of gold, silver, ivory, polystyrene, hydrocellulose, polyvinyl	s.c.	–	500/1469	500 sarcomas
1958	*Schmähl*	Implantation of asbestos fibers	s.c.	810± 120	–	3 spindle cell sarcomas
		Mineralized asbestos	s.c.	630± 120	–	2 spindle cell sarcomas
1959	*Haddow* and *Horning*	Imferon 1 ml^3/week for 6 m D = 10 mg Fe.	s.c.	180	24/30	19 sarcomas or fibrosarcomas 5 histiocytomas
1961	*Druckrey* and *Bannasch*	"Erdschwarz" M 90 80% suspension in glycerol	s.c.	706± 180	9/16	9 sarcomas
1961	*Gilman* and *Ruckenbauer*	Metal powder 30 mg = d = D	i.m.	525	8/20	8 mainly spindle cell sarcomas
		Metal powder 20 mg = d = D	i.m.	388	27/35	27 mainly spindle cell sarcomas
		Metal powder, washed refinery sample 20 mg = d = D	i.m.	389	28/31	38 mainly spindle cell sarcomas

Table 3 (continued)

Year	Author	Substance and dosage	Appl.	t	n	Tumors
		Penicillin G 120 000 I.U.	i.m.	519	0/30	---
		Cobaltoxyde d = D = 10 mg	i.m.	489	5/10	5 sarcomas
		Thorotrast d = D = 30 mg	i.m.	402	0/10	---
1961	*Gilman*	Copper oxide d = D = 20 mg 2 injection sites/animal	i.m.	595	0/32	---
		Ferrous oxide d = D = 20 mg 2 injection sites/animal	i.m.	601	0/32	---
		Nickel oxide d = D = 20 mg partly 2 injection sites/animal	i.m.	595	21/32	26 rhabdo- myosarcomas
		Nickel disulphide d = D = 20 mg partly 2 injection sites/animal	i.m.	365	36/32	36 rhabdo- myosarcomas
		Cobalt oxide d = D = 20 mg partly 2 injection sites/animal	i.m.	–	13/32	13 tumors
		Cobalt sulfite d = D = 20 mg partly 2 injection sites/animal	i.m.	–	28/30	35 tumors
1961	*Gross*	Triphenylmethyne "Patent blue AE" Brillant blue FCF 2 x/week 1-2% solution identical injection site	s.c.	285	–	high sarcoma rate
1961	*Ikeda*	Thorotrast	s.c.	365	7/40	7 spindle cell sarcomas
1961	*Schramm*	20-methylcholanthrene d = D = 0.5 mg and Linoleic acid ethylester 10 times 0.2 ml	s.c.	300	18/25	18 sarcomas, generally spindle- cell type
		20-methylcholanthrene d = D = 0.5 mg	s.c.	300	18/20	18 sarcomas, generally spindle- cell type
1962	*Zollinger*	Ferri-hydroxide-poly- maltose 1 ml/week for 10 weeks	i.m.	545	45/50	45 malignant histiocytomas (reticulosarcoma)
1963	*Brown*	Trypan blue				
		2.5 mg/100 g b.w.	s.c.	450	21/25	1 tumor
		5.0 mg/100 g b.w.	s.c.	420-570	17/25	5 tumors
		10.0 mg/100 g b.w.	s.c.	330-510	18/25	9 tumors
		15.0 mg/100 g b.w.	s.c.	360-510	15/27	3 tumors
		20.0 mg/100 g b.w. 5-8 times twice a week	s.c.	300-510	5/25	4 tumors

Table 3 (continued)

Year	Author	Substance and dosage	Appl.	t	n	Tumors
1963	*Dasler* and *Milliser*	Polyvinyl alcohol implantation	s.c.	540	5/12	5 fibrosarcomas
1963	*Shulman* et al.	Polyethylene film	s.c.	720	4/55	4 fibrosarcomas
1963	*Shulman* et al.	Polyethylene net	s.c.	720	1/52	1 fibrosarcoma
1963	*Toth* and *Shubik*	7,10-dimethylbenzanthracene d = D = 1 mg	s.c.	210	9/119	9 spindle cell sarcoma
		d = D = 0.1 mg	s.c.	700	8/112	8 spindle cell sarcoma
		d = D = 0.075 mg	s.c.	700	3/58	3 spindle cell sarcoma
		d = D = 0.050 mg	s.c.	700	0/45	---
		d = D = 0.025 mg	s.c.	700	2/88	2 spindle cell sarcoma
		d = D = 0.010 mg	s.c.	700	0/116	---
1964	*Heath* and *Daniel*	Cadmium d = D = 14 mg powder	i.m.	392	9/10	9 malignant tumors, generally rhabdomysarcoma with fibrosarcoma component
		Cadmium d = D = 28 mg powder	i.m.	322	6/10	6 malignant tumors, generally rhabdomyosarcoma with fibrosarcoma component
1964	*Heath* and *Daniel*	Nickel, pure d = D = 28.3 mg	i.m.	280	10/10	10 tumors of the striated musculature
1964	*Hueper*	Polyurethane pieces				
		"Control-Mobaysheet"	s.c.	720	1/30	1 fibrosarcoma
		"Goodrich granules"	s.c.	720	3/35	3 fibrosarcomas
		films	s.c.	720	4/35	4 fibrosarcomas
		"Metrell Ostamer"	s.c.	720	1/35	1 fibrosarcoma
1964	*Oppenheimer* et al.	Cellophane film changed every month	s.c.	800	9/70	9 tumors
		In comparison with: Polystyrene	s.c.	600	16/252	16 tumors
		Cellophane film film and "bag" changed every month	s.c.	800	0/35	---
		In comparison with: Polystyrene	s.c.	650	---	---
1964	*Haddow* et al.	Cadmium inital dose 20 mg, again after 46 days; afterwards 8 times 2 mg a week of a cadmium-rat-ferritin precipitate (56 mg precipitate ≙ 0.95 mg cadmium)	s.c.	810	7/20	7 spindle cell sarcomas

Table 3 (continued)

Year	Author	Substance and dosage	Appl.	t	n	Tumors
		Cadmium sulfate d=0.5 mg, 10 times a week D=5 mg Cd-sulfate, Cd	s.c.	480	14/20	14 spindle cell sarcomas
1964	*Pollard* et al.	20-methylcholanthrene				
		d = D = 1 mg sterile	s.c.	230	13/13	13 tumors
		d = D = 1 mg usual	s.c.	286	21/25	21 tumors
		d = D = 1 mg sterile	s.c.	180	13/14	13 tumors
		d = D = 1 mg sterile	s.c.	240	13/141	13 tumors
		d = D = 1 mg sterile	s.c.	224	11/14	11 tumors
		d = D = 1 mg usual	s.c.	204	9/11	9 tumors
1964	*Roe* et al.	Imferon (iron-dextran complex) d = 0.5 ml, 24 times a week D = 12 ml & 50 mg Fe/ml				fibroma fibrosarcomas polymorph. cell sarcoma
		1 injection site	s.c.	500	14/24	14
		2 injection sites	s.c.	600	7/24	7
		4 injection sites	s.c.	600	12/24	12
		6 injection sites	s.c.	600	10/24	10
		6 injection sites	–	600	0/32	---
1964	*Undritz* and *Fraenkel*	Iron d = 20 mg 78 times applicated D = 7 mg Fe/100 g b.w./week	i.m.	857	169/212	169 sarcomas
		d = 0.35 mg Fe/100 g b.w./ week, total 148 times	i.m.	1095	19/252	19 sarcomas
		d = 0.07 mg Fe/100 g b.w./ week, total 141 times	i.m.	1095	2/164	2 sarcomas
1965	*Roe* and *Haddow*	Imferon (iron-dextran complex) d = 1 mg/50 g b.w. total 79 times D = 830 mg (average)	s.c.	365	3/24	3 sarcomas
		Jectofer (iron-sorbite-citric-acid complex) d = 1 mg/50 g b.w. total 79 times D = 830 mg (average)	s.c.	–	0/24	---
1966	*Chester* et al.	Polyethylene (different forms)	s.c.	–	–	fibromas, fibrosarcomas
		Polyethylene + barium sulfate (different forms)	s.c.	–	–	fibroma, fibrosarcomas
		Stainless steel (different forms)	s.c.	–	–	---
1966	*Kazantzis* and *Hanbury*	Cadmium sulfide d = D = 25 mg	s.c.	365	6/10	6 sarcomas
		Cadmium sulfide d = D = 50 mg	i.m.	510	5/14	5 spindle cell sarcomas
		Cadmium oxide d = D = 25 mg	s.c.	240	8/10	8 spindle cell of polymorph. cell sarcomas

References

Badger, G.M., Cook, J.W., Hewett, C.L., Kennaway, E.L., Kennaway, N.M., Martin, H.R.: The production of cancer by pure hydrocarbons.–Part IV. Proc. roy. Soc. B **131**, 170-191 (1942).

Barry, G., Cook, J.W., Haslewood, G.A.D., Hewett, C.L., Hieger, I., Kennaway, E.L.: The production of cancer by pure hydrocarbons. – Part III. Proc. roy. Soc. B **117**, 318-332 (1935).

Berenblum, I.: Carcinogenesis as a biological problem. New York: North-Holland American Elsevier 1974.

Berenblum, I., Kendal, L.P.: The production of tumours in the fowl with a colloidal solution of 1,2,5,6-dibenzanthracene. Brit. J. exp. Path. **15**, 366-371 (1934).

Berenblum, J.: A speculative review: The probable nature of promoting action and its significance in the understanding of the mechanism of carcinogenesis. Cancer Res. **14**, 471-477 (1954).

Bogliolo, L.: Sopra i blastomi sperimentali da ossido di torio. Pathologica **30**, 422-430 (1938).

Bonser, G.M., Clayson, D.B., Jull, J.W., Pyrah, L.N.: The carcinogenic properties of 2-amino-1-naphtholhydrochloride and its parent amine 2-naphthylamine. Brit. J. Cancer **6**, 412-424 (1952).

Boyland, E., Brues, A.M.: The carcinogenic action of dibenzcarbazoles. Proc. roy. Soc. B **122**, 429-441 (1937).

Boyland, E., Roe, F.J.C., Gorrod, J.W.: Induction of pulmonary tumours in mice by Nitrosonornicotine, a possible constituent of tobacco smoke. Nature (Lond.) **202**, 1126-1131 (1964).

Brethauer, E.W., Black, S.C.: Polonium-210: Removal from smoke by Resin Filters. Science**156**, 1375-1376 (1967).

Brown, D.V.: Subcutaneous neoplasms induced by trypan blue. Lab. Invest. **12**, 1221-1227 (1963).

Brunschwig, A., Tschetter, D.: Age factor and latent period in production of sarcoma by methylcholanthrene in rats. Proc. Soc. exp. Biol. (N.Y.) **36**, 439-441 (1937).

Bücheler, J., Thomas, C.: Experimentell erzeugte Drüsenmagentumoren bei Meerschweinchen und Ratte. Beitr. Path. **142**, 194-209 (1971).

Bürkle de la Camp: quoted by Struppler (1959).

Bullock, F.D., Rohdenburg, G.L.: Spontaneous tumors of the rat. J. Cancer Res. **2**, 39-60 (1917).

Bullock, F.D., Curtis, M.R.: Spontaneous tumors of the rat. J. Cancer Res. **14**, 1-115 (1930).

Burns, W.A., Kanhouwa, S., Tillman, L., Saini, N., Herrmann, J.B.: Fibrosarcoma occurring at the site of plastic vascular graft. Cancer (Philad.) **29**, 66-72 (1972).

Chester, M., Southam, M.D., Virginia, I., Babcock, B.S.: Induction of subcutaneous tumors in rats by plastic loops and spirals. Amer. J. Obstet. Gynec. **96**, 134-140 (1966).

Contzen, H.: Die lokale Gewebereaktion auf implantierte Kunststoffe in Abhängigkeit von deren Form. Langenbecks Arch. klin. Chir. **304**, 922 (1963).

Contzen, H., Straumann, F., Paschke, E.: Grundlagen der Alloplastik mit Metallen und Kunststoffen. Stuttgart: Thieme 1967.

Cook, J.W., Schoenthal, R.: Carcinogenic activity of metabolic products of 3,4-Benzpyrene: Applicatic to rats and mice. Brit. J. Cancer **6**, 400-406 (1952).

Cooper, D.A., Crane, R., Boncot, K.R.: Primary carcinoma of the lung in nonsmokers. Arch. environm. Hlth. **16**, 398-400 (1968).

Crain, R.C.: Spontaneous tumors in the Rochester strain of the Wistar rats. Amer. J. Path. **34**, 311-335 (1958).

Curtis, M.R., Bullock, F.D., Dunning, W.F.: A statistical study of the occurrence of spontaneous tumors in a large colony of rats. Amer. J. Cancer **15**, 67-121 (1931).

Dansi quoted by Eichler, O., et al. 1966.

Dasler, W., Milliser, R.V.: Induction of tumors in rats by subcutaneous implants of surgical sponges. Experientia (Basel) **19**, 424-428 (1963).

Datta, R.K., Datta, B.: Role of methylated nucleic acids in carcinogenesis. Exp. molec. Path. **10**, 129-140 (1969).

Davenport, H.A., Savage, J.L., Durstine, M.J., Queen, F.B.: Induction of tumors in rats by carcinogens in various lipids. Cancer Res. **1**, 821-824 (1941).
Davis, R.K., Stevenson, G.T., Busch, K.A.: Tumor incidence in normal Sprague Dawley female rats. Cancer Res. **16**, 194-197 (1956).
Dickens, F.: Carcinogenic lactones and related substances. Brit. med. Bull. **20**, 96-101 (1964).
Dickens, F., Jones, H.E.H.: Carcinogenic activity of a series of reactive lactones and related substances. Brit. J. Cancer **15**, 85-100 (1961).
Dickens, F., Jones, H.E.J.: Further studies on the carcinogenic action of certain lactones and related substances in the rat and the mouse. Brit. J. Cancer **19**, 392-403 (1965).
Dietz, W.: Ösophaguskarzinom und Tabakabusus. Krebsarzt **6**, 352-354 (1951).
Domagk, G.: Die synthetisch hergestellten carcinogenen Substanzen und ihre Beziehungen zu physiologischen Produkten. Z. Krebsforsch. **44**, 160-186 (1936).
Druckrey, H.: Die Entstehung von Krebs. Mkurse ärztl. Fortbild. **10**, 199 (1960).
Druckrey, H., Schmähl, D.: Cancerogene Wirkung von 4-Dimethylamino-triphenylmethan bei subkutaner Gabe an Ratten. Naturwissenschaften **42**, 215 (1955).
Druckrey, H., Schmähl, D.: Cancerogene Wirkung von Kunststoff-Folien. Z. Naturforsch. **7b**, 353-356 (1952).
Druckrey, H., Nieper, H.A., Lo, H.W.: Carcinogene Wirkung von Parafuchsin im Injektionsversuch an Ratten. Naturwissenschaften **43**, 543 (1956).
Druckrey, H., Schmähl, D., Preussmann, R.: Fluoreszierende Verunreinigungen beim Paraffinum liquidum und organischen Lösungsmitteln. Arzneimittel-Forsch. **9**, 600-604 (1959).
Druckrey, H., Preussmann, R., Ivankovic, S., So, B.T., Schmidt, C.H., Bücheler, J.: Zur Erzeugung subcutaner Sarkome an Ratten. Carcinogene Wirkung von Hydrazodicarbonsäure-bis-(methylnitrosamid), N-Nitroso-N-n-butyl-harnstoff, N.Methyl-n-nitroso-nitroguanidin und B-Nitroso-imidazolidon. Z. Krebsforsch. **68**, 87-102 (1966).
Druckrey, H., Kruse, H., Preussmann, R., Ivankovic, S., Landschütz, Ch.: Cancerogene alkylierende Substanzen. III. Alkyl-halogenide, -sulfate, sulfonate und ringgespannte Heterocyclen. Z. Krebsforsch. **74**, 241-270 (1970).
Druckrey, H., Preussmann, R.: Zur Entstehung carcinogener Nitrosamine am Beispiel des Tabakrauchs. Naturwissenschaften **49**, 498-499 (1962).
Dunning, W.F., Curtis, M.R., Bullock, F.D.: Respective roles of heredity and somatic mutation in origin of malignancy. Amer. J. Cancer **28**, 681-712 (1936).
Dunning, W.F., Curtis, M.R., Eisen, M.J.: The carcinogenic activity of methylcholanthrene in rats. Amer. J. Cancer **40**, 85-127 (1940).
Duuren, B.L. van, Sivak, A., Langseth, L.: The tumor-promoting activity of tobacco leaf extract and whole cigarette tar. Brit. J. Cancer **21**, 460-463 (1967).
Dyad'Kova quoted by Shubik, Ph., Hartwell, J.L., 1957.
Eichler, O., Farah, A., Herken, H., Welch, A.D. (Hrsg.): Handbuch der experimentellen Pharmakologie, Bd. XVI/12. Berlin-Heidelberg-New York: Springer 1966.
Eiselsberg, v.: Über einen Fall von erfolgreicher Transplantation eines Fibrosarkoms bei Ratten. Wien. klin. Wschr. **3**, 927-931 (1890).
Eisen, M.J.: Betel chewing among natives of Southwest Pacific islands; lack of carcinogenic action. Cancer Res. **6**, 139-141 (1946).
Eldredge, N.T., Luck, J.M.: Electrophoretic studies on watersoluble proteins of liver during azo dye carcinogenesis in the rat. Cancer Res. **12**, 801-806 (1952).
Esmarch, O.: Deposition of methylcholanthrene in some organs in the rat. Acta path. microbiol. scand. **19**, 79-99 (1942).
Ezeyza, S.: Neosalvarsán, sulfato de atropina y Tripaflavina en ratas inyectadas subcutanéamente: carencia de poder cancerígeno de los dos primeros y producción de un sarcoma en el lugar de la injección de Tripaflavina. Sem. méd. (B. Aires) **100**, 778-780 (1952).
Fitzhugh, A.F.: Malignant tumours and high polymers. Science **118**, 783 (1953).
Friedrich-Freska, H.: Grenzdosis, Kombinations- und Spätwirkungen von Methylcholanthren und Benzpyren an der Haut von Mäusen, ein Beitrag zum Wirkungsmechanismus von cancerogenen Kohlenwasserstoffen. Biol. Zbl. **60**, 498 (1940).
Gaylord, H.R.: 6. Vers. Amer. Ass. Path. a. Bact. in Baltimore 1906. Ref. Z. Krebsforsch. **4**, 676 (1906).
Ghiringheli, C., Parola, P.L., Zanaboni, A. (1953): quoted by Shubik and Hartwell, 1957.
Gilman, J.P.W.: Metal carcinogenesis. II. A study on the carcinogenic activity of cobalt, cooper, iron, and nickel compounds. Cancer Res. **22**, 158-162 (1961).

Gilman, J.P.W., Ruckerbauer, G.M.: Metal carcinogenesis I. Observations on the carcinogenicity of a refinery dust, cobalt oxide, and colloidal thorium dioxide. Cancer Res. **22**, 152-157 (1961).
Gross, E.: Über die Erzeugung von Sarkomen durch die besonders gereinigten Triphenylmethanfarbstoffe Lichtgrün SF und Patentblau AE bei der wiederholten subcutanen Injektion an der Ratte. Z. Krebsforsch. **64**, 287-304 (1961).
Haagensen, C.D., Krehbiel, O.F.: The morphology of the sarcomas produced by 1,2,5,6-dibenzanthracene. Amer. J. Cancer **26**, 368-377 (1936).
Haddow, A., Roe, F.J.C., Duckes, C.E., Mitchley, B.C.V.: Cadmium neoplasia: Sarcomata at the site of injection of cadmium sulphate in rats and mice. Brit. J. Cancer **18**, 667-673 (1964).
Haddow, A., Horning, E.S.: On the carcinogenicity of a iron-dextran complex. J. nat. Cancer Inst. **24**, 109-147 (1960).
Haerem, A.T.: Carcinogenic effect of sulfonamides. Proc. Soc. exp. Biol. (N.Y.) **68**, 330-332 (1948).
Hammond, E.C., Horn, D.: Smoking and death rates—report on forty-four months of follow-up of 187,783 men. J. Amer. med. Ass. **166**, 1159-1172, 1294-1308 (1958).
Hansen, P.B., Bichel, J.: Carcinogenic effect of sulfonamides. Acta radiol. (Stockh.) **37**, 258-265 (1952).
Harrold: Quoted by Schmähl (1970).
Hartmann, W.D.: Die Wirkung von hohen Außentemperaturen auf das Wachstum von 20-Methylcholanthren-Tumoren der Ratte und die Wirkung niedriger Außentemperaturen auf das Wachstum von 1, 2, 5, 6-Dibenzanthrazen-Tumoren bei Mäusen. Krebsarzt **2**, 496-501 (1947).
Heath, J.C., Daniel, M.R.: The production of malignant tumours by nickel in the rat. Brit. J. Cancer **18**, 261-264 (1964).
Heath, J.C., Daniel, M.R.: The production of malignant tumours by cadmium in the rat. Brit. J. Cancer **18**, 124-129 (1964).
Hendry, J.A., Rose, F.L., Walpole, A.L.: Cytotoxic agents. I. Tethylolamides with tumour inhibitory activity and related inactive compounds. Brit. J. Pharmacol. **6**, 201-234 (1951).
Hendry, J.A., Homer, R.F., Rose, F.L., Walpole, A.L.: Cytotoxic agents: II. Bis-Exposides and related compounds. Brit. J. Pharmacol. **6**, 235-255 (1951).
Hill, R.M., Rutledge, E.K., Ware, A.G., Schultz, F.H., Livingston, W.H.: The effects of lowered body temperatures on methylcholanthrene fibrosarcomas in rats. Cancer Res. **3**, 841-848 (1943).
Holland, R.H., Acevedo, A.R.: Current status of arsenic in American cigarettes. Cancer (Philad.) **19**, 1248-1250 (1966).
Hueper, W.C.: Surgical use of polyurethane plastics questioned on basis of animal experiment. Amer. J. clin. Path. **34**, 46 (1961).
Hueper, W.C.: Cancer induction by polyurethan and polysilicone plastics. J. nat. Cancer Inst. **33**, 1005-1027 (1964).
Hueper, W.C.: Are sugars carcinogens? Cancer Res. **25**, 440-443 (1965).
Ikeda, M.: Experimental studies on sarcoma formation with thorium dioxide sol (thorotrast) (Japanese). Fukuoka Acta med. **52**, 481-493 (1961).
Jaffé, W.G.: The response of rats to the simultaneous application of two different carcinogenic agents. Cancer Res. **7**, 113-116 (1947).
Jasmin, G.: Influence of age, sex and glandular extirpation on muscle carcinogenesis in rats. Experimentia (Basel) **21**, 149 (1965).
Kazantzis, G., Hanbury, W.J.: The induction of sarcoma in the rat by cadmium sulphide and by cadmium oxide. Brit. J. Cancer **20**, 190-199 (1966).
Keen, P., de Moor, N.G., Shapiro, M.P., Cohen, L., Cooper, R.L., Campbell, J.M.: The aetiology of respiratory tract cancer in the South African Bantu. Brit. J. Cancer **9**, 528-538 (1955).
Khanolkar, V.R.: Cancer in India in relation to race, nutrition and customs. J. nat. Cancer Inst. **11**, 640-642 (1950).
Khanolkar, V.R.: Oral cancer in India. Acta Un. int. Cancr. **15**, 67-77 (1959).
Klinke: quoted by Eichler, O., et al. 1966.
Kolb, W., Baumgärtel, K., Dreschhoff, G.: quoted by Schmähl (1970).
Kupperman, H.S., Greenblatt, R.B.: The effect of steroid hormones on carcinogenic activity of benzpyrene. Exp. Med. Surg. **6**, 156-166 (1948).

Lazar, P., Chouroulinkov, J., Liberman, C., Guerin, M.: Benzo(a)pyrene content and carcinogenicity of cigarette smoke condensate–results of short-term and long-term tests. J. nat. Cancer Inst. **37**, 573-579 (1966).
Lacassagne, A., Buu-Hoi, N.P., Royer, R., Zajdela, F.: Contribution à l'etude du pouvoir cancérigène du 2-aminofluorène. C.R. Soc. Biol. (Paris) **142**, 481-483 (1948).
Legene, Esmonet: Volumineux fibrome sous-cutané chez un rat. Bull. Soc. Anat. Paris **80**, 38-45 (1905).
Lindner, A., Müller, A.: Über die krebserzeugende Wirksamkeit von 10-Chlor-1,2-benzanthracen und 10-Brom-1,2-benzanthracen. Sci. pharm. (Wien) **22**, 224-228 (1954).
Loveless, A.: Possible relevance of 0-6 alkylation of deoxyguanosine to the mutagenicity and carcinogenicity of nitrosamines and nitrosamides. Nature (Lond.) **223**, 206-207 (1969).
Magat, M.A., Ambromowitsch, I.A., Lebenson, E.G.: Dependence of growth and metastases of experimental tumors on method of administration of carcinogenic hydrocarbons. Med. Zhur. Akad. Nauk. UdSSR **11**, 53-66 (1941).
McCoy, G.W.: A preliminary report on tumors found in wild rats. J. med. Res. **21**, 285-296 (1909).
Meier, A.: Gibt es einen Zuckerkrebs? Schweiz. Z. Path. **5**, 226 (1942).
McDougall, I.R.: Malignant tumours at site of bone plasting. J. Bone J. Surg. **38**, 709-713 (1956).
Mohr, H.J., Nothdurft, H.: Bindegewebskapseln um subcutan eingeheilte Fremdkörper und ihre Entartung zu Sarkomen. Klin. Wschr. **36**, 493-494 (1958).
Muir, C.S., Kirk, R.: Betel, tobacco, and cancer of the mouth. Brit. J. Cancer **14**, 597-608 (1960).
Nakahara, W., Fukuoka, F.: Summation of carcinogenic effect of chemically unrelated carcinogenes, 4-nitroquinoline,N-oxide, and 20-methylcholanthrene. Gann **51**, 125-137 (1960).
Neurath, G.: Zur Frage des Vorkommens von N-Nitroso-Verbindungen im Tabakrauch. Experientia (Basel) **23**, 400-404 (1967).
Nishiyama, Y.: Experimental production of sarcoma in rats by repeated injection of glucose solution. Gann **32**, 85 (1938).
Nolte, D.: Über Sarkome nach Kriegsverletzung. Mschr. Unfallheilk. **69**, 124 (1966).
Nothdurft, H.: Die experimentelle Erzeugung von Sarkomen bei Ratten und Mäusen durch Implantation von Rundscheiben aus Gold, Silber, Platin und Elfenbein. Naturwissenschaften **42**, 75-76 (1955).
Nothdurft, H.: Experimentelle Sarkomauslösung durch eingeheilte Fremdkörper. Strahlentherapie **100/102**, 192-210 (1956).
Nothdurft, H.: Tumorerzeugung durch Fremdkörperimplantation. Abhandl. dtsch. Akad. Wiss. Berlin, 79-89 (1960).
Nothdurft, H. (1961): quoted by Ott (1970).
Oberling, C., Sannie, C., Guerin, M., Guerin, P.: Recherches sur l'action cancérigène du 1,2-benzopyrène. Bull. Ass. franç. Cancer **25**, 156-180 (1936).
Oettel, H., Frohberg, H., Nothdurft, H., Wilhelm, G.: Die Prüfung einiger synthetischer Farbstoffe auf ihre Eignung zur Lebensmittelfärbung. Arch. Toxikol. **21**, 9 (1965).
Oppenheimer, B.S., Oppenheimer, E.T., Stout, A.P.: Sarcomas induced in rodents by imbedding various plastic films. Proc. Soc. exp. Biol. (N.Y.) **79**, 366-369 (1952).
Oppenheimer, E.T., Wilhite, M., Stout, A.P., Danishefsky, I., Fishman, M.M.: A comparative study of the effects of imbedding cellophane and polystyrene films in rats. Cancer Res. **24**, 379-387 (1964).
Ott, G.: Fremdkörpersarkome. Experimentelle Medizin, Pathologie und Klinik, Vol. 32. Berlin-Heidelberg-New York: Springer 1970.
Ott, G., Volmar, J., Hieronymi, G.: Krebsgefährdung nach Implantation von Kunststoffen. Langenbecks Arch. klin. Chir. **302**, 608-628 (1963).
Ott, G., Jansen, H.H.: Sarkomentstehung nach Homio- und Heterotransplantation von Knochen. Langenbecks Arch. klin. Chir. **314**, 1-14 (1966).
Pollard, M., Matsuzawa, T., Salomon, J.C.: Induction of neoplasms in germfree rodents by 3-methylcholanthrene. J. nat. Cancer Inst. **33**, 93-97 (1964).
Pott, P.: Chirurgical observations relative to the cataract, the polyps of the nose, the cancer of the scrotum, the different kinds of ruptures, and the mortification of the toes and feet. London 1775.

Radford, E.P., Hunt, V.R.: Polonium-210: A volatile radioelement in cigarettes. Science **143**, 247-249 (1964).
Ratcliffe, H.L.: Spontaneous tumors in two colonies of rats of the Wistar Institute of Anatomy and Biology. Amer. J. Path. **16**, 237-254 (1940).
Roe, F.J.C.: The role of 3,4-benzopyrene in experimental tobacco carcinogenesis. Acta Un. int. Cancr. **19**, 730-732 (1963).
Roe, F.J.C., Haddow, A., Dukes, C.E., Mitchley, B.C.V.: Iron-dextran carcinogenesis in rats: effects of distributing injected material between one, two, four, or six sites. Brit. J. Cancer **18**, 801-808 (1964).
Roe, F.J.C., Haddow, A.: Test of an iron sorbitrol-citric acid complex (Jectofer) for carcinogenicity in rats. Brit. J. Cancer **19**, 855-859 (1965).
Roffo, A.H.: Durch Tabak beim Kaninchen entwickeltes Carcinom. Z. Krebsforsch. **33**, 321-332 (1931).
Ross, W.C.J.: Biological alkylating agents. London: Butterworth, 1962.
Roussy, G., Oberling, C., Guerin, M.: Action cancérigène du dioxyde du thorium chez le rat blanc. Bull. Acad. Méd. (Paris) **112**, 809-816 (1934).
Roussy, G., Oberling, C., Guerin, M.: A propos de l'action sarcomatogène du dioxyde do thorium colloidal; résultats globaux des nos expériences. Bull. Ass. franç. Cancer **25**, 716-721 (1936).
Sanghvi, L.D., Rao, K.C., Khanolkar, V.R.: Smoking and chewing of tobacco in relation to cancer of the upper alimentary tract. Brit. med. J. **1955 II**, 1111-1114.
Schairer, E., Rechenberger, J.: Der Einfluß von d,l-Leucyl-Glycin auf Bildung und Wachstum von Benzpyrengeschwülsten. Klin. Wschr. **21**, 64 (1942).
Schairer, E.: Die Entstehung von Benzpyrentumoren bei Ratten und ihre Beeinflussung durch d,l-Leucylglyzin sowie saure und alkalische Kost. Virchows Arch. path. Anat. **311**, 149-164 (1943).
Schmähl, D.: Fluoreszensuntersuchungen an Zigarettenrauch. Z. Aerosol-Forsch. **4**, 1-11 (1955).
Schmähl, D.: Cancerogene Wirkung von 4-Amino-propiophenon nach Fütterung an Ratten. Naturwissenschaften **21**, 564 (1957).
Schmähl, D.: Cancerogene Wirkung von Asbest bei Implantation an Ratten. Z. Krebsforsch. **62**, 561-567 (1958).
Schmähl, D.: Vergleichende Untersuchungen an Ratten über die carcinogene Wirksamkeit verschiedener Tabakextrakte und Tabakrauchkondensate. Arzneimittel-Forsch. (Drug Res.) **18**, 814-817 (1968).
Schmähl, D.: Prüfung von Polyvinylpyridin-N-oxid auf carcinogene Wirkung by Ratten und Mäusen. Arzneimittel-Forsch. (Drug. Res.) **19**, 1313-1314 (1969).
Schmähl, D.: Entstehung, Wachstum und Chemotherapie maligner Tumoren. Aulendorf i.W.: Editio Cantor 1970.
Schmähl, D., Consbruch, U., Druckrey, H.: Fluoreszenzmessungen an Zigarettenrauch. Arzneimittel-Forsch. (Drug Res.) **4**, 71-75 (1954).
Schmähl, D., Thomas, C.: Experimentelle Untersuchungen zur Synkarzinogenese. II. Versuche zur Krebserzeugung bei Mäusen bei gleichzeitiger Applikation von Urethan und 9,10-Dimethyl-1,2-benzanthracen. Z. Krebsforsch. **66**, 291-302 (1964a).
Schmähl, D., Thomas, C.: Vergleichende Prüfung von Tabakrauchkondensaten bei subcutaner und oraler Applikation auf cancerogene Wirkung bei Ratten. Z. Krebsforsch. **66**, 291-296 (1964b).
Schramm, T.: Einfluß von Fettsäuren auf die Carcinogenese. Acta biol. med. germ. **7**, 486-501 (1961).
Schueren, G. van der, Debie, F.: Comportement specifique de differents organes vis-à-vis d'un hydrocarbüre cancérigène. Acta Un. int. Cancr. **7**, 178-183 (1950).
Secher, K.: Kasuistischer Beitrag zur Kenntnis der Geschwülste bei Tieren. Z. Krebsforsch. **16**, 297 (1919).
Selbie, F.R.: Experimental production of sarcoma with thorotrast. Lancet **1936 II**, 847-848.
Shanta, V., Krishnamurthi, S.: Further study in aetiology in carcinomas of the upper alimentary tract. Brit. J. Cancer Inst. **17**, 8-23 (1923).
Shubik, Ph., Hartwell, J.L.: Survey of compounds which have been tested for carcinogenic activity. Publ. Health Serv. Publ. No 149, Suppl. 1. U.S. Gov. Print. Off. Washington 1957.
Shulman, J., Wiznitzer, Th., Neuman, Z.: A comparative study of sarcoma formation by implanted polyethylene film and mesh in white rats. Brit. J. plast. Surg. **16**, 336-339 (1963).

Stout and Eirich: quoted by Ott (1970).
Struppler, V.: Sarkome nach Knochennagelung. Mschr. Unfallheilk. **62**, 121-127 (1959).
Suguira, K,: Studies upon a new transplantable rat tumor. J. Canc. Res. **12**, 2 (1928).
Sunderman, F.W., Sunderman, F.W., jr.: Studies on nickel carcinogenesis. The subcellular partition of nickel in lung and liver following inhalation of nickel carbonyl. Amer. J. clin. Path. **40**, 563-575 (1963).
Sunderman, F.W., Donelly, A.J.: Studies of nickel carcinogenesis. Metastasing pulmonary tumors in rats induced by the inhalation of nickel carbonyl. Amer. J. Path. **46**, 1027-1041 (1965).
Sunderman, F.W., Sunderman, F.W., jr.: Nickel poisoning. XI. Implications of nickel as a pulmonary carcinogen in tobacco smoke. Amer. J. clin. Path. **35**, 203-209 (1961).
Sutton, B.: On fatty tumors. Lancet 1083 (1885).
Takizawa, N.: Experimental production of sarcoma in the mouse by injection of glucose, fructose and galactose. Gann **34**, 1 (1940).
Thomas, J.A.: Innocuité du méthylcholanthrène inséré sous for de granule dans la chambre antérieure de l'oeil chez le rat. C.R. Soc. Biol. (Paris) **126**, 1176-1178 (1937).
Thomas, C., Schmähl, D.: Vergleichende morphologische und toxikologische Untersuchungen an Maus, Ratte, Meerschweinchen, Kaninchen und Hund zur Prüfung der cancerogenen Wirkung und Organotrophie des Diäthylnitrosamins. Virchows Arch. path. Anat. **340**, 122-136 (1965).
Thompson, S.W., Huseby, R.A., Fox, M.A., Davis, C.L., Hunt, R.D.: Spontaneous tumors in the Sprague Dawley Rat. J. nat. Cancer Inst. **27**, 1037-1057 (1961).
Toth, B., Shubik, P.: Carcinogenesis in Lewisrats injected at birth with 7,12-dimethylbenanthracene. Brit. J. Cancer **17**, 540-545 (1963).
Umeda, M.: Experimental production of sarcoma in rats by injection of Rhodamine B. Gann **43**, 120-122 (1952).
Undritz, E., Fraenkel, B.: The production of localized sarcomas in the rat by long-term intramuscular injection of organic iron preparations in three dosages. In: Iron Metabolism, ed. by F. Gross. Berlin-Göttingen-Heidelberg: Springer 1964.
Velich, A.: Übertragbarkeit des Sarkoms. Wien. med. Bl. 711 (1898).
Wallenius, K.: Experimental oral cancer in the rat. With special reference to the influence of saliva. Acta path. microbiol. scand. Suppl. 180. Copenhagen: Munksgaart 1966.
Walpole, A.L., Williams, H.C., Robert, D.C.: The carcinogenic action of 4-amino-diphenyl and 3,2-dimethyl-4-aminodiphenyl. Brit. J. industr. Med. **9**, 255-263 (1952).
Walpole, A.L., Roberts, D.C., Rose, F.L., Henry, J.A., Homer, R.F.: Cytotoxic agents. IV. Carcinogenic actions of some monofunctional ethyleneimine derivates. Brit. J. Pharmacol. **9**, 306-323 (1954).
Wheeler, G.P.: Studies related to the mechanisms of action of cytotoxic alkylating agents. A review. Cancer Res. **22**, 651-688 (1962).
Woolley, P.G., Wherry, W.B.: On twenty-tweo spontaneous tumors in wild rats. J. med. Res. **25**, 205-215 (1911/12).
Wynder, E.L., Hoffmann, D.: A study of tobacco carcinogenesis. VII.The role of higher polycyclic hydrocarbones. Cancer **12**, 1079-1086 (1959).
Wynder, E.L., Hoffmann, D.: Tobacco and tobacco smoke. New York: Academic Press 1967.
Zackheim, H.S.: Tumours of the skin. In: Pathology of tumours in laboratory animals. Ed. V.S. Turusov, I.A.R.C. Lyon 1973.
Zollinger, H.U.: Weichteiltumoren bei Ratten nach sehr massiven Eiseninjektionen. Path. Microbiol. (Basel) **25**, 296-300 (1962).

Visceral Candidosis *

Anatomic Study of 34 Cases

K. SALFELDER**, K. UEDA, E.L. QUIROGA, and J. SCHWARZ

I. Introduction

Candidosis, formerly regarded as a mycotic infection of mucous membranes with little or no clinical significance, has become an important, often generalized and fatal fungal infection in debilitated patients. Review of routine material in Cincinnati produced surprising results with reference to organ involvement, portal of entry, tissue reactions, and structure of fungi in tissues, in part not described in recent reviews (*Fetter* et al., 1967; *Winner*, 1971).

Furthermore, details of candidosis superimposed upon underlying diseases, granulomatous cutaneous candidosis, so-called monilial granuloma, and the role of candidosis in "granulomatous disease of childhood" are discussed in this report.

---

* From the Department of Pathology, College of Medicine, University of Cincinnati, Cincinnati, Ohio, USA.

** Visiting Professor in the Department, 1973-1974. Reprint requests to Dr. *K. Salfelder*, Apartado 75, Merida, Venezuela, South America.

It is felt that the study of tissues in the deep mycoses—in contrast to many other infectious diseases— is an especially rewarding task for the morphologist. With proper techniques and thorough investigation for fungal elements numerous findings are revealed.

II. Material and Methods

In several Pathology Services of Cincinnati (Childrens Hospital, Shriners Burns Institute, General Hospital, Jewish Hospital, and Coroner's Office)*, cases with the diagnosis of visceral, deep-seated systemic or generalized candidosis were collected in a nonsystematic manner. The number of available cases, therefore, is not representative of the true incidence of the disease in this locality.

Thirty-four cases with involvement of at least one viscus were included in this study. History, clinical information, autopsy protocols, biopsy reports, slides, and paraffin blocks were available in each case, although in many cases, blocks were recut and/or slides restained. Almost all cases were restudied with the hematoxylin and eosin (H and E) stain and the Grocott method. In addition, Gridley, PAS, and Gram stains were used in several cases.

Most of the cases with burns (from the Shriners Institute) were included in the report of *Law* et al. (1972).

Some diagnostically dubious cases are included and discussed in this study, i.e., cases with possible infection by species of fungi other than *Candida* or possibly by mixed fungal infections.

III. Results

1. General Data

As shown in Table 1, a complete autopsy was performed in 33 cases; in one case the specimen was the result of a lobectomy.

Twelve cases were from Childrens Hospital, ten from Shriners Burns Institute, seven from General Hospital, four from Jewish Hospital, and one from the Coroner's Office.

The material was accumulated from 1960-1974; the largest number of cases were collected in 1970 (five cases) and in 1973 (six cases). The youngest patient was 6 days old, and the eldest 83 years old. Five patients were of perinatal age, three patients from 3 months to 1 1/2 years of age, 19 from 3-15 years, and 7 patients were older than 24 years.

The patients included 19 females and 15 males. Twenty-eight patients were white; five were black, and in one case race was not recorded.

The underlying diseases were burns (12 cases), diabetes (4 cases), leukemia (4 cases), hepatopathies (4 cases), perinatal conditions (4 cases), renal transplants because of chronic glomerulonephritis, granulomatous disease of childhood, and bowel diseases with surgery (2 cases each).

---

* We thank Drs. *R. Smith, J. McAdams, O.J. Kim,* and *B.G. MacMillan*, who gave permission and/or made available the material for this study.

Table 1. Clinical data and cultures

Hospital and No.	Year	Age-Sex Race	Underlying disease(s) and duration	Associated disease(s)	Treatment	Cultures during life: Bacteria	Cultures during life: Fungi	Cultures at necropsy: Bacteria	Cultures at necropsy: Fungi
1 CH	1960	5y-W-F	Acute lymphoblastic leukemia 2 years		Cytotoxins Steroids				Bowel, lung: *Candida*
2 CH	1960	3y-W-F	Acute lymphoblastic leukemia 1 month	Bowel resection, sepsis	Cytotoxins Steroids	Blood, group B: *Salmonella*		Lung: *Pseudomonas* Liver: Gram neg. bacilli	
3 CH	1960	2d-W-M	Prematurity		Insulin shock Steroids venous catheter				Spinal fluid: *Candida*
4 CH	1962	6d-W-M	Kernicterus Hyaline membrane disease		Venous catheter			Trachea: α-*streptococci*	Trachea, lung: *Candida*
5 CH	1964	3y-W-M	Acute lymphoblastic leukemia 16 months		Cytotoxins Steroids Transfusions	Blood: *E. coli, P. aeruginosa*		Blood, lung: *P. aeruginosa*	Lung, liver: *Candida*
6 CH	1964	26d-W-F	Prematurity; duodenal atresia (surgery)		Venous catheter	Skin: *Pseudomonas*	Skin: *Candida*		
7 CH	1965	3m-W-F	Malabsorption syndrome Volvulus (surgery) 3 months		Venous catheter		Skin, mouth: *Candida*	Blood: *Pseudomonas*	
8 CH	1965	11y-W-M	Acute myelogenous leukemia 8 months	Sepsis		Skin: *Pseudomonas*		Skin, blood, lung: *Pseudomonas*, etc.	
9 JH	1966	41y-B-F	Esophageal varices Liver cirrhosis, 5 months					Kidney: coliform bacilli	
10 JH	1967	3m-W-M	Granulomatous disease Childhood					Spinal fluid, blood: coliform bacilli	

Table 1 (continued)

Hospital and No.	Year	Age-Sex Race	Underlying disease(s) and duration	Associated disease(s)	Treatment	Cultures during life: Bacteria	Cultures during life: Fungi	Cultures at necropsy: Bacteria	Cultures at necropsy: Fungi
11 JH	1967	12y-W-F	Burns (40%), 6 weeks		Venous catheter; Transfusions	Skin, urine: *Pseudomonas*	Urine: *Candida*		
12 CO	1967	3y-W-F	Burns (40%), 7 weeks		Venous catheter	Blood: *Pseudomonas*	Skin, urine: *Candida*		Lung, liver, kidney: *Candida*
13 CH	1968	9y-M	Infectious hepatitis 8 weeks		Venous catheter Steroids	Pleural fluid: *Staphylococci*	Pleural fluid, urine: *Candida*	Lung: *Staphylococci*	
14 SHR	1968	3y-W-F	Burns (56%), 8 weeks	Cytomegalic inclusion disease		Skin: *Pseudomonas*			
15 SHR	1969	1y-W-F	Burns (75%), 4 weeks	Duodenal ulcer	Venous catheter				
16 SHR	1969	3y-W-M	Burns (55%), 3 weeks	Duodenal ulcer	Venous catheter				
17 SHR	1969	8y-W-F	Burns (93%), 3 weeks			Skin: *Pseudomonas*			
18 JH	1969	39y-W-F	Regional ileitis (surgery) Peritonitis		Venous catheter Steroids				
19 SHR	1970	13y-W-F	Burns (75%), 9 weeks	Sepsis	Venous catheter	Catheter: *Staphylococci* *Klebsiella* *Peudomonas*			
20 SHR	1970	9y-W-M	Burns (60%), 3 weeks	Duodenal ulcer Sepsis	Venous catheter			Vein, lung, endocardium: *Staphylococci*	Vein, lung: *Candida*

Table 1 (continued)

Hospital and No.	Year	Age-Sex Race	Underlying disease(s) and duration	Associated disease(s)	Treatment	Cultures during life Bacteria	Cultures during life Fungi	Cultures at necropsy Bacteria	Cultures at necropsy Fungi
21 SHR	1970	5y-B-F	Burns (70%), 7 weeks	Gastric and duodenal ulcer				Vein: Gram neg. bacilli Ulcer: Gram neg. bacilli	Vein, stomach, dura: *Candida*
22 SHR	1970	3y-W-F	Burns (70%), 9 weeks	Sepsis					
23 SHR	1970	5y-W-M	Burns (60%), 3 weeks		Venous catheter				
24 GH	1971	13d-B-M	Prematurity		Venous catheter				
25 CH	1972	26d-W-F	Neonatal hepatitis	Cytomegalic inclusion disease	Transfusions		Spinal fluid: *Candida*		Blood, lung: *Candida*
26 CH	1973	3y-W-F	Infectious hepatitis Aplastic anemia, 2 months			Blood: *E. coli*		Lung: *E. coli*	
27 GH	1973	43y-W-M	Chronic glomerulonephritis Renal transplant, 18 months	Osteomyelitis spine	Steroids	Spine: Gram neg. bacilli	Spine: *Aspergillus*	Blood, group D: *Streptococci*	Lung: *Candida* Spinal fluid: *Aspergillus*
28 GH	1973	15y-W-F	Renal transplant, 10 weeks Chronic glomerulonephritis	Cytomegalic inclusion disease	Steroids			Lung, skin: *Pseudomonas* Skin: β-hemolytic *strep.*	Lung: *Candida*
29 GH	1973	83y-B-F	Diabetes Gangrene (surgery)	Ulcerative colitis, Sepsis, myocardial infarct					
30 GH	1973	24y-W-M	Juvine diabetes–Type IV Hyperlipoprotinemia 14 years					Lung: *Pseudomonas* *Klebsiella*	Lung: *Candida*

Table 1 (continued)

Hospital and No.	Year	Age-Sex Race	Underlying disease(s) and duration	Associated disease(s)	Treatment	Cultures during life		Cultures at necropsy	
						Bacteria	Fungi	Bacteria	Fungi
31 SHR	1973	3y-B-F	Burns (60%), 5 weeks	Duodenal ulcer					
32 CH	1974	6y-B-M	Granulomatous disease Childhood				Lung: *Aspergillus*		
33 JH	1974	78y-W-M	Bilateral amputation Diabetes, 3 years	Carcinoma colon, surgery, peritonitis Pulmonary tuberculosis	Venous catheter			Blood, lung, bladder: *Klebsiella*	Blood, bladder: yeast
34 GH	1974	59y-W-M	Carcinoma pancreas Diabetes	Sepsis	Venous catheter		Mouth, sputum, trachea, urine: *Candida*		Catheter: *Candida*

Abbreviations: y = years, w = weeks, m = months, d = days; CH = Children's Hospital, SHR = Shriner's Burns Insitute, GH = General Hospital, JH = Jewish Hospital, CO = Coroner's Office.

Complicating diseases were usually other infections. In 25 cases, infection was demonstrated by cultural methods and/or anatomic findings. Bacterial infection was found in 24 cases. Cytomegalic virus (CMV) infection was noted in three cases, two of which demonstrated an additional bacterial infection.

Cultures for fungi were recovered in 26 cases. Cultures of six cases were taken during life, ten were recovered at post-mortem examination, and ten cases were cultured both during life and at post-mortem examination.

Fungi were isolated in 18 cases, 16 of which showed *Candida.* Four of the five cultures obtained during life contained *Candida*, the other showing *Aspergillus. Candida* was present in eight of nine cultures from post-mortem examination, the last culture revealed "yeasts" without specification. Each of the four cases with cultures obtained both during life and at post-mortem examination contained *Candida*, but one of these cases also contained *Aspergillus.*

Complete data about the treatment could not be obtained in all the cases. In 17 cases, venous catheters had been inserted, in some of the cases at different sites (often in one patient). In three cases, transfusions had been given. Steroids had been used in eight cases and cytotoxins in four cases. Frequently, these treatments were given concurrently. Antibiotics had been given in the majority of cases (not mentioned in Table 1).

Table 2 provides data about organ involvement with indication of organs histologically examined in each case. Exclusive fungemia at one site is also mentioned. Peritoneum, endocardium, and meninges are not counted separately when seen in tissue sections in continuity with the respective organs. Salivary glands, thymus, and parathyroid were seldom available and were omitted, since there was no positive case.

In order of frequency, the lungs were the organs most frequently involved (19 cases), followed by the kidneys (18), brain (13), digestive tract (12), myocardium (8), liver and skin (5 each), upper respiratory tract (4), and spleen, adrenal glands, and bladder (3 cases each). This order of frequency is not greatly altered by the instances when only fungemia (without any tissue lesions) was found in an organ.

Thrombi found at gross examination were not always examined histologically.

The involvement of dura mater, spinal cord, bone, skeletal muscle, lymph node, thyroid, gall bladder, tongue, and cartilage are rare occurrences.

Genital organs (both male and female), pancreas, and pituitary were not found with *Candida* lesions.

Twenty-one cases were found to have *Candida* lesions at more than two sites and 13 cases were found to have lesions at only one or two sites (including one case with involvement of lungs, brain, and meninges [case 27]).

The cases with involvement of one site refer to the lungs on three occasions (cases 10, 11, and 32, the latter was a surgical specimen); and in one instance to the brain (case 6, in whom there was no detectable portal of entry).

In two cases, involvement of two sites was found, by direct extension of the lesion in one and lymphogenous spread in the other: in case 14, a cutaneous candidosis extended into the underlying bone; and in case 30, with lung lesions due to *Candida*, there was a *Candida* lymphadenitis in a hilar lymph node.

In the remaining seven cases, a limited hematogenous dissemination apparently took place; four cases also showed *Candida* lesions at the other site (case 2 from stomach to liver, case 29 from colon to brain, case 9 from skin to kidney, and case 27 from lungs to

Table 2. Fungus lesions found in tissue sections

Organs	1	2	3	4	5	6	7	8	9	10	11	12	13	14	15	16	17	18	19	20	21	22	23	24	25	26	27	28	29	30	31	32	33	34	Total Fungus lesions	Fungemia
Skin		–		–	–	–		–	+		–	–		+	–		+		–		–	–	–	+	–			–			+				5/19	
Upper respiratory tract	–	–	–	+		–	–	–					–						+					+	–	–		+				●	–		4/14	
Lungs	+	*	+	+	–	–	–	–		+	+	+	+	–	–	+		+	+	*	*	+	+	+	–	+	+	+	–	+	–	+	–	+	19/32	3
Lymph nodes	–				–		–	–	–				–			–					–				–	–	–		–	+				–	1/14	
Vein thrombus				+			+								+				+	+	+	+						–					+		8/9	
Myocardium	+	–	–	+	–	–	+	–	–	–	–		+	–	–	–	–	*	–	–		+	–	+	+	+	–	–	–	–	*		*	–	8/31	3
Endocardium																				+	+				+										3	
Digestive tract	–	+	+	+	–	–	–	+	–	–	–	–	–		+	+		–	+	–		–	–	+	–	–	–	+	+	–	+		–	+	12/30	
Peritoneum																		+			+														2	
Liver	–	+	+	+	–	–	–	+	–	–	–	–	–	–	*	*	–	–	*			*	*	+	–	–	–		+	–	–		–	–	5/30	5
Spleen	–	–	–	–	+	–	–	–	–	–	–	*	–	–	*	*	*	–	*	*		+	–	*	–	+			–	–	–		–	–	3/30	7
Pancreas	–	–	–	–	–	–		–	–		–	–	–	–	–	–	–	–		–	–	–	–		–	–	–		–	–	–		–	–	–/28	
Kidney	+	–	–	+	+	–	+	+	+	–	–	+	+	–	*	+	–	+	+	+	+	+	–	+	+	+	–		–	–	+		–	–	18/32	1
Bladder	–	–	–	–	–	–	–		–			–		–						–		+	–		–	–	–	–			+			+	3/19	
Genitalia, female	–	–					–		–									–	–						–			–			–				–/9	
Genitalia, male			–	–	–	–		–					–			–											–			–			–	–	–/11	
Brain	–	–	+	+	+	+	+	–		–		–	+		–			+	–	–	–	+	–	+	+	–	+	–	+	–	+		–	–	13/28	
Leptomeninges			+	+	+								+									+		+	+		+								8	
Dura					+																														1	
Spinal cord	+		+	–	–	–	–	–					–													–		–							2/10	
Pituitary		–	–	–	–	–	–	–			–		–			–										–	–	–						–	–/14	
Thyroid	–	–		+	–	–	–	–			–	–	+	–		–		–	–					–	–	–	–	–	–		–		–	–	2/23	
Adrenal gland	–	–	–	–	–	–	–	–	–	–		+	–	–	–	+	–	–		–	–	*	–	+	–	–	–	–	–	–	–		–	–	3/31	1
Bone and bone marrow	–	–	–	–	–	–	–	–	–	–	–		–	+		*	*	–	–	–		*		–	–	–	–	–	–	–			–	–	1/28	3
Skeletal muscle	–	–	–	+	–	–	–						–												–	–	–			–					1/12	
Gall bladder																																		+	1	
Total –	13	14	11	9	15	17	14	14	10	8	10	7	12	9	6	6	5	8	6	7	5	4	9	2	14	13	13	11	10	11	7		12	12		
+	4	2	6	11	5	1	4	3	2	1	1	3	6	2	2	4	1	4	5	3	4	8	1	10	5	4	3	3	2	2	5	1	1	4		
*		1										1			3	3	2	1	2	2	1	3	1	1							1		1			

● = Surgical specimen (Pulm. Lobectomy); – = examined, negative; + = fungal lesion; * = fungemia brain and meninges). In the three other patients, the only manifestation of spread was a fungemia (case 17, skin candidosis with fungemia in spleen and bone marrow; case 23 with pulmonary candidosis and fungemia in liver; case 33 with candidosis of a venous thrombus and fungemia in the myocardium) (Fig. 1).

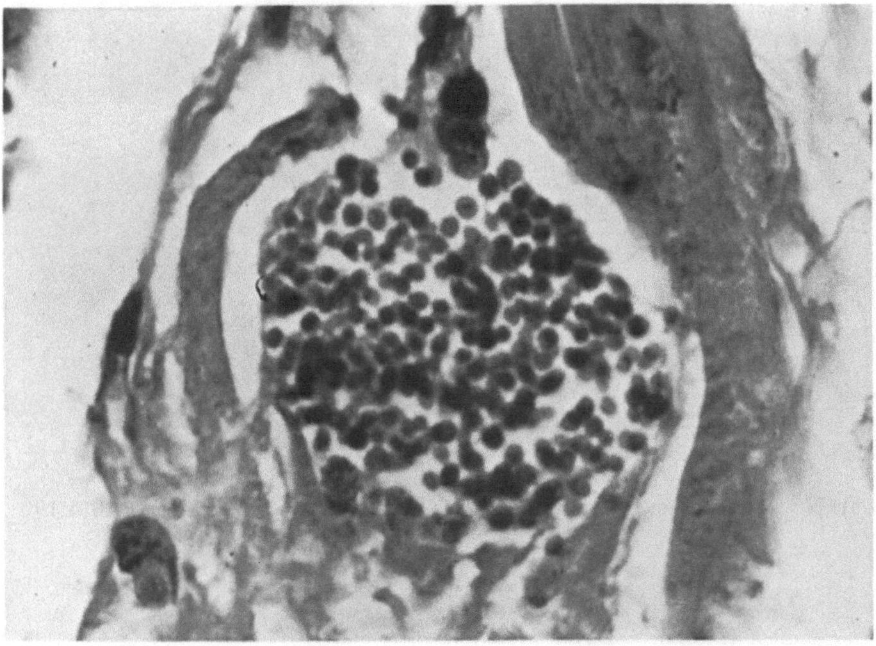

Fig. 1. Case No. 33: Myocardium. Nest of spherical small Grocott-positive yeast cells of *Candida* in a distended interstitial capillary. Could be confused with intrafibral cysts or pseudocysts of *Toxoplasma gondii* or *Trypanosoma cruzi.* H and E, x 1150

2. Special Pathology

a) Respiratory Organs

The *upper respiratory tract* was examined in only 14 cases. In four, *Candida tracheitis* was found. Two were from the perinatal group. One was a child with burns and one an adolescent with renal transplant. In all of these cases, the lung and the digestive tract also showed lesions due to *Candida* (in two cases the esophagus and in another two the stomach). In two cases (24 and 28) the cartilage was invaded by numerous hyphae (Fig. 2) and in case 24 the tongue showed *Candida* lesions with granulomatous reaction and intact mucosa.

The *lungs* were examined histologically in 32 of the 34 cases. The two cases in which the lungs were not examined were: case 9 with *Candida* involvement of skin and kidneys in a patient with liver cirrhosis and case 17 with candidosis of skin and fungemia in a patient with burns.

In 19 cases, pulmonary lesions were found due to *Candida* and in three additional cases only fungemia in the lungs was found. The ten cases in which *Candida* was not found

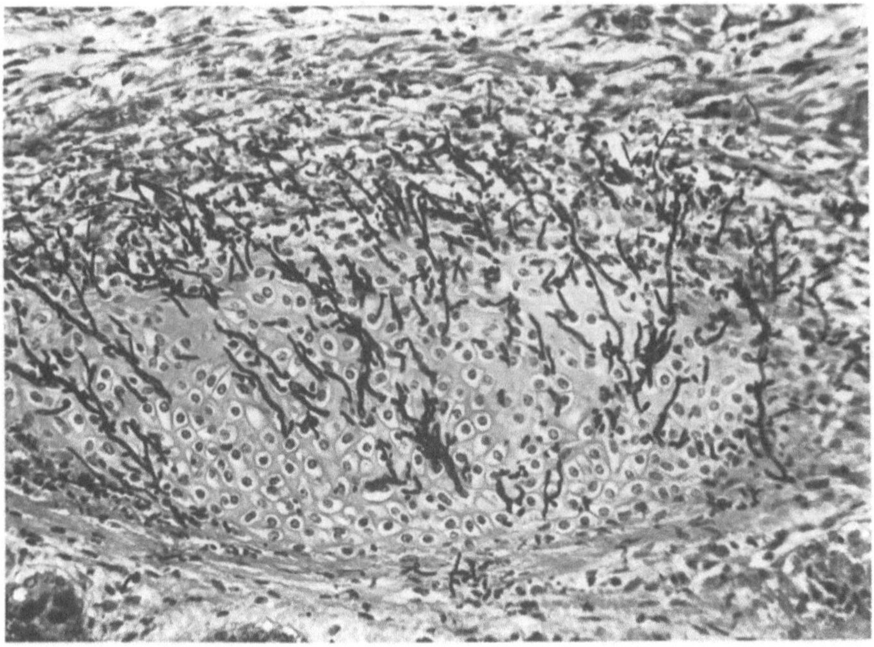

Fig. 2. Case No. 24: Trachea. Invasion of cartilage by thin hyphae of *Candida*. PAS, x 190

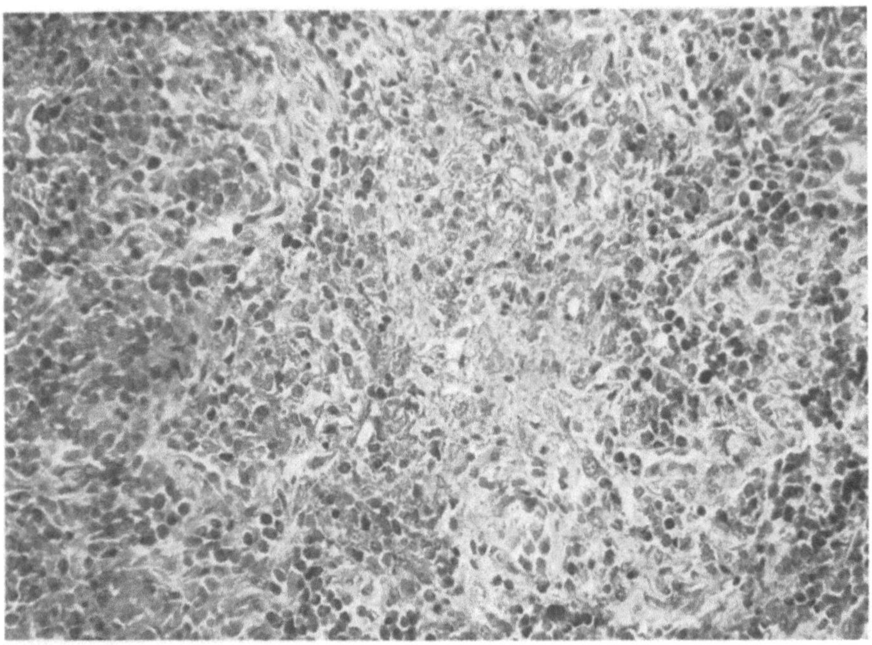

Fig. 3. Case No. 26: Lung. *Candida* lesion. Fungi not visible. Mononuclear and plasma cell infiltrates. H and E, x 130

in the lungs were from all age groups. The basic diseases were burns (3 cases), leukemia (2), diabetes (2), neonatal hepatitis (1), prematurity and repaired duodenal atresia (1), and volvulus with bowel resection (1).

In ten cases (cases 4, 10, 11, 19, 23, 24, 27, 28, 30, and 32) infection was considered to have occurred by inhalation. In three of these only the lungs were found to be infected by *Candida*. In one, there was, in addition a satellite *Candida* lymphadenitis and in another a hematogenous spread had occurred to the liver. In the sixth case, there was pulmonary candidosis and hematogenous dissemination with *Candida* lesions in brain and meninges (case 27). In the remaining four cases, candidosis of the upper respiratory tract was present. Underlying diseases in this group were burns (3 cases), granulomatous disease of childhood (2 cases) prematurity and Kernicterus (2 cases), renal transplant (2 cases), and juvenile diabetes (1 case).

Hyaline membranes were only found three times: in two cases with abscesses containing fungus cells, and in one case with an additional granulomatous reaction. In two cases the heavy leukocytic intra-alveolar pneumonia was thought to be due mainly to bacterial infection, and in one case only minimal hemorrhages were found (Figs. 3, 4, and 5).

In the other nine cases (cases 1, 3, 12, 13, 16, 18, 22, 26, and 34) the pulmonary *Candida* infection was thought to be the result of hematogenous dissemination. Fungal elements were found mostly inside blood vessels, often within emboli and with vasculitis (Figs. 6 and 7). Tissue reaction considered to be due to *Candida* consisted in four cases of septic, often hemorrhagic infarcts. In addition, in two, a granulomatous reaction was present. In two other cases, only perivascular interstitial cell nodules were seen, and in addition, in one

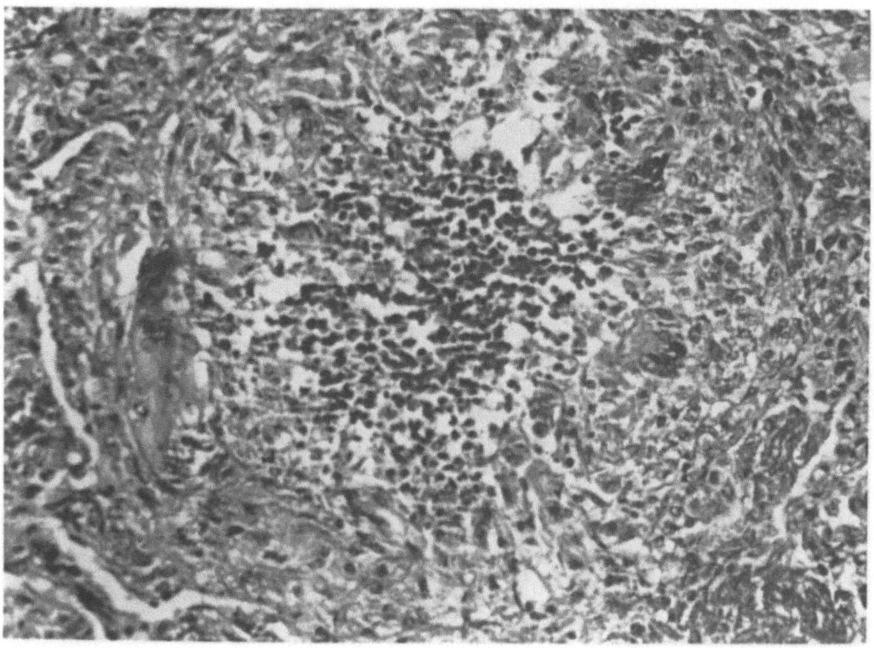

Fig. 4. Case No. 12: Lung. Abscess with rim of epithelioid cells. Fungus cells not visible with this stain. H and E, x 190

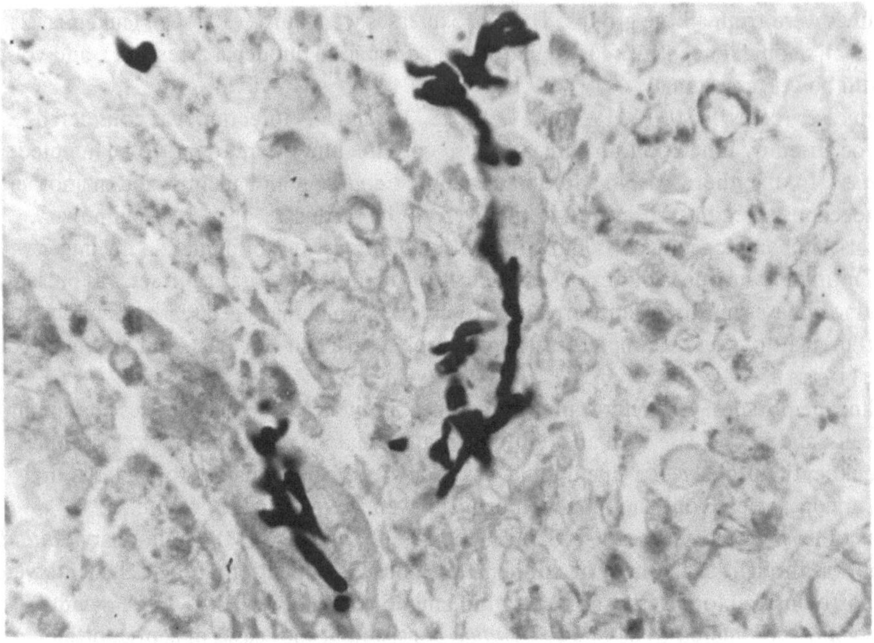

Fig. 5. Case No. 32: Lung. Atypical fungus structures in the abscesses with granulomatous reaction. Grocott, x 400

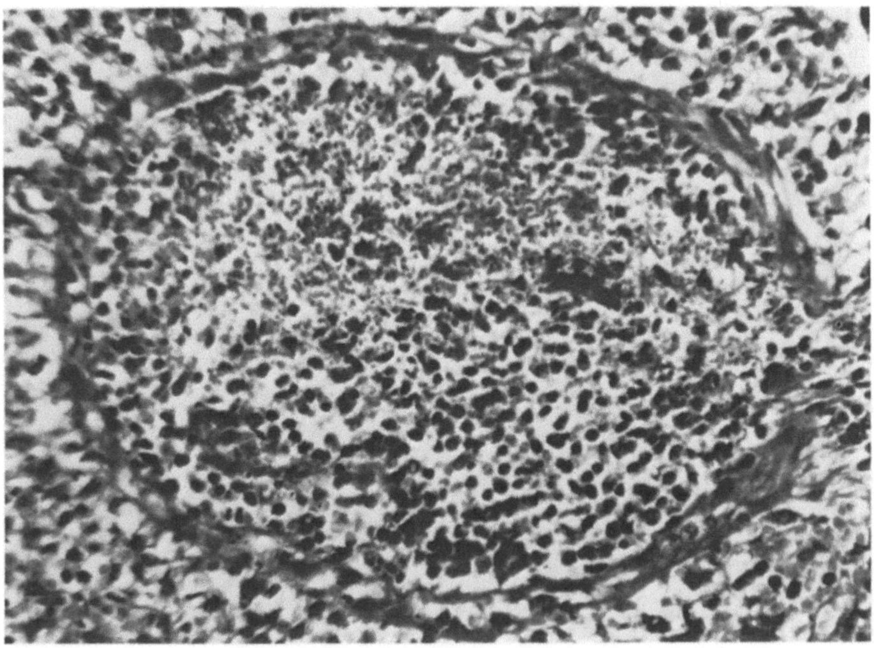

Fig. 6. Case No. 22: Lung. Vasculitis. Fungus cells not clearly visible; hyphae not at all. Compare with Fig. 7. H and E, x 300

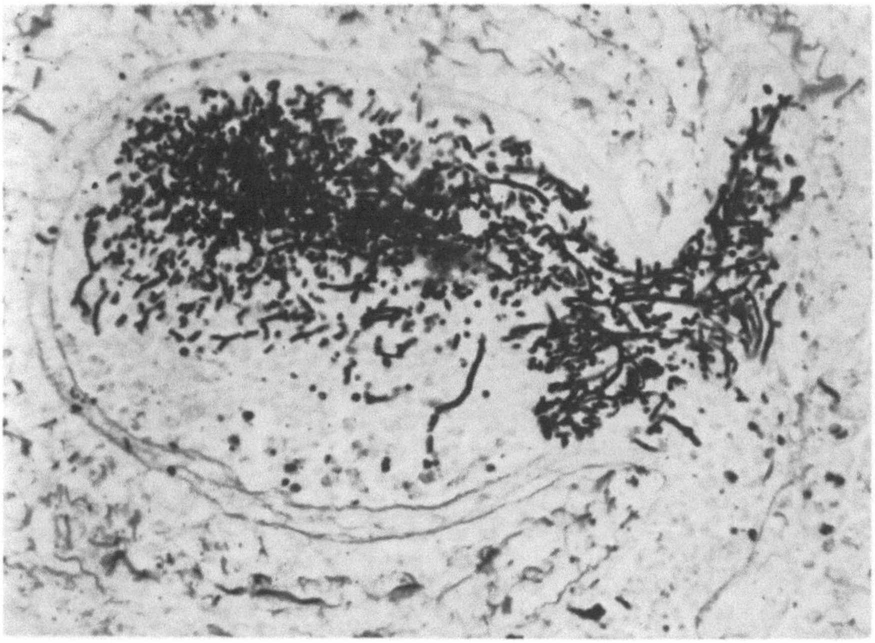

Fig. 7. Same blood vessel as in Fig. 6. Grocott, x 260

instance, hyaline membranes. In one case, there were microabscesses, in another pneumonic foci of different sizes, and in another emboli with fungi and minimal inflammatory reaction.

Associated diseases in this group were: burns in three cases, infectious hepatitis in two, and prematurity, enterocolitis with surgery, leukemia and renal transplant, in one case each.

There were neither fungus balls in the lungs, nor invasion of the pleura by fungal elements.

b) Urogenital System

In 19 cases, fungal elements were found in the kidneys: in 18 with tissue reactions and in one case only a fungemia with minimal interstitial-cell infiltrates, possibly independent of the fungi present in the blood vessels.

Renal involvement occurred in all age groups and in patients with all types of underlying diseases. In three of four leukemic patients, and in eight of the 12 patients with burns, renal involvement was noted. But no renal candidosis was found in the four diabetic patients, or in one transplanted kidney (the organs of the other transplant patient were not examined).

In one case pure bacterial pyemic abscesses were found without fungal elements in the abscesses. In another one (case 25) a marked CMV infection associated with the renal candidosis was noted (Fig. 8).

Multiple bilateral abscesses were the rule, but in some instances no gross lesions were noted at all.

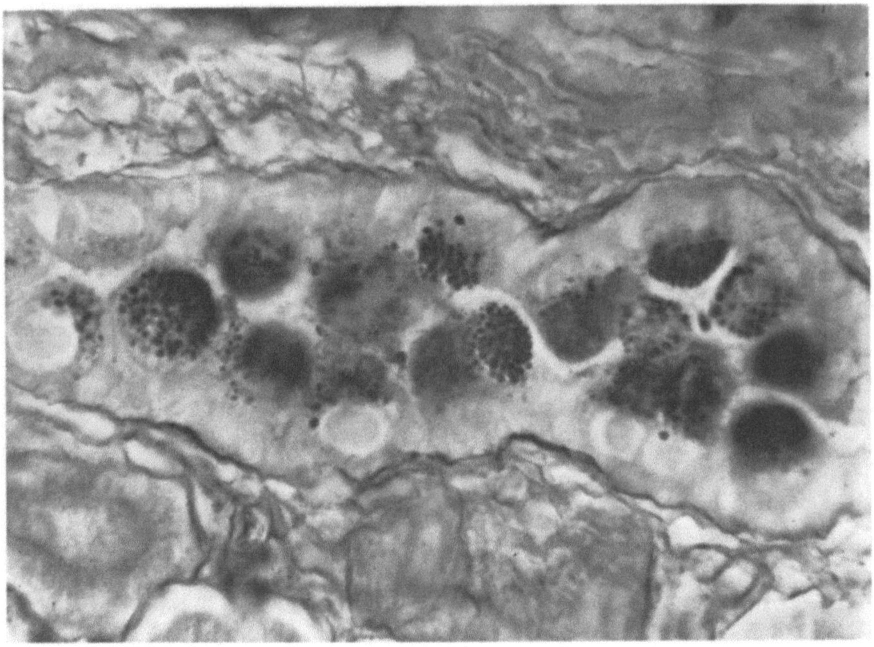

Fig. 8. Case No. 25: Kidney. Cytomegalic cells in tubule with Grocott-positive cytoplasmic granules. Grocott, x 400

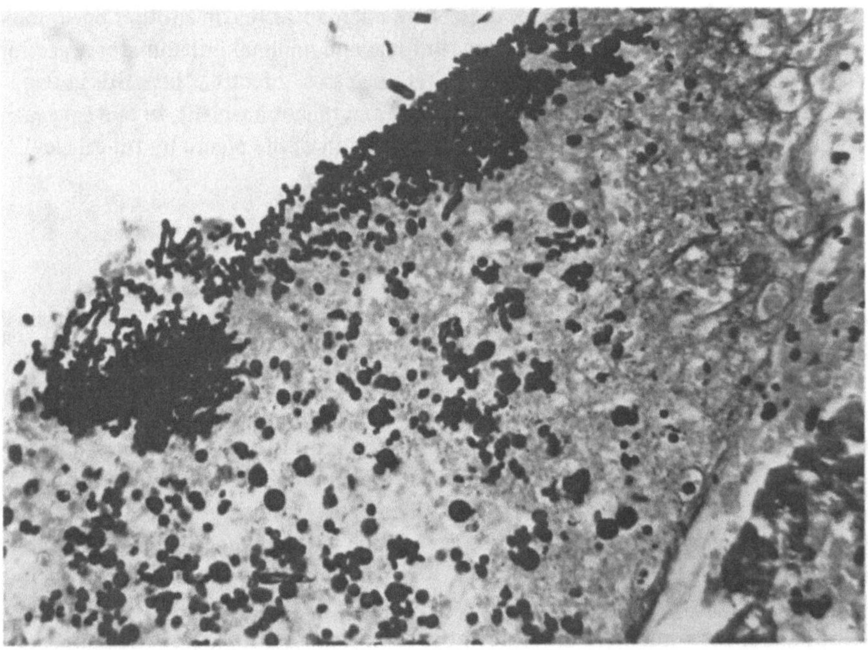

Fig. 9. Case No. 21: Kidney. Numerous large and small yeast cells and hyphae at the tip of the papilla. Grocott, x 300

The predominant tissue reaction was the presence of multiple abscesses of different sizes (12 cases) with surrounding hemorrhages in two. In nine cases a necrotizing papillitis (Fig. 9) was also noted. In three of these cases, large numbers of fungal elements forming a fungus ball were present in the caliceal cavities (cases 5, 9, and 16). In two cases with leukemia, necrotic foci were found showing yeast cells without tissue reaction.

In two cases minimal tissue reaction with accumulation of plasma cells and lymphocytes with few leukocytes was found.

Four cases, in addition to abscesses, showed a granulomatous reaction (Figs. 10 and 11).

In addition to acute *Candida* glomerulitis (Figs. 12, 13, 14, and 15) in ten of the 19 positive cases, signs of membranous-proliferative, focal glomeruolnephritis were found with proliferation of endothelial and epithelial cells of Bowman's capsule and thickening of the basal membranes (Figs. 16 and 17). These lesions were present only in a few glomeruli in the respective cases with renal candidosis. In three cases, thrombi were present in glomerular capillaries (2) and in veins (1), and in one case (12) a necrotizing arteritis was encountered.

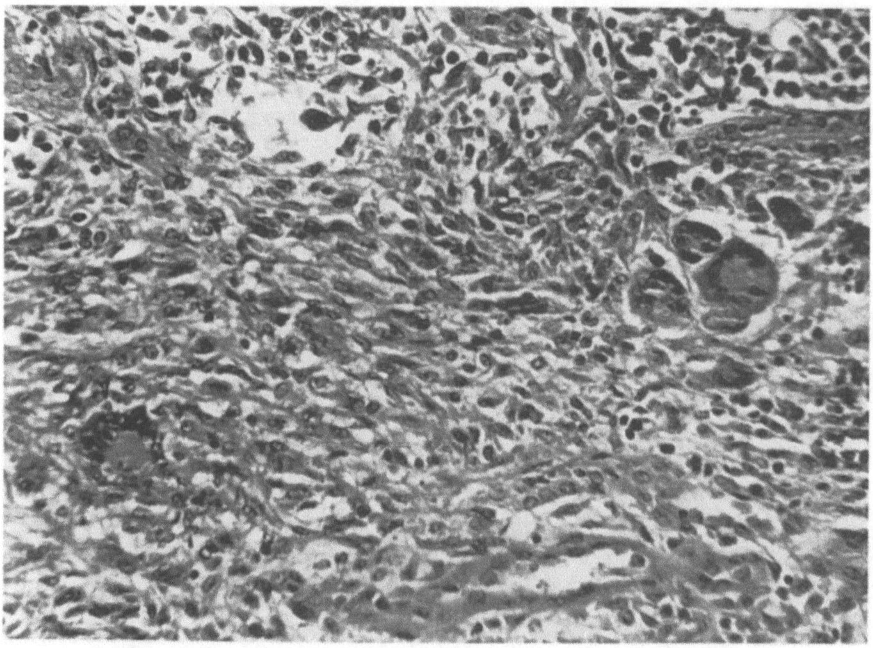

Fig. 10. Case No. 12: Kidney. Granulomatous reaction with epithelioid and giant cells. H and E, x 130

In this case of a burned patient, a *Pseudomonas* infection existed and a granulomatous reaction was also present in the kidneys (Fig. 18). In this group of 10 cases, all basic diseases were represented; associated infection was missing in only two of them.

Atypical fungal elements (see Chapter "Fungi in Tissues, pp. 212) were seen in six kidneys (of the 19 cases with renal involvement).

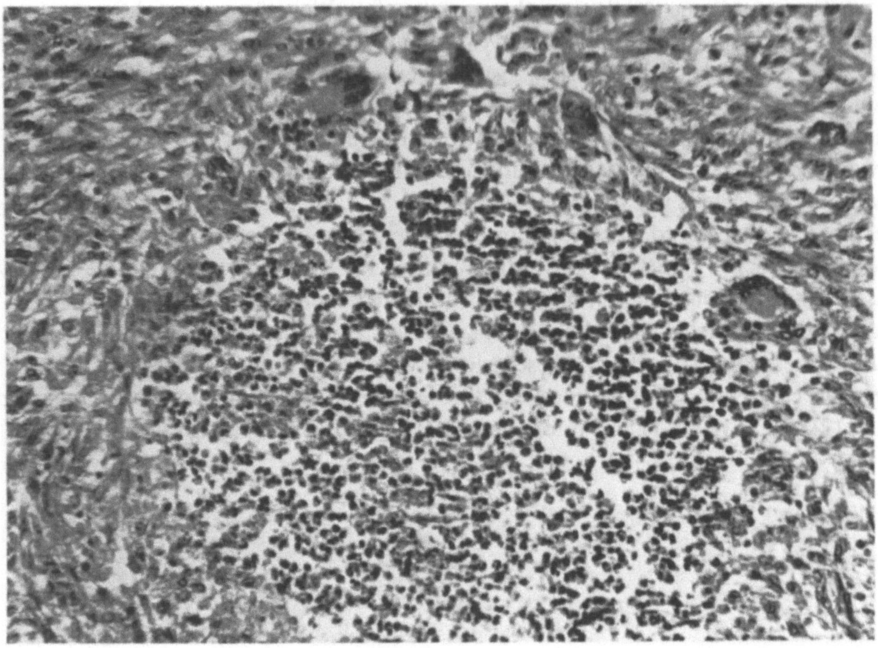

Fig. 11. Same case and organ as in Fig. 10. Abscess with granulomatous reaction. H and E, x 190

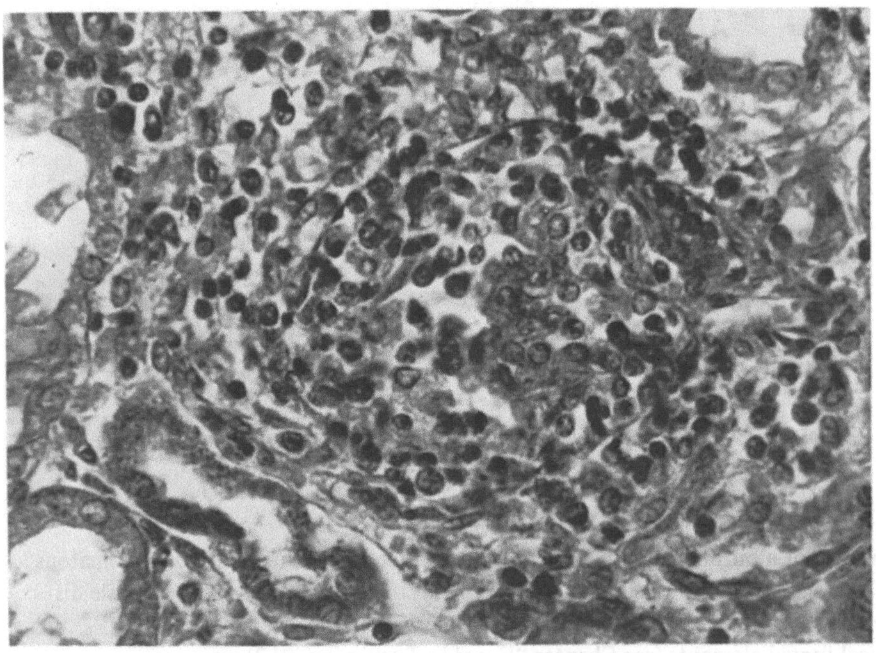

Fig. 12. Case No. 7: Kidney. *Candida* glomerulitis. Fungus cells not visible in this stain. H and E, x 450

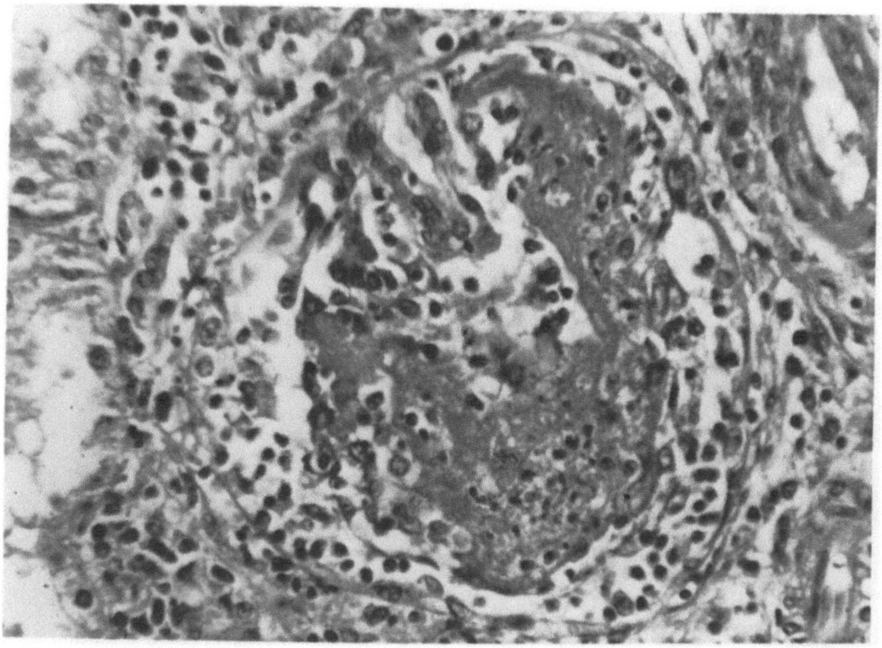

Fig. 13. Case No. 20: Kidney: Leukocytic and fibrinous necrotizing *Candida* glomerulitis. Fungus cells not visible. H and E, x 350

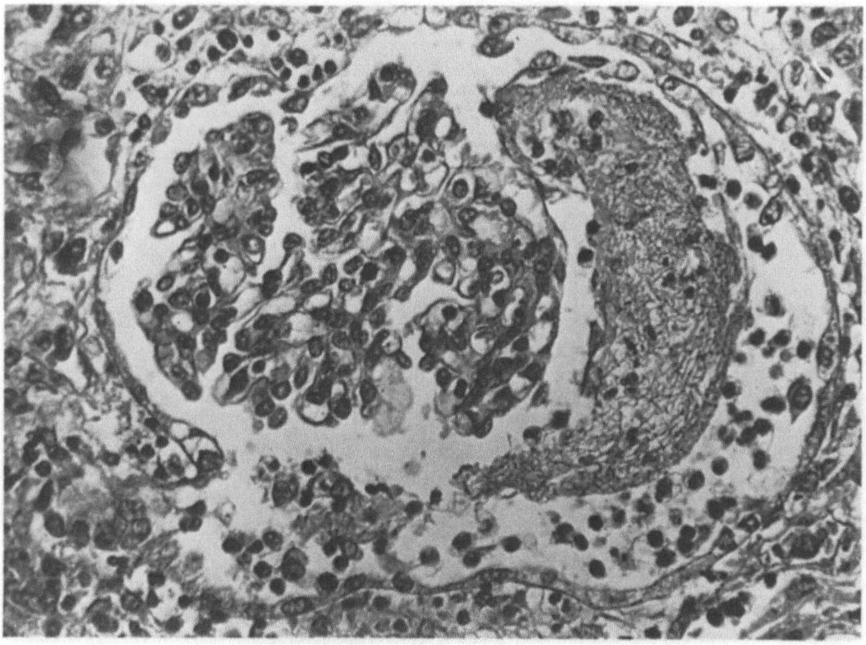

Fig. 14. Same case and organ as in Fig. 13. Another glomerulitis with fibrinous and leukocytic inflammation. H and E, x 350

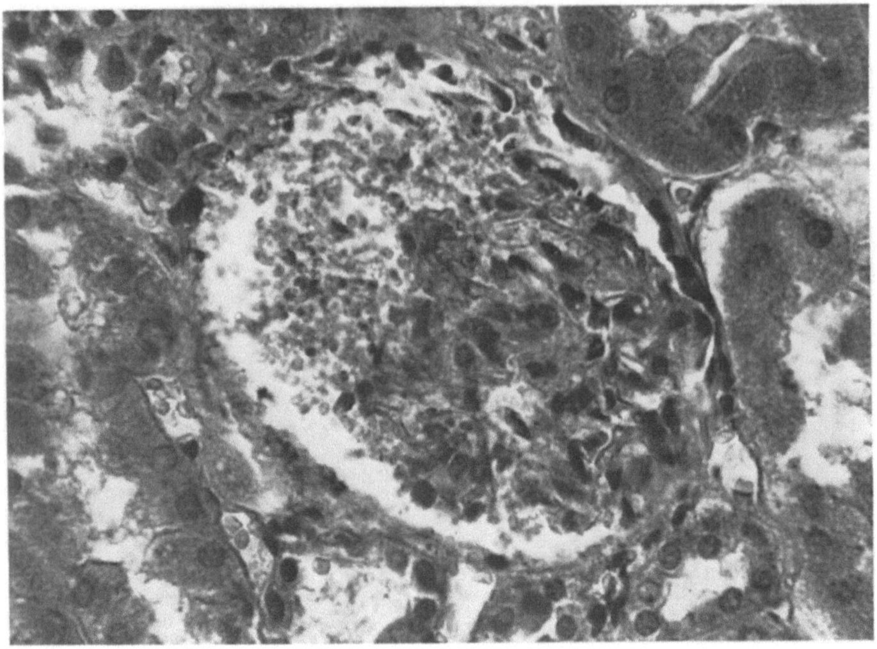

Fig. 15. Case No. 1: Kidney. Necrotizing *Candida* glomerulitis with numerous yeast cells. H and E, x 450

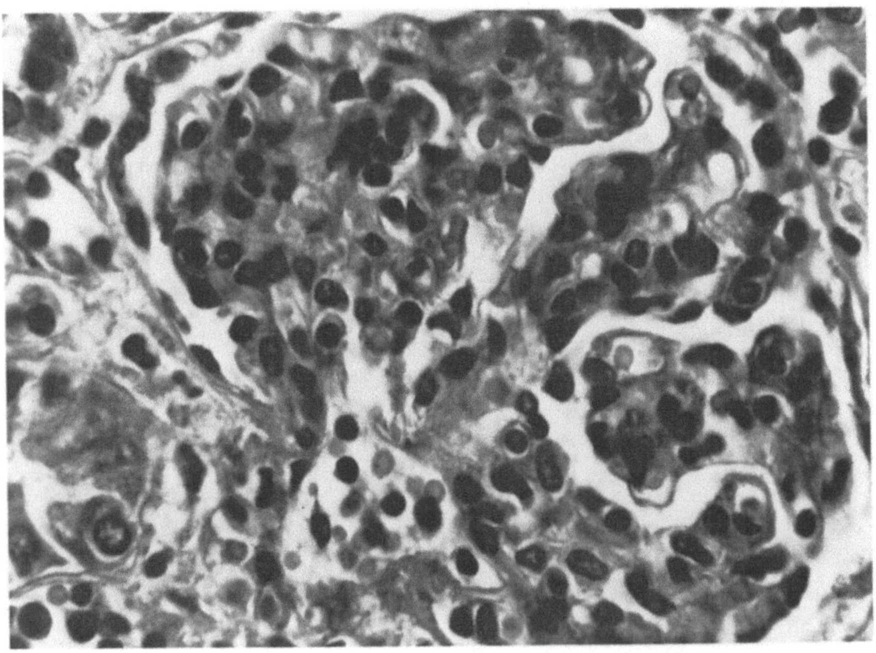

Fig. 16. Case No. 19: Kidney. Focal membranous glomerulitis. H and E, x 600

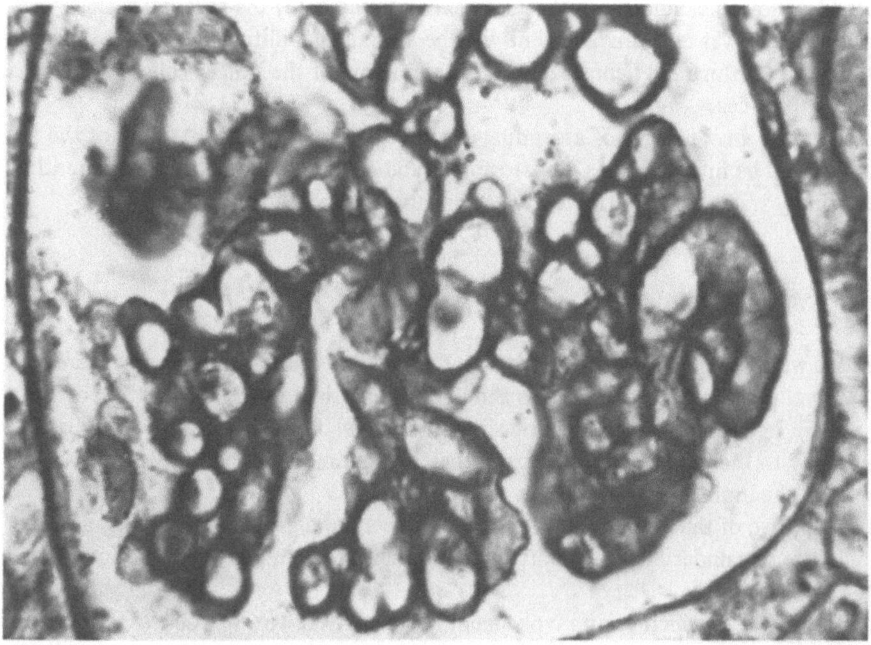

Fig. 17. Case No. 20: Kidney. Thickening of basement membranes of glomerulus in candidosis. Grocott, x 750

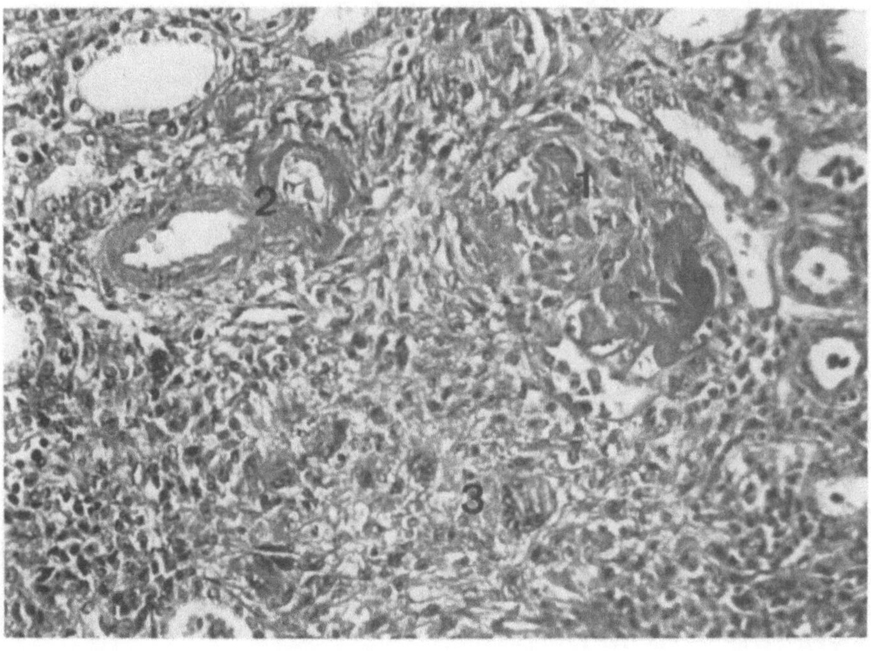

Fig. 18. Case No. 12: Kidney. Fibrinoid, necrotizing glomerulitis *(1)* and arteritis *(2)* and granulomatous reaction *(3)* in candidosis. H and E, x 190

The *bladder* showed candidosis in three cases. In one (case 22) the mucosal surface was intact. In two, there was renal involvement with necrotizing papillitis but without fungus balls in the calyces. A marked hematogenous dissemination of the fungus cells was present in all of these three cases.

No case studied showed signs of ascending (pyelonephritic) *Candida* infection of the kidneys. All could be interpreted as hematogenous and descending infections. Apparently the lower urinary tract was not affected by secondary descending infection.

c) Central Nervous System

Tissue lesions were observed in all 13 cases with *Candida* infection of the *brain*. There was no pure cerebral fungemia. Only once was cerebral candidosis the only localization of the disease (case 6). In another case, a diabetic showed *Candida* colitis apparently with dissemination only to the brain. In another case, there was pulmonary candidosis with isolated spread to the brain.

Surprisingly, five of the 13 patients were from the perinatal group and a sixth was three months old. The remainder of the 13 patients with brain involvement were older children and adults.

The basic diseases in which brain involvement occurred were two with burns, one with leukemia, and one with diabetes, in addition to the hepatopathies and pathologic conditions of the perinatal group.

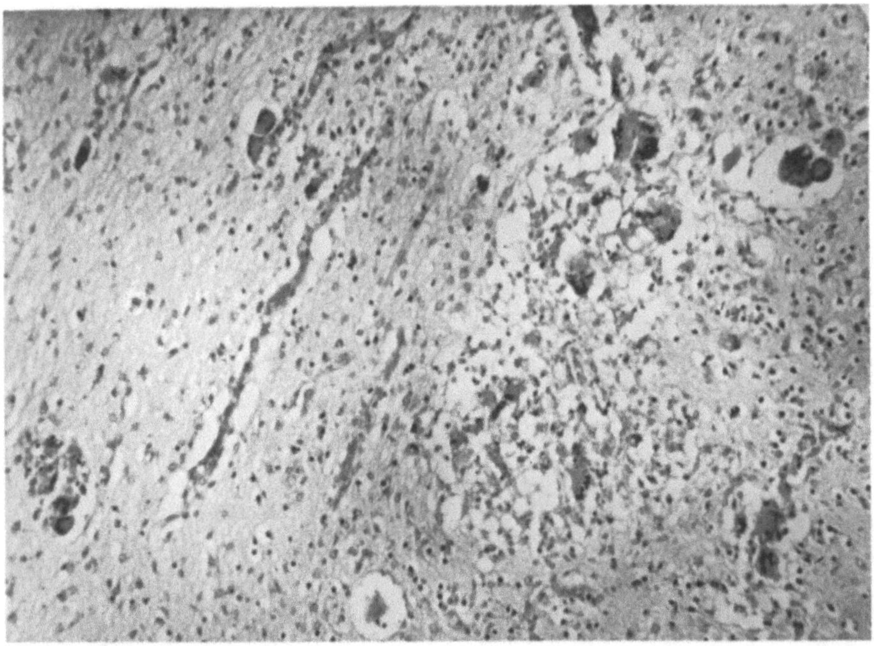

Fig. 19. Case No. 25: Brain. Necrosis with granulomatous reaction. Fungus cells not seen with this stain and magnification. Clear elongated spaces in giant cells are fragments of hyphae (arrow). H and E, x 140

In nine cases, there was a bacterial infection and in one of these, there was, in addition, a massive extracerebral CMV infection.

Tissue reactions varied considerably: in five cases, the main feature was the presence of abscesses of different sizes. In addition, infarcts, hemorrhages, perivascular infiltrates, and glial nodules were observed. In four cases, only cellular infiltrates of different intensity were observed, mostly consisting of mononuclear cells. In two cases, only small cellular infiltrates with a few leukocytes were seen. In four cases (cases 6, 25, 27, and 31) a granulomatous reaction was found (Figs. 19 and 20). Three of these showed almost exclusive involvement by granulomatous reaction. The fourth case, however, also had abscesses. This fourth case is included in the group of five with abscesses (mentioned above). Inflammation extended in five cases toward the surface of the ventricles (ependymitis) (Fig. 21) with formation of a fungus ball in case 3 and a granulomatous ependymitis and granulomatous reaction in the choroid plexus in case 27. Glial nodules without fungal elements were also seen in two cases of generalized candidosis but without *Candida* involvement of the brain.

Atypical fungus cells were seen in the brain in two cases (13 and 31).

Candida leptomeningitis was observed in eight cases, associated with candidosis of the brain. Tissue reaction was variable including a purulent reaction. One case had granulomatous reaction in the leptomeninges and in the brain. In case 5, candidosis was found in the brain, leptomeninges, and *dura mater.*

Involvement of the *spinal cord* was found twice (cases 1 and 3) (Fig. 22). In one case without involvement of the brain; in the other (3) in the presence of candidosis of brain and meninges. In this case, the meninges of the spinal cord were also involved. In case 3,

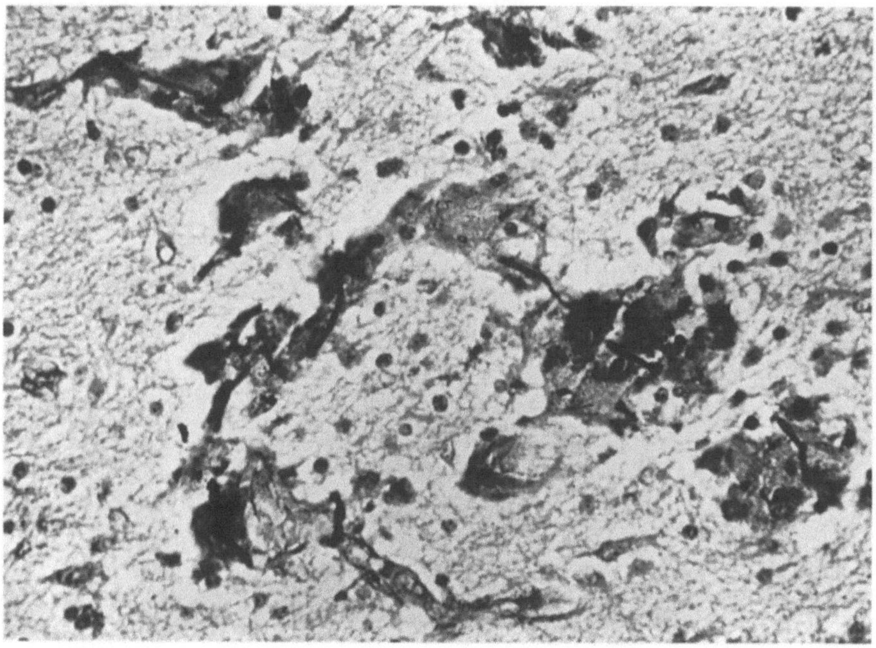

Fig. 20. Same case and organ as in Fig. 19. Fragments of hyphae in giant cells. H and E and Grocott, x 300

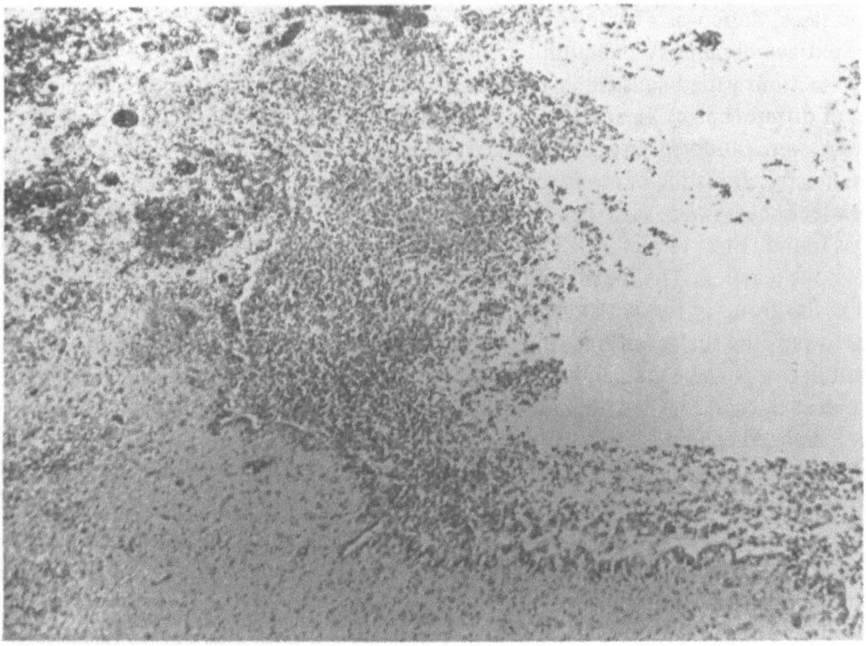

Fig. 21. Case No. 3: Brain. *Candida* ependymitis with numerous fungus cells in the ventricle (fungus ball). H and E, x 72

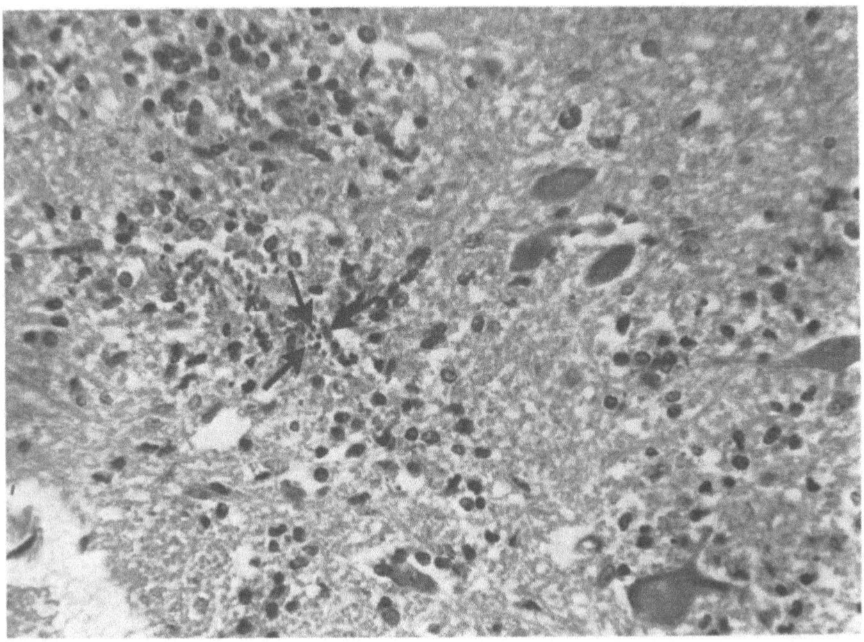

Fig. 22. Case No. 1: Spinal cord. *Candida* myelitis. Focus in the gray matter. The small granules are yeast cells (arrow). H and E, x 300

fungus elements were observed in the cerebral ventricles (fungus ball) and meninges. Therefore, the possibility of a directly descending extension exists, while in case 1, infection of the spinal cord apparently occurred by hematogenous dissemination.

d) Heart and Blood Vessels

The *myocardium* showed *Candida* lesions in eight cases; fungemia was present in three.

Lesions were seen exclusively in the perinatal group and in children up to 9 years of age. Basic diseases were burns and leukemia, each occurring once, otherwise the children had perinatal pathology and hepatitis.

In four cases abscesses were found (Fig. 23); in two there was granulomatous reaction—one associated with abscesses; two cases had necrotic foci loaded with yeast cells without cellular reaction (one a leukemia) (Fig. 24); in one case the fungi were found in cellular infiltrates consisting of round cells and leukocytes. In case 13, abscesses and a granulomatous reaction were found with fungi, but abscesses showing exclusively bacteria and recent infarcts were also noted. In case 19 an eosinophilic myocarditis with abscesses was found (without fungi). In this same case, a marked eosinophilic pancreatitis with abscesses coexisted (also without fungus cells).

Candida endocarditis was observed three times (Fig. 25), in one case associated with *Candida* myocarditis with abscesses. The endocarditis being a direct extension from myocardial lesions showed a fibrinopurulent exudate. In the other two cases, bacteria were

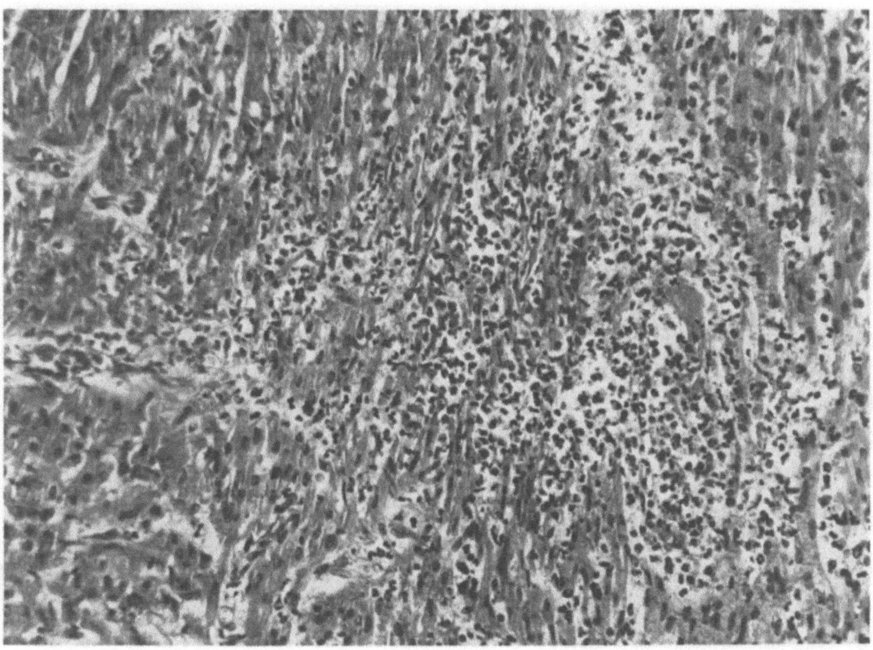

Fig. 23. Case No. 4: Myocardium. Purulent *Candida* myocarditis. Formation of abscess. Fungus cells not visible. H and E, x 190

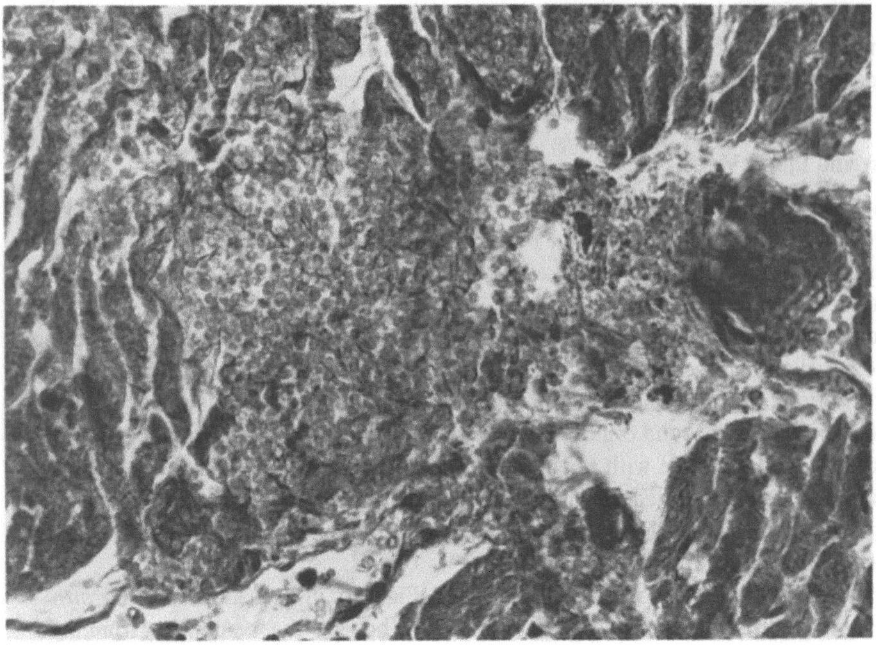

Fig. 24. Case No. 1.: Myocardium. Necrotic focus with numerous yeast cells. H and E, x 275

found in the endocardial lesions in addition to the yeast cells. In both cases, partly organized venous thrombi with fungi could be found (see below).

Fungi in thrombi within *large veins* (e.g., jugular, subclavian, iliac, inferior cava) were observed in eight cases (Figs. 26 and 27); in one other case in which thrombi were examined histologically fungi were not seen. In two cases, in addition to fungi, bacteria were also observed in the thrombi. Five of these eight cases were burn patients, one was a diabetic, one had bowel pathology with surgery, and the last had Kernicterus.

In case 24 with candidosis of the umbilical region, fungi were also found in a venous thrombus of this area.

These data on thrombi represent the minimal incidence, since numerous thrombi were observed only grossly.

Smaller and larger blood vessels in many organs showed *Candida* vasculitis partly with thromboembolism. These findings were not tabulated.

e) Reticulo-Endothelial System

In the *liver Candida* lesions were found in five cases; in five further cases there was fungemia (all in burn cases). From the five cases with lesions, three were from the perinatal group. The remaining two were children with leukemia. Tissue reaction was rather uniform; in the perinatal group nodular mononuclear infiltrates were seen and there were only small numbers of leukocytes; no clear-cut abscess formation was seen (Figs. 28 and 29). In the leukemic

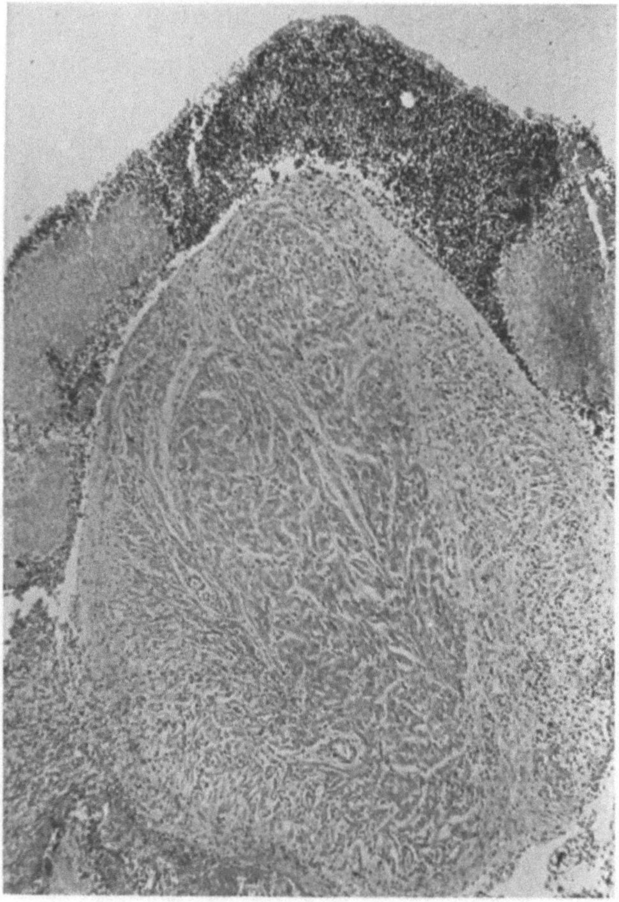

Fig. 25. Case No. 25: Endocardium. *Candida* endocarditis. Numerous yeast cells in the pale areas. Also candida myocarditis. H and E, x 50

cases, however, rather nonreactive necrotic foci were found with an accumulation of yeast cells or hyphae (Fig. 30). The gall-bladder showed involvement only once in case 34. This was a patient with carcinoma of the pancreas, obstructive jaundice, and terminal diabetes. Other organs with *Candida* lesions were lungs, digestive tract, and bladder. Candidosis of the gall-bladder may have developed by intraductal ascension or by hematogenous spread.

In the *spleen*, three cases showed *Candida* lesions; fungemia was detected in seven further cases (six in burn cases). There were no cases of the perinatal group in this small series, all were children. One case had leukemia, another burns, and the third had infectious hepatitis as basic diseases. Granulomatous tissue reactions were noted twice in a case with leukemia; nonreactive foci were present in one case (Figs. 31 and 32).

In the *bone marrow* fungemia was seen in three cases, all were burn patients. Lesions of the bone were present in one case, see below (Skeletal System).

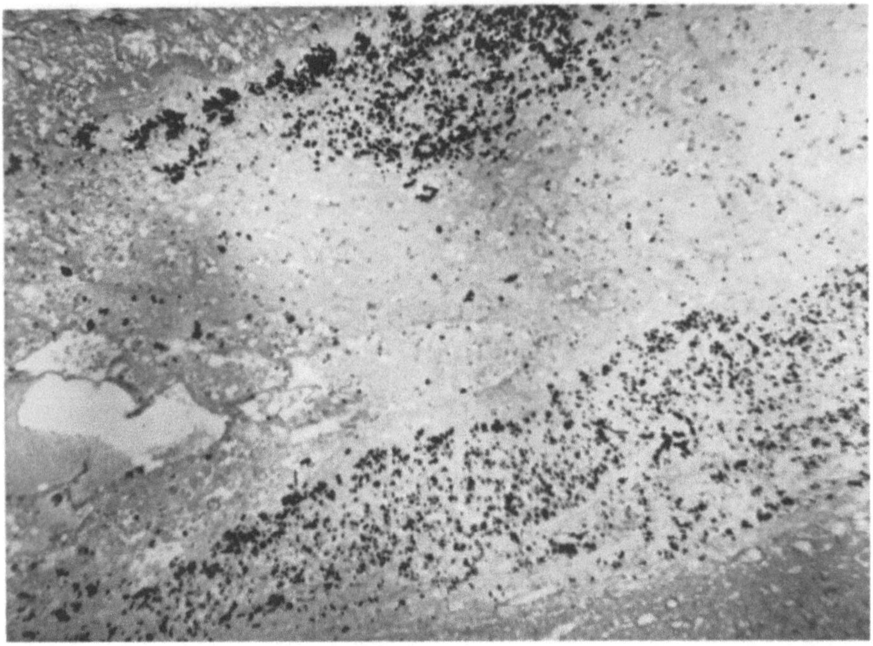

Fig. 26. Case No. 20: Jugular vein. The vein is cut longitudinally. Numerous fungus cells in thrombus. Grocott, x 50

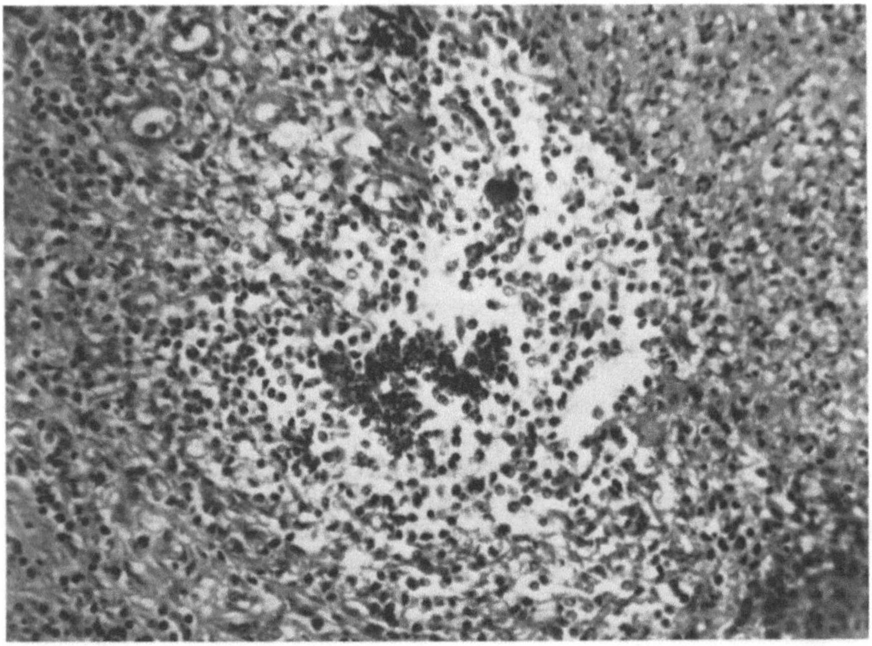

Fig. 27. Case No. 19: Subclavian vein. Organized thrombus with purulent softening (abscess formation). Bacteria and fungi not visible with this stain. H and E, x 190

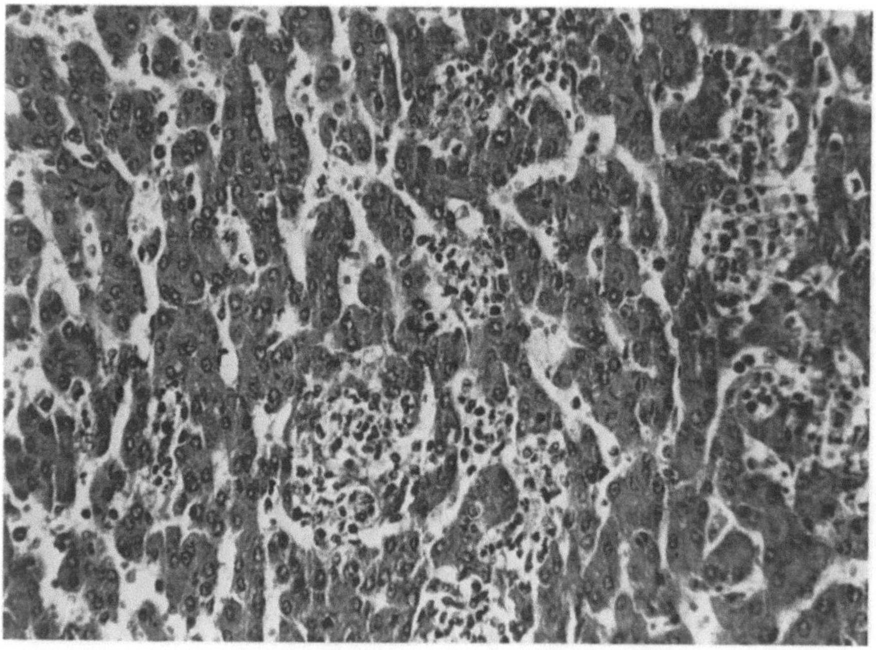

Fig. 28. Case No. 24: Liver. Multiple nodular infiltrates with mononuclear cells and few leukocytes. H and E, x 190

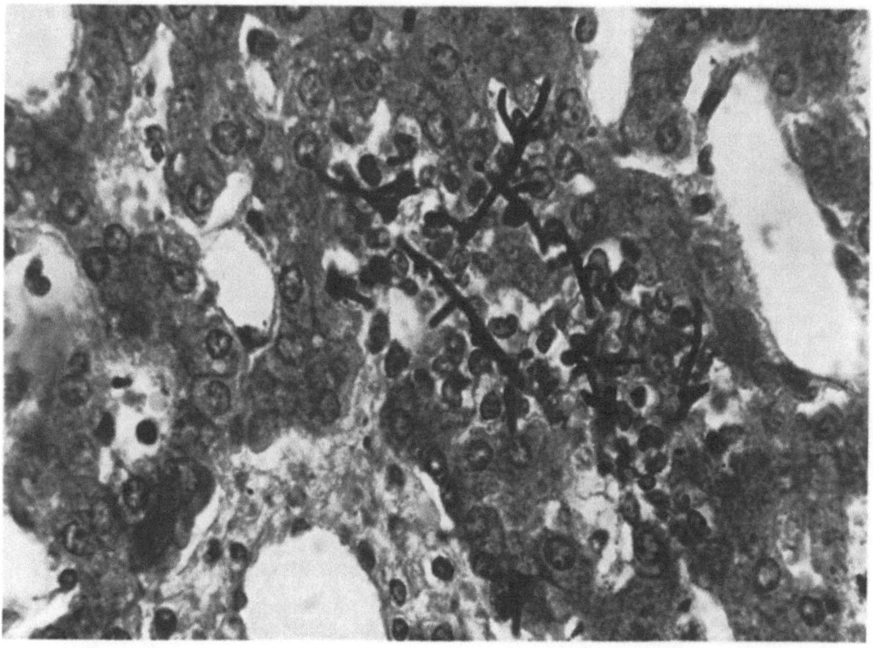

Fig. 29. Same lesion as in Fig. 28. *Candida* hyphae in the cellular infiltration clearly visible. PAS, x 450

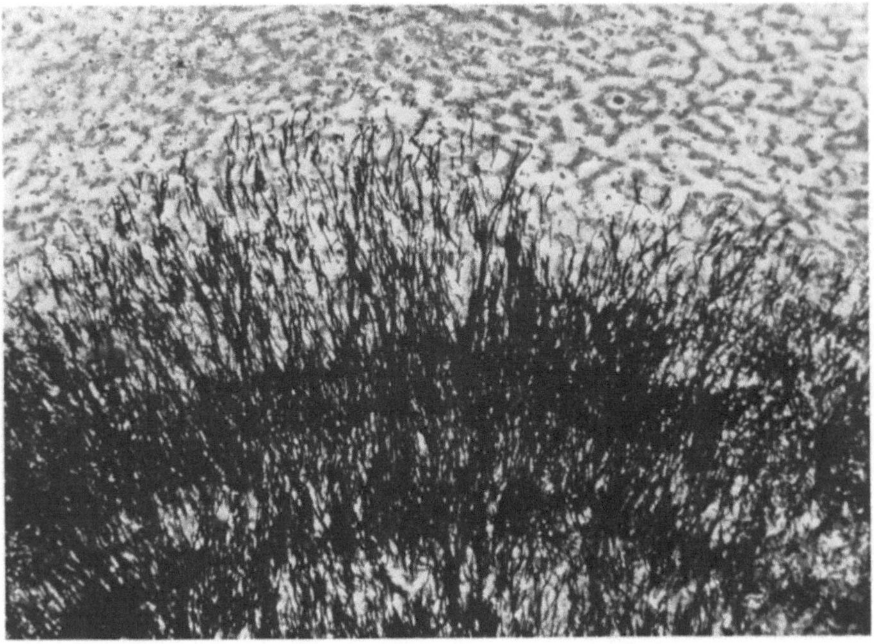

Fig. 30. Case No. 2: Liver. Large necrotic area with numerous hyphae of *Candida*. PAS, x 95

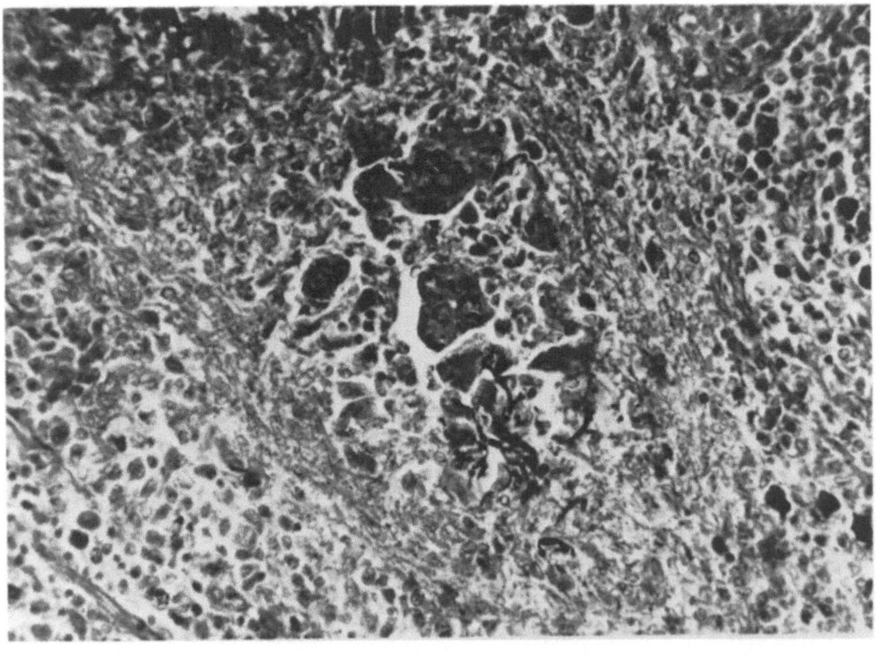

Fig. 31. Case No. 5: Spleen. Granuloma with giant cells containing fragments of hyphae. PAS, x 250

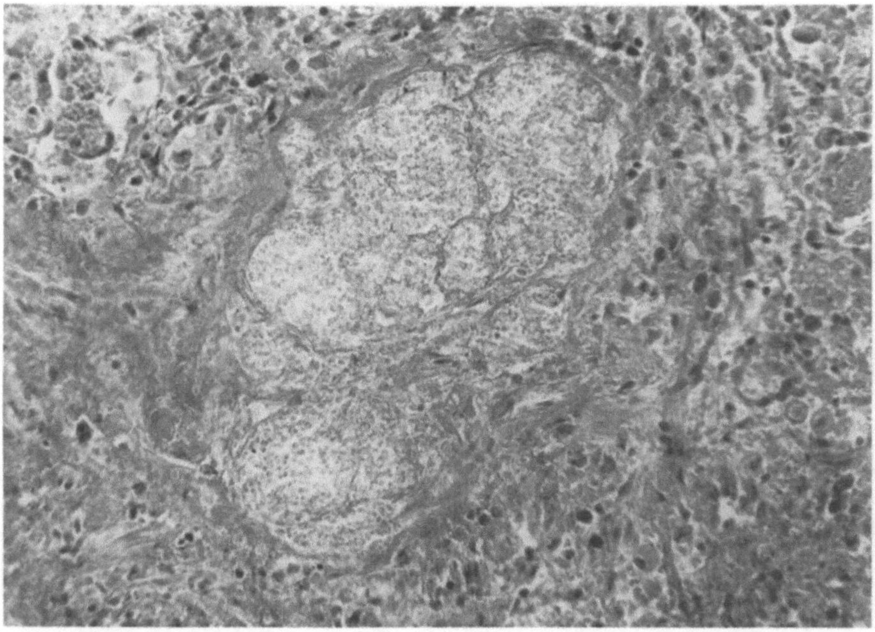

Fig. 32. Case No. 26: Spleen. Necrotic focus with numerous faintly visible yeast cells. H and E, x 190

Candida lesions were observed only once in *lymph nodes*. This was a regional lymphadenitis in a case with isolated pulmonary candidosis. Fungal elements were found in a small poorly circumscribed accumulation of inflammatory cells with a few leukocytes without formation of a true abscess.

When the findings of the organs of the reticulo-endothelial system are summarized and compared a rather uniform pattern is revealed. The total number of cases with fungi in the liver and spleen were the same, each organ showing ten cases. Lesions did not occur simultaneously at these sites. Fungemia was present simultaneously in liver, spleen, and bone marrow quite frequently, mostly in burn patients. In three of the four leukemic patients hepatic and/or splenic lesions were found. Tissue reactions in the liver were similar in the perinatal group and in liver and spleen of patients with leukemia. Granulomatous reaction was also found in one of these patients. Bone marrow did not show lesions after hematogenous spread of the fungi, only one case of *Candida* lymphadenitis was seen. Finally, atypical fungus cells were not found in these organs.

f) Endocrine Organs

In the *adrenal gland* lesions were seen in three cases. Two of these were children with burns and one a premature infant. Tissue reactions consisted of mononuclear nodules in two (one of them being the premature infant) (Figs. 33 and 34), and a granulomatous reaction in the third case (Fig. 35). The fungemia in a fourth case was found in a burn patient.

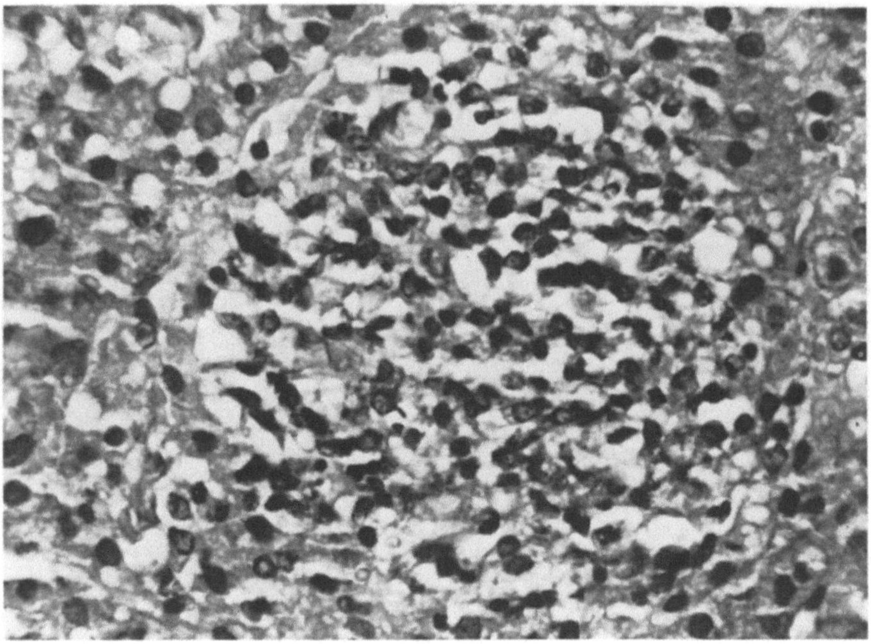

Fig. 33. Case No. 16: Adrenal gland. Circumscribed cellular nodule (mononuclear elements). Fungus cells not visible. H and E, x 450

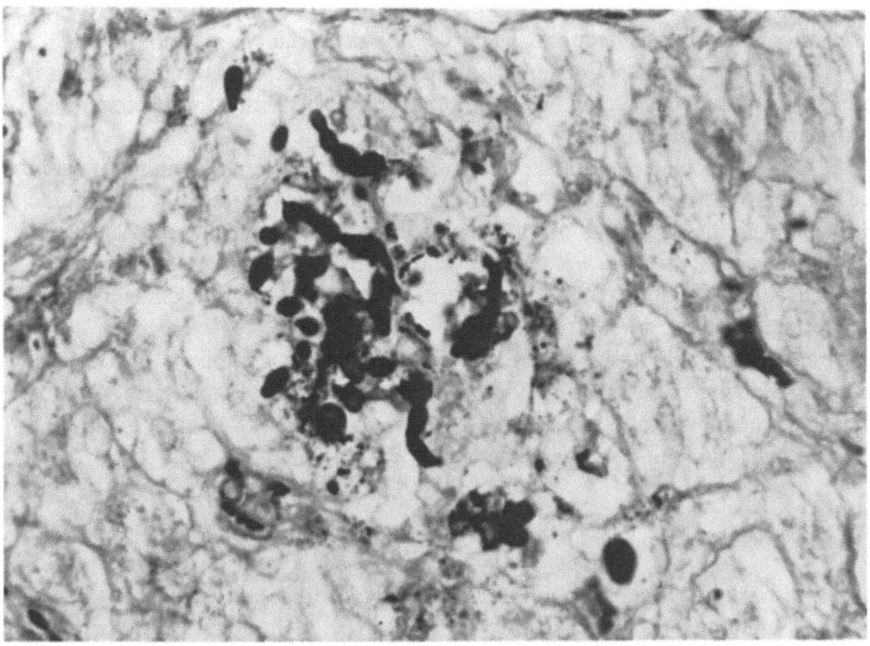

Fig. 34. Same lesion as in Fig. 33. Fungus cells clearly visible. Grocott, x 450

Fig. 35. Case No. 12: Adrenal gland. Giant cell containing fungus cells. H and E, x 450

The *thyroid* showed *Candida* lesions in two cases. The first was a perinatal patient with Kernicterus and hyaline-membrane disease, and the second a nine-day-old infant with post-hepatitic bone-marrow failure. There were abscesses in both these cases and, in additon, one showed a granulomatous reaction. In this latter case, atypical fungus cells were also seen.

In the four cases with *diabetes* no involvement of endocrine organs was found. Symptomatology of endocrine disorders (*Candida* Endocrine Syndrome–*Whitaker* et al., 1956) was not noted, to our knowledge, in any of the 34 cases.

g) Skin and Mucosa of Digestive Tract

Cutaneous candidosis was seen in five cases (skin was examined histologically 19 times). One case of *Candida* omphalitis is included (24). In this case, in addition to candidosis of the surface, fungus cells were found in a venous thrombus. Basic diseases in this group were three burn cases (from a total of 12), one case of prematurity, and one case of liver cirrhosis. In one burn case, atypical fungus cells were present in the tissue.

Candidosis of the *mucosa of the alimentary tract* was seen in 12 cases. Five cases showed involvement of the esophagus, four showed involvement of the stomach (Fig. 36), two of the duodenum, three of the small bowel, and two of the large bowel. Candidosis was localized to one organ in nine cases, to two organs in one case, and to three organs in two cases.

Basic diseases were burns in four cases, diabetes in three cases, leukemias in three cases, perinatal conditions in two cases (a total of three including a diabetic case), and one renal transplant case.

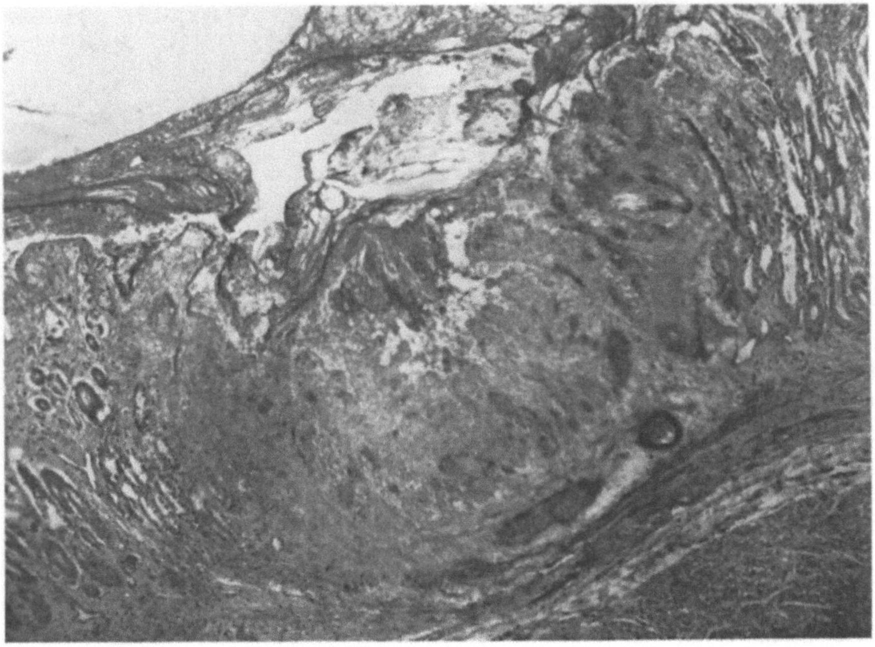

Fig. 36. Case No. 12: Stomach. Necrotic area in mucosa with numerous yeast cells (not visible at this magnification). H and E, x 48

Peritonitis wit fungus cells in the granulation tissue was observed in cases 18 and 21; in the first, following resection because of enterocolitis and in the second, after perforation of a duodenal ulcer in a burn case. In one of these cases there was a granulomatous reaction in the peritoneum. Mucosal candidosis was not seen in these cases.

No granulomatous reaction was found in the skin or in the mucosae.

h) Skeletal System

Candida osteomyelitis due to hematogenous spread was not found, but fungemia in the bone marrow in three cases was noted. In case 14, a candidosis of the cranial skin showed a deep infiltration and extension into bone with marked granulomatous reaction and numerous atypical fungus cells (Figs. 37, 38, and 39). This was a burn patient. Apparently the underlying bone was damaged by the burns and was therefore more susceptible to fungal invasion.

Skeletal muscle was examined randomly in only 11 cases. While 10 cases did not show lesions, in one case a *Candida* lesion of the diaphragm (Fig. 40) was found incidentally, apparently due to hematogenous spread, since in this case (case 4) fungal lesions were seen at 11 other sites.

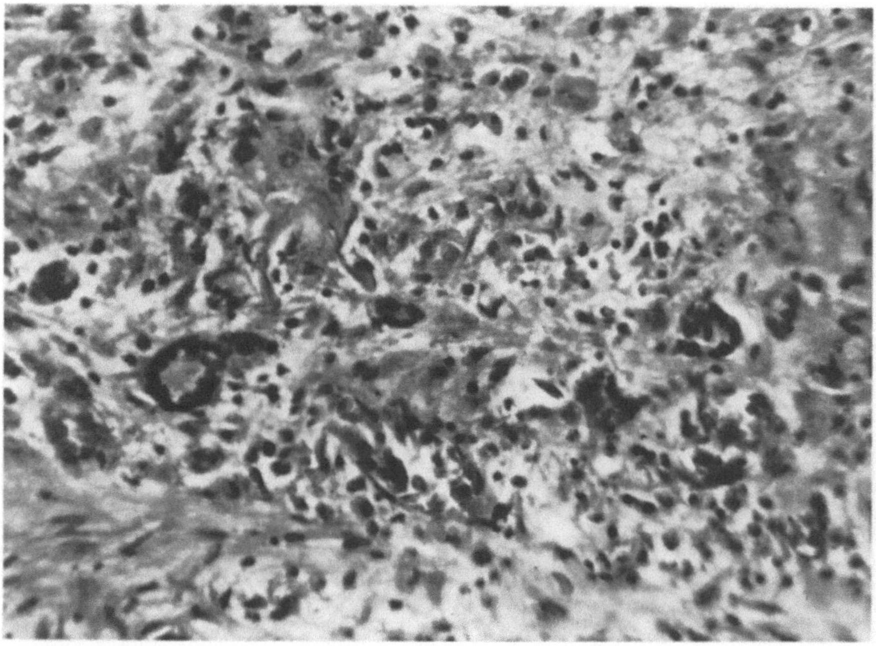

Fig. 37. Case No. 14: Bone. Granulomatous osteomyelitis *(Candida).* Fungal elements not visible. H and E, x 300

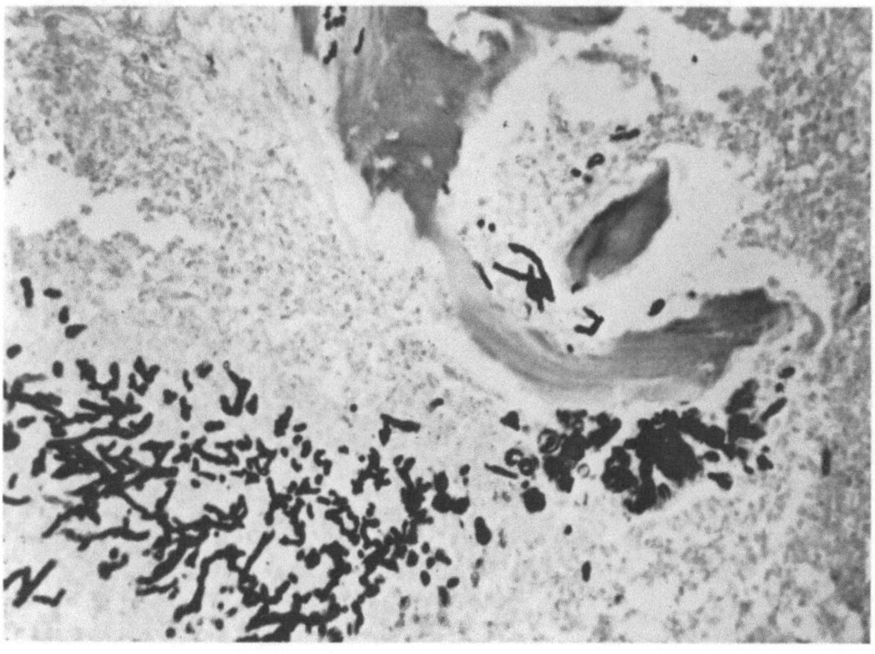

Fig. 38. Same case and organ as in Fig. 37. *Candida* osteomyelitis with destruction of osseus trabeculae. Numerous partly atypical fungus cells. Grocott, x 190

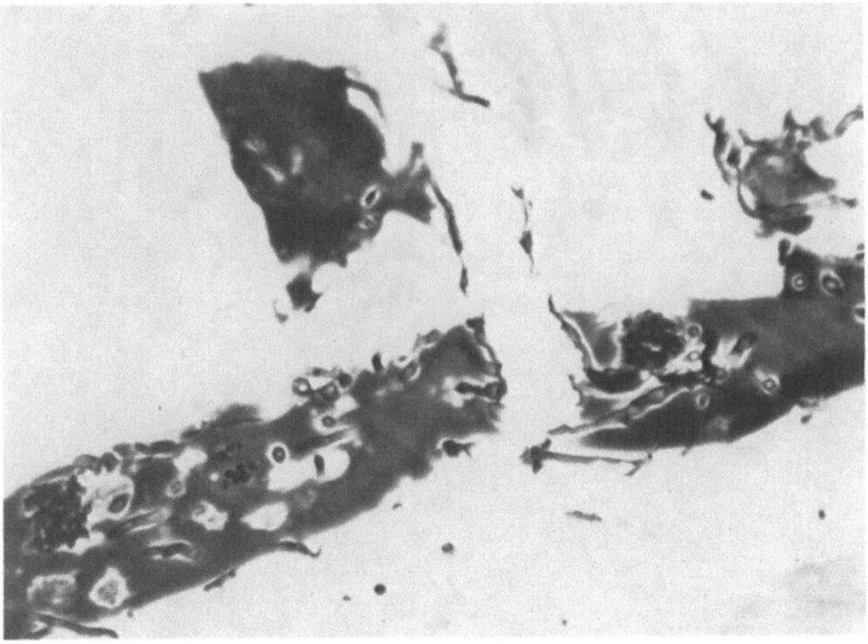

Fig. 39. Same case and organ as in Fig. 37. Invasion of osseus trabeculae by fungus cells. Grocott, x 275

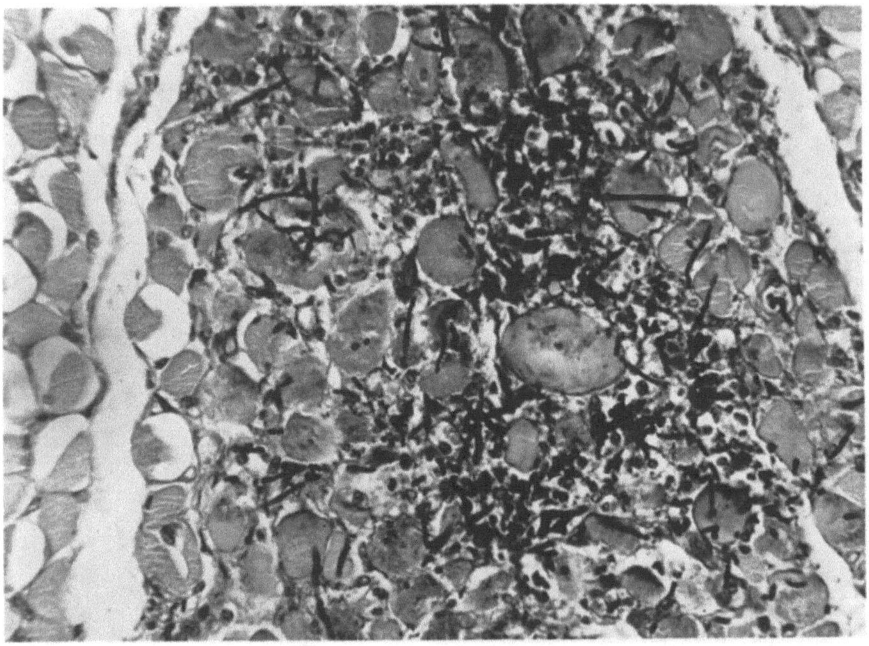

Fig. 40. Case No. 4: Diaphragm. Circumscribed inflammation with numerous fungal elements of *Candida*. PAS, x 300

3. Tissue Reactions

The majority of cases showed bacterial infection in addition to fungal infection. Tissue reactions due to the bacterial infection must be distinguished from those due solely to fungi. This was not possible in all lesions. We considered tissue reaction due to fungi, if lesions contained fungus cells in the Grocott stain.

Three main types of tissue reaction were found:

1. A nonspecific inflammation was noted with cellular infiltrates of variable density showing mononuclear cells, round cells, variable numbers of leukocytes, and/or abscesses. In some cases, this cellular response was minimal. These nonspecific reactions were either present alone or in combination with the following.

2. An inundation of tissues with fungus cells in smaller or larger areas with circumscribed necrosis (breakdown) of the tissue and lack of or minimal cellular infiltrates was recorded. This was observed mostly in leukemic patients in liver, spleen, myocardium, kidneys, and gastric mucosa. It was also present in case 26, with infectious hepatitis and aplastic anemia. In these patients, however, granulomatous reaction and abscesses were observed repeatedly at the same site or at other sites, i.e., the necrosis was not a reaction occurring exclusively in a given case. Necrotic foci with fungi were found also in leukemic infiltrates.

3. A granulomatous reaction due to fungus cells was seen. In addition to this reaction, nonspecific inflammation was found at the same or different sites in most cases. This reaction was seen in 16 cases and at 23 sites. Table 3 shows the case numbers, organ involvement, and the presence of atypical fungus cells. One organ was involved in 11 cases, two organs were involved in three cases, and three organs were involved in two cases. The lungs showed the reaction in five cases, kidneys in four cases, brain in four cases, myocardium and spleen in two cases, and meninges, peritoneum, bone, thyroid, and adrenal gland in one case each.

Table 3. Granulomatous reaction. Case numbers, organs involved, and atypical fungus cells

No. of Case	Organs	Atypical Fungi
5	Spleen	
6	Brain	
9	Kidney	
10	Lung	
12	Lung, adrenal gland, kidney	
13	Thyroid, myocardium	Thyroid, myocardium, lung, brain
14	Bone	Bone
16	Kidney	
18	Peritoneum	
19	Lung	
22	Lung, spleen	Kidney
24	Myocardium, tongue	
25	Kidney, brain, leptomeninges	
27	Brain	
31	Brain	Brain, skin, kidney
32	Lung	

Three cases involving five different organs with granulomatous reaction showed atypical fungus cells. In addition, atypical fungus cells were seen in one more case of the total 16 cases, in an organ without granulomatous reaction and in two of the three cases mentioned above in organs without granulomatous reaction.

The granulomatous reaction occurred in all age groups, in both sexes, regardless of race, and regardless of the basic diseases, surprisingly, however, granulomatous reaction was absent in the four diabetics. The reaction consisted of the presence of giant cells and formation of poorly circumscribed nodular structures, i.e., real tubercular granulomas were rarely observed. Few and scattered giant cells in otherwise nonspecific inflammation were seen in three cases. In most cases there were abscesses with giant cells in the lining granulation tissue and in four cases epithelioid cells were also seen. Coagulation necrosis was not observed. Tubercular granulomas were seen in two cases, but in the majority of them no fungi were observed. In case 10, they were interpreted as granulomatous disease of childhood.

In the giant cells of 11 cases (13 lesions) fragments of hyphae and/or yeast cells were found.

The nodular mononuclear cell infiltrate in the liver (occurring mostly in the perinatal age group) and glial nodules and cuffing in the brain must be mentioned as particular tissue reactions in certain organs. The latter reaction occurred in combination with others. Glomerular lesions are discussed above.

4. Fungi in Tissues

Hyphae, pseudohyphae, and yeast cells were found in tissues in variable numbers, both alone or together, with no relation to tissue reaction, basic diseases, or any other pattern. It is extremely difficult, if not impossible to differentiate between true hyphae and pseudohyphae of *Candida* in every instance. If the classical picture of pseudohyphae formation is present, it consists of elongated, stretched-out yeast cells, with constrictions and bud formation—commonly occurring at the point of constriction—between two segments of the elongated yeast cells. However, the segments can become quite long and may not lie at the same level in the morphologic preparations; such formations then will be clearly interpreted differently by various observers. Our material does not offer support for the assumption that hyphal elements indicate colonization and invasion as opposed to yeast cells alone—which would be saprophytic. Yeast-like cells occur sometimes in clusters, mostly extracellularly, preferentially within blood vessels (fungemia), and are occasionally similar in appearance to *Histoplasma capsulatum*. Grocott-positive, dust-like particles (destroyed fungus cells) were seen rarely (Fig. 41).

In H and E stained sections small or large numbers of yeast cells or hyphae were frequently not recognized. Without the Grocott method, diagnosis of the fungal etiology could not be made in a certain number of lesions (lymph node, certain cerebral foci, hepatic, and pulmonary lesions). Gram's method (fibrin-Weigert) did not stain fungal organisms regularly.

In twelve cases swollen hyphae with or without septa (in fragments) and with cyst-like swellings and/or large yeast cells in tissue, were morphologically similar, in some respects, to the fungus cells of *Geotrichum, Aspergillus, Phycomycetes, Blastomyces*, or looked

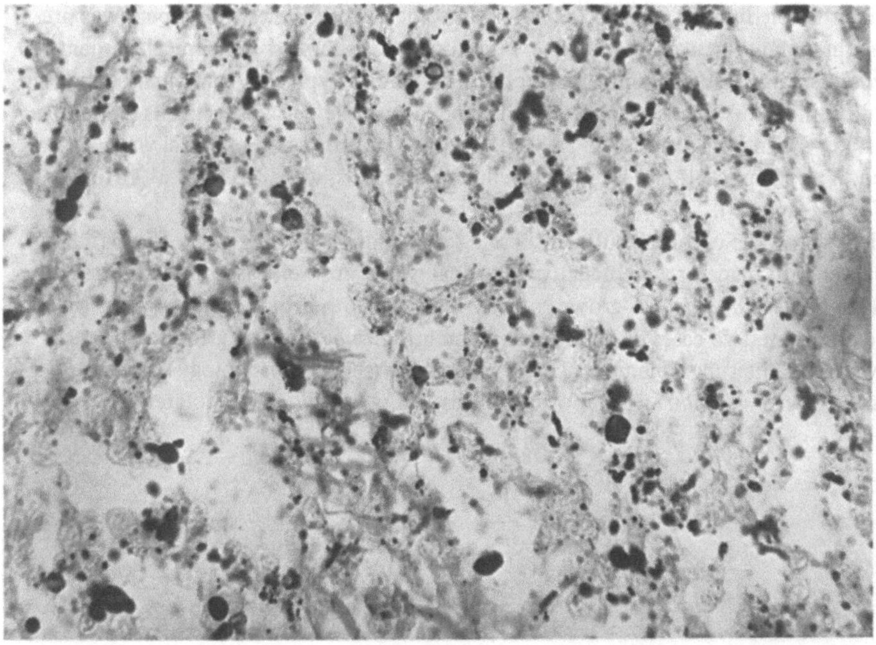

Fig. 41. Case No. 16: Kidney. In this necrotic area, partly preserved fungus cells and numerous Grocott-positive (black) particles are seen which represent debris of fungus cells. Grocott, x 450

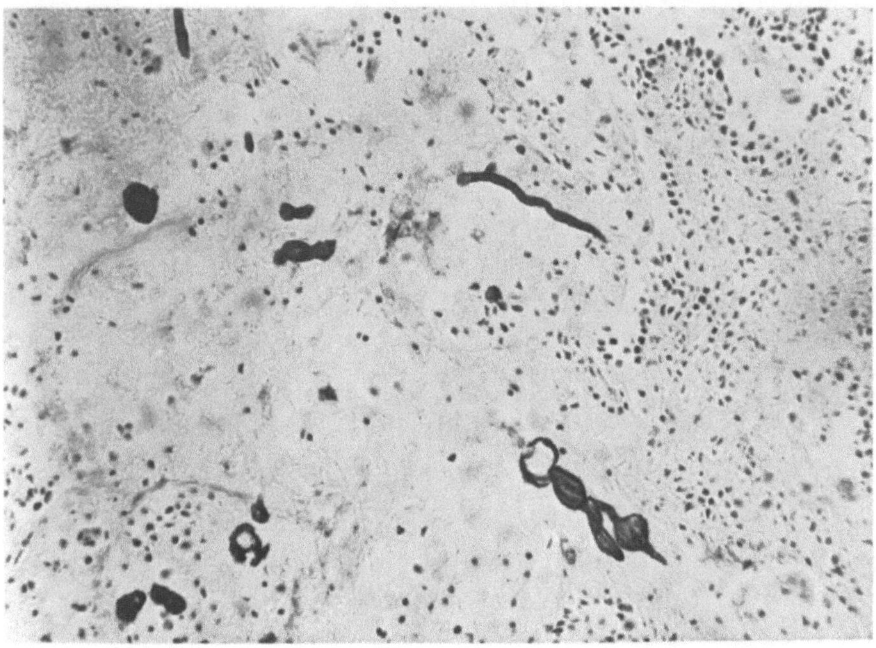

Fig. 42. Case No. 21: Lung. Atypical hyphal elements; wall defective cells? Grocott, x 300

like unidentifiable hyphae (Figs. 42 and 43). In some of these cases, this aspect was present only in certain organs while in others the fungus cells looked like typical elements of *Candida* (small yeast cells and thin hyphae). Distended hyphae in transverse sections can be difficult to distinguish from large (distended) wall defective yeast cells (spheroblasts) (Figs. 44 and 45). In case 13, atypical hyphae with invasion of vessel walls, typically seen in phycomycosis and large intracellular yeast cells, were observed at different sites (Figs. 46 and 47).

Table 4 shows the culture results in these 12 cases (also see Table 1). In five patients cultures were not made or were negative for fungi, while in seven cases with atypical fungus cells in tissues *Candida* was grown, mostly from sites other than seen in tissues (in four cases postmortem, two cases during life, and one instance before and after death).

In two cases, the fungal elements in the tissues showed the morphology of *Candida*, while in cultures *Aspergillus* was reported: in case 27, *Aspergillus* was isolated from the

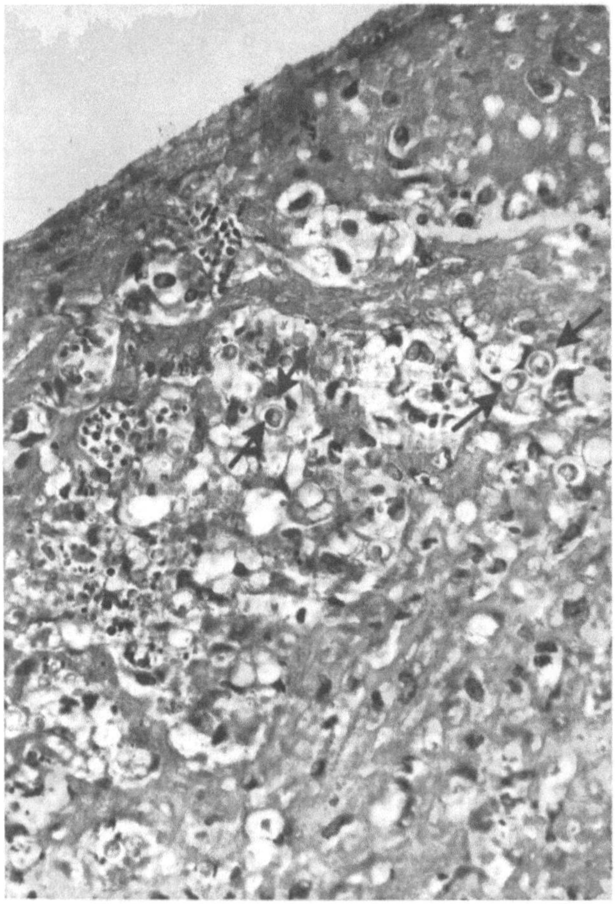

Fig. 43. Case No. 21: Subclavian vein. In a partly organized thrombus, small and larger yeast cells are seen. The large yeast cells resemble *Blastomyces dermatitidis* (arrow). H and E, x 300

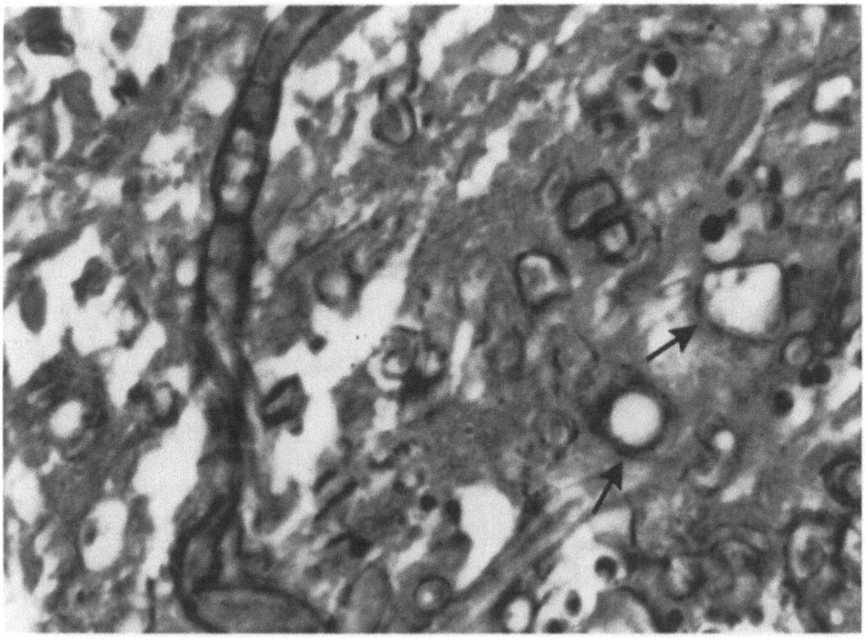

Fig. 44. Case No. 31: Skin. Broad hyphae with septa (longitudinal) and transversally cut (arrow) distended hyphae, difficult to distinguish from large distorted yeast cells. H and E, x 1150

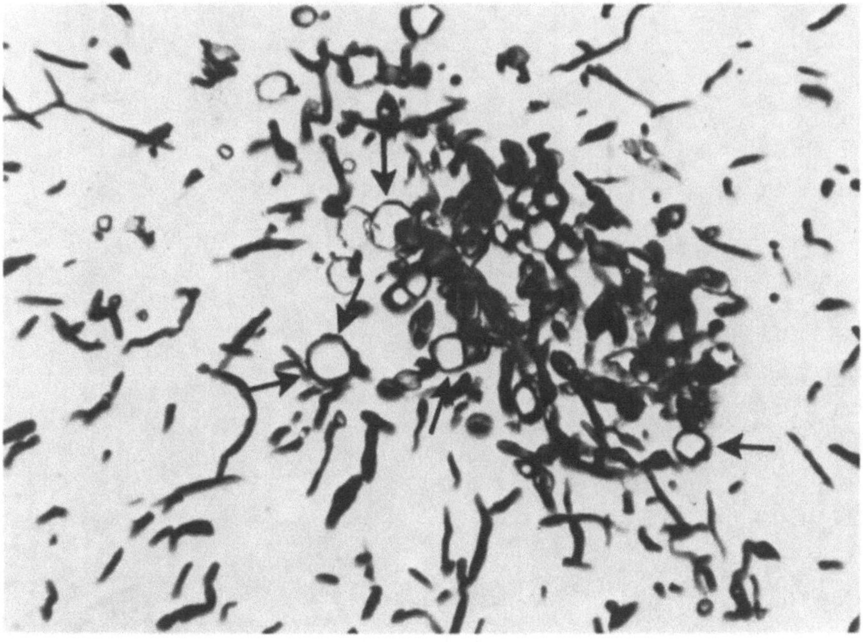

Fig. 45. Case No. 31: Brain. Partly distended (swollen) hyphae (transversally cut) or distorded yeast cells (spheroplasts) (arrow). Grocott, x 190

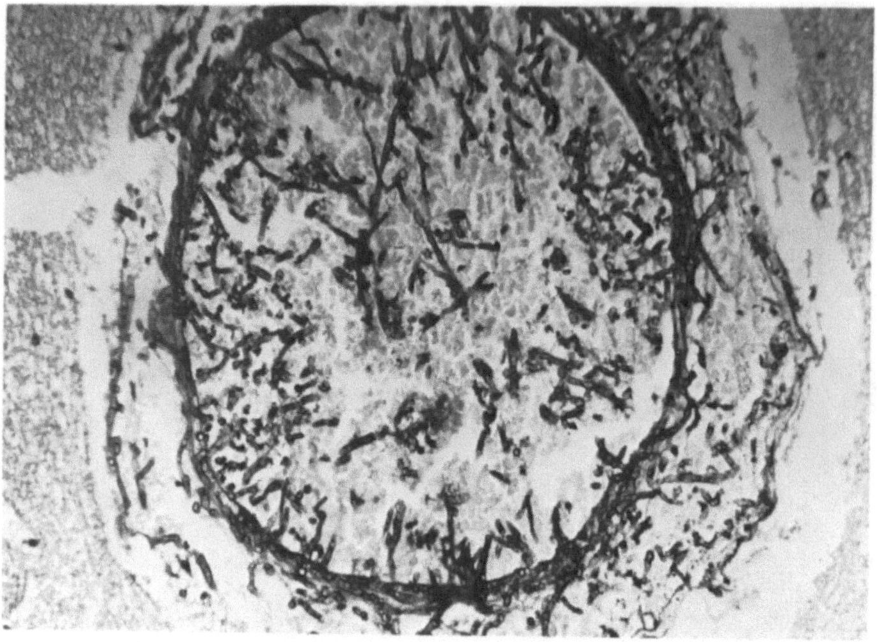

Fig. 46. Case No. 13: Brain. Broad hyphae in blood vessel wall, within and without the lumen of the vessel. Grocott, x 190

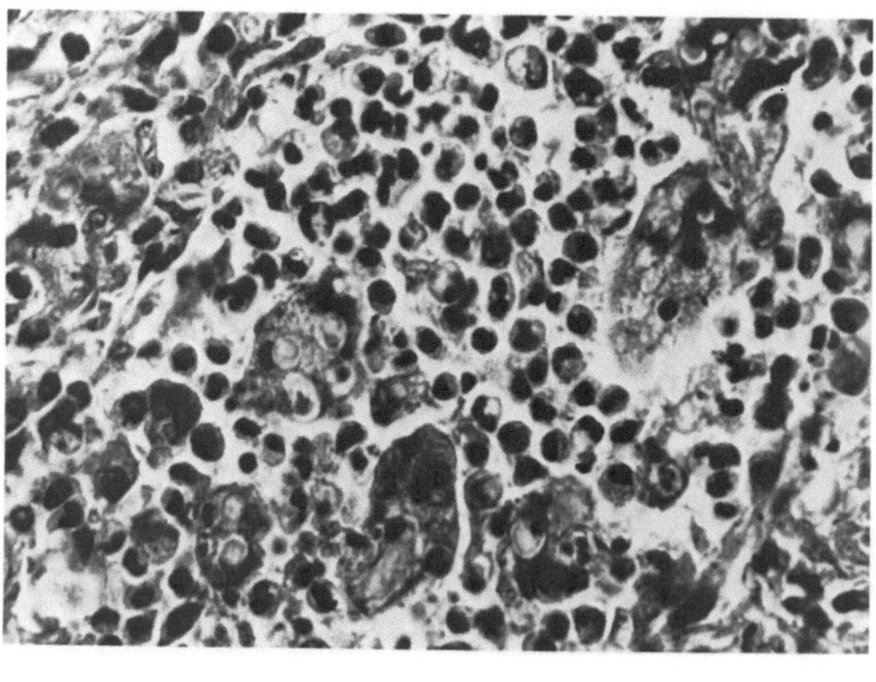

Fig. 47. Same case as in Fig. 46: Myocardium. Granulomatous reaction with large yeast cells in giant cells. H and E, x 600

spine in vivo and from the spinal fluid at post-mortem examination. *Candida* was isolated from the lungs. In case 32, *Aspergillus* was grown from the surgical specimen.

Table 4. Atypical fungus in tissues–results of cultures

	Atypical morphology in tissues			*Candida*-positive cultures	
Case No.	Organs	Hyphae	Yeast cells	Intravital	Postmortem
1	Lung	+			++ (Bowel, lung)
7	Myocardium		+	+++ (Skin, perineum, mouth)	
	Kidney		+		
8	Esophagus	+	+		
13	Thyroid		+		
	Lung		+	++ (Urine, pleural fluid)	
	Myocardium	+	+		
	Brain		+		
14	Bone	+			
20	Kidney		+		++ (Vein, lung)
21	Lung	+	+		
	Vein thrombus		+		++ (Vein, duodenal and gastric ulcer
	Kidney		+		
22	Kidney		+		
26	Lung	+	+		
	Myocardium		+		
28	Lung	+			+ (Lung)
31	Skin	+			
	Kidney	+			
	Brain	+			
34	Lung		+	++++ (Mouth, sputum, urine, trachea)	+ (Catheter)

5. Underlying Diseases

Debilitating conditions were noted in all cases, often several per patient.

In the 12 *burn* cases (11, 12, 14, 15, 16, 17, 19, 20, 21, 22, 23, and 31) 11 were from Shriners Burns Institute and one from General Hospital. All were children from 1 1/2-13 years of age. Bacterial infections were noted in eight cases, especially with *Pseudomonas*. There was CMV infection in one case. Gastroduodenal ulcers were noted in five cases. Positive *Candida* cultures were obtained in only four cases.

In seven cases, more than three organs showed *Candida* lesions. Fungemia was found in 10 cases. Skin candidosis was seen in only three out of ten cases examined. Adrenal glands were involved in three (of the four positive) cases, in two with minute lesions and in one with fungemia.

Granulomatous reaction was seen in seven cases and focal membranous glomerulitis in five. Atypical fungus cells were found in five instances. Case 14 was included in this series because the extension of cutaneous candidosis to the underlying bone seemed outstanding.

The group of cases with *diabetes* (29, 30, 33, and 34) had marked complications. Two of these cases had malignant tumors, two had undergone surgery, and one patient died of an acute myocardial infarct. No kidney involvement was observed in this group, there was no granulomatous reaction, and there were no atypical fungal elements.

The four *leukemic* patients were children (the eldest was 11 years old) (cases 1, 2, 5, and 8). Three had acute lymphoblastic and one acute myelogenous leukemia. Lesions consisted mainly of necrotic foci loaded with fungus cells. These lesions were also found in leukemic infiltrates, granulomatous reaction was also seen in one instance. *Candida* lesions in the liver were seen in two cases and in the spleen in one case. In two cases, there was also candidosis of the digestive tract. Atypical fungus cells were noted in two cases.

Hepatopathies were present in four cases of candidosis (9, 13, 25, and 26); one case of neonatal hepatitis, two cases of infectious hepatitis in children, and one case of liver cirrhosis in a middle-aged woman. It is very important to note that the two patients with infectious hepatitis developed posthepatitic bone-marrow failure (aplastic anemia) with marked neutropenia. In three cases, granulomatous reactions was observed and in two, atypical fungus cells.

Pathologic conditions of the *perinatal age group* were present in four cases (3, 4, 6, and 24). In three cases, (one had also undergone surgery because of duodenal atresia) prematurity was noted and in one case Kernicterus and hyaline-membrane disease were present. One case tabulated in another group (hepatopathies) also belongs to this perinatal age group (25).

In three of these four cases, a heavy dissemination was present with lesions in more than five organs and in all cases the brain was involved (the leptomeninges were also involved in three cases). Myocardium, liver, and digestive tract were involved in three cases each. Granulomatous reaction was present in three cases. Atypical fungus cells were not found. Mononuclear cell infiltrates with fungus cells in the three cases of liver involvement in this age group are also noteworthy.

In the two cases with *renal transplant* (27 and 28) in addition to generalized candidosis a bacterial infection was seen. In one case a massive CMV infection was observed. The lungs were involved in both these cases. In one case a granulomatous reaction was seen and in the other atypical fungus cells were noted.

In two cases (10 and 32) *chronic granulomatous disease of childhood* was thought to be the basic disease. In case 10 (a 3-month-old child) tubercular granulomas with necrosis were observed in lungs, liver, spleen, and bone marrow. Acid-fast bacilli were not found in tissues and were not grown in cultures. In the lungs, fungal elements resembling *Candida* were found in some granulomas.

In case 32, only a surgical specimen (lobectomy) was examined. In this 6-year-old boy, who had been hospitalized several times for infectious episodes and who had a family history of chronic granulomatous disease of childhood, extensive lung lesions were observed. Chronic fibrosis with plasma cells and eosinophils and abscesses with granulomatous reaction were also seen. Some of these foci contained hyphal elements morphologically resembling *Candida*, but *Aspergillus* grew on the cultures.

Finally, *gastro-intestinal abnormalities* were present in two patients (7 and 18). Bowel resection was performed because of volvulus in a 3-month-old child and in the adult patient several resections were performed because of chronic "regional" ileocolitis. In one case, atypical fungus cells were seen and in the other patient a granulomatous reaction was found in the peritoneum. Candidosis of the bowel mucosa was absent in both patients.

6. Portals of Entry

Possible portals of entry are listed in Table 5. Burn wounds, in which the presence of cutaneous candidosis had not been proved, were also considered in this table since candidosis may have been present at sites other than those examined. The genitourinary tract is not mentioned, since anatomic evidence excluded this portal of entry in our series.

Table 5. Possible portals of entry

Case No.	Skin	Upper respiratory tract	Digestive tract	Vein thrombus	Lungs
1					+
2			+		
3			+		+
4		+	+	+	+
5					
6					
7				+	
8			+		
9	+				
10					+
11	B				+
12	B				+
13					+
14	+B				
15	B		+	+	
16	B		+		+
17	+B				
18					+
19	B	+	+	+	+
20	B			+	
21	B			+	
22	B			+	+
23	B				+
24	+	+	+	(+)	+
25					
26					+
27					+
28		+	+		+
29			+		
30					+
31	+B		+		
32					+
33				+	
34			+		+

Based on anatomic evidence *no portal of entry* was found in three cases (5, 6, and 25). In one case only the brain showed *Candida* lesions. In two cases, candidosis was found at five different sites. One case had leukemia, in another case surgery had been performed, (anastomosis of a duodenal atresia), and in the third case transfusions had been given to an infant of the perinatal age group with neonatal hepatitis.

Only *one possible portal of entry* was noted in 16 cases. Skin was indicated in three cases (9, 14, and 17) (two of the three in burn patients). The mucosa of the digestive tract was indicated in three cases (2, 8, and 29), venous thrombi in two cases (7 and 33), and the lungs in eight cases (1, 10, 13, 18, 26, 27, 30, and 32).

Two possible portals of entry were noted in eight cases: digestive tract and lungs (3 and 34); skin (burn lesions) and lungs (11, 12, and 23); skin (burn lesions) and venous thrombi (20 and 21); and skin and digestive tract (31).

Three possible portals of entry were found in four cases: Skin (burn lesions) digestive tract and venous thrombus (15); skin (burn lesions), digestive tract and lungs (16); skin (burn lesions), lungs and venous thrombus (22); upper respiratory tract, digestive tract, and lungs (28).

Four possible portals of entry were observed in only one case (4): upper respiratory tract, digestive tract, lungs, and venous thrombus.

Five possible portals of entry were seen in two cases (19 and 24). The (+) in case 24 of Table 5 refers to the finding of fungus cells in a thrombus of a vein near the umbilicus.

IV. Discussion

Almost all of the 34 cases of our series belong to the third clinical variety of systemic candidosis, which *Symmers* (1964) referred to as "septicaemic candidosis occurring as a complication of other diseases or of their treatment." In a series of this author, septicemic candidosis, without predisposing factors, had occurred in four out of 10 cases (*Symmers*, 1966).

Our series presents all the disadvantages of a retrospective study in reference to completeness of anatomic material and culture studies. However, in most cases numerous organs could be examined histologically. *Candida* was isolated in almost half of the organs examined. Speciation of *Candida* was not carried out.

Involvement of organs was ascertained and details about all organs examined histologically were also given. Therefore, the results fairly reflect the true involvement of principal organs, whereas in other studies, the total of examined organs was not sufficient for assessment of involvement. Absence of lesions in pancreas and internal genitalia may reflect reality, since these organs were examined frequently. In a few other organs, such as pituitary, parathyroid, and others, not enough material was reviewed to rule out involvement with certainty. Fungemia, in addition to lesions elicited by the fungus, could be found in numerous tissues by applying the Grocott method as systematically as possible, giving additional evidence of the magnitude of hematogenous dissemination.

Lung involvement was slightly higher than that of the kidneys, particularly when compared with figures in other series (*Winner*, 1971). However, it must be emphasized that the frequent presence of *Candida* organisms in the lungs, as seen in the autopsy material of one of the present authors (K.S.), who used the Grocott method systematically, in the great majority of cases does not imply lung disease. This is also emphasized by the "fungemias" described in this paper–some without any tissue reaction. In mice, injection of strains of *Candida* other than *C. albicans* often produces fungemia with high colony counts in a variety of organs, but without tissue reaction; with *C. albicans* discrepancies between number of organisms recovered by quantitative cultural methods and tissue reaction also exist in certain organs (*Oblak* and *Schwarz*, unpublished data).

There was no anatomic evidence of ascending (pyelonephritic) *Candida* infection in our series. *Candida* pyelitis, *Candida* papillary necrosis, and fungus balls in the renal calyces were always thought to be the result of a hematogenous dissemination to the kidneys with secondary descending spread.

The central nervous system was involved in fourteen cases. *Fetter* et al. (1967) collected only 45 such cases from the literature up until 1967. The involvement of the brain in all of our cases of the perinatal group points to a special susceptibility of this tissue at this age. The involvement of the ependyma (*Candida* ependymitis) may lead to the formation of veritable "fungus balls" in the ventricles. *Candida* meningitis was found in our series only associated to or by propagation from brain candidosis. Clinical reports of *Candida* meningitis refer mostly to isolated meningitis (*Kuehner* and *Stroeder*, 1975), since it is not possible to prove brain involvement. In our series, isolated meningitis was not found.

In the cases of *Candida* involvement of the skin in our series, granulomatous reaction was not seen, and was likewise absent in the series of skin lesions associated with disseminated candidosis (*Bodey* and *Luna*, 1974). The so-called "*Candida* granuloma" (monilial granuloma) is a well-defined clinical entity of *Candida* skin infection in both children and adults (*Hauser* and *Rothman*, 1950; *Goldman* and *Schwarz*, 1966). From the histologic point of view as a rule no granulomatous reaction is present (see granulomatous reactions in our series), but only a nonspecific granulation tissue; the name "*Candida* granuloma", therefore, should be modified, perhaps to "chronic relapsing mucocutaneous candidosis."

Lesions were rarely seen in our series in bone, thyroid, and skeletal muscle, as also found by *Winner* (1971). The involvement of the dura, spinal cord (*Candida* myelitis), adrenal gland, gall-bladder, lymph node, cartilage and hematogenous spread to the tongue, were not mentioned in the more recent reviews (*Fetter* et al., 1967; *Winner*, 1971). This may be partly due to the systematic examination with the superior Grocott method in our series. The assumed yeast cells of *Candida* in tissue sections of the gall-bladder in numerous cases (*Hermanek*, 1965) do not represent fungus cells, but Grocott-positive tissue particles or artefacts.

In numerous cases and organs of our series a granulomatous reaction was found, although this feature has been reported in candidosis only rarely (*Sauerteig* et al., 1971; *Fetter* et al., 1967). Often there were fragments of hyphae or yeast cells in giant cells. This reaction was neither found regularly, nor in all lesions of a given case, and was independent of the presence of typical or atypical fungus cells, and of the basic diseases. There is no simple explanation for this type of tissue response. On the basis of these findings, candidosis must now be added to the group of deep mycoses in which granulomatous reaction may be seen in a certain percentage of cases. This seems to be of some importance, because morphologic diagnosis of deep mycosis relies at least partly on tissue reactions. The presence of granulomatous reaction, therefore, should not exclude candidosis from consideration. Occasionally, *Candida* yeast cells can be found intracellularly, adding one more differential diagnostic problem.

The focal membranous proliferative glomerulitis described in a certain number of cases with *Candida* lesions in the kidneys, is either an immunologic phenomenon or may be the result of direct action of fungus cells circulating in the blood stream.

The fibrinoid necrotizing glomerulitis and arteritis in one case could be an allergic reaction. Allergic vasculitis in candidosis has been described in the skin by *Tomsikova* et al.

(1974). Immunologic and allergic conditions in candidosis and the methods to confirm them have been reviewed recently by *Seeliger* et al. (1975).

Signs of disseminated intravascular coagulation (DIC) were present in a number of our series, a feature common in infectious diseases and also in some deep mycoses (*Robboy* et al., 1972). However, since associated bacterial infection was present in most of the cases in our series, it is difficult to attribute these findings (immunologic reactions and DIC) exclusively to *Candida*.

In ideal conditions the exact diagnosis of deep mycoses should rely on morphology of fungus cells in tissues, tissue reactions, and cultures. However, in many circumstances cultures are not available or culture results alone do not permit etiologic conclusions of a given tissue lesion to be drawn. The presence of organisms in culture and failure to observe fungi in tissue and/or failure of tissue reaction, questions the validity of the cultural findings. A great number of deep mycoses has been originally described by pathologists on morphologic grounds and the efforts of the pioneers in the field of mycologic pathology such as *Binford, Baker, Symmers*, and others gave us sufficient morphologic data for diagnosis in numerous deep mycoses, at least of typical cases. In mycoses caused by yeast cells with yeast cells in the tissues, and in some other cases where typical fungal structures are present, we feel that diagnosis may be made based upon histology, often alone. A growing number of pathologists show good diagnostic abilities in seminars and short courses and there are always experts in certain fields who can help to solve diagnostic puzzles, although microbiologists and clinicians may object. Morphologic species diagnosis, on the other hand, is difficult or impossible, when only hyphal elements are found in tissues, without other typical fungus cell structures.

After these introductory remarks, it should be stated clearly that we cannot be sure whether all of our 34 cases are *Candida* infections. While morphology of fungus cells was typical in many cases and organs, *Candida* was isolated in cultures in less than 50% of the cases of our series. There were positive *Aspergillus* cultures in two cases, that seemed to have *Candida* in the tissues and in many cases atypical fungus cells were found in the tissues. Therefore, infection by other species (e.g., *Geotrichum, Aspergillus, Phycomycetes, Blastomycetes*, etc.) or mixed infections in the cases with positive *Candida* culture, cannot be excluded in a number of cases. Double infections by *Candida* and *Aspergillus* are frequent (*Salfelder* et al., 1973; *Bader* and *Bader*, 1974; *Peña*, 1971).

Fungus cells in tissues, especially in necrotic lesions, may show degenerative changes as swelling and deformities (spheroplasts) (*Goldman* and *Schwarz*, 1962). In many of our cases, these atypical structures may thus be explained and the fungus cells may represent *Candida* species. This interpretation is favored by the fact that in more than 50% of these cases with atypical fungus cells, cultures were positive for *Candida* and the fungus cells at other sites in these cases were often typical for *Candida*, with apparent transitional forms from small ovoid yeast cells to the "cyst-like" structures interpreted as spheroplasts.

The particular features of candidosis in each of the basic diseases, described in the respective sections, may be fortuitous or are difficult to explain. The lack of reaction in the necrotic lesions with abundant fungi is characteristic and may be attributed to immunodepression in terminal leukemias and aplastic anemia. Only the cases of chronic granulomatous disease of childhood (*Bridges* et al., 1959) require comment. This entity was diagnosed clinically in one case and assumed to exist in an other, since no etiology of the granulomas was found. In both cases hyphal elements, presumably of *Candida*, were observed only in a certain number of granulomas. Hence, the possibility that *Candida* infection occurred secondarily.

The portal of entry is obscure in many cases of systemic candidosis (*Winner*, 1971). Particularly in reasonably complete post-mortem examinations, as in our series, candidosis is often widespread and in the majority of cases no single portal of entry can be specifically determined. On the basis of anatomic evidence, no portal of entry could be found in three cases and one possible portal of entry was found in 16 cases. In the remainder, there were from two to five possible portals of entry. It can be assumed but not proven that the portal of entry in quite a few cases of our series was by i.v. inoculation (by catheters, infusions, transfusions, including one case of omphalitis with catheterization at this site). No case of candidosis of the umbilical cord (*Aterman*, 1968) was observed in our series. The findings in our series seem to support the statement of *Symmers* (1966), that there is no good evidence that lungs are important as the portal of entry of *Candida* in cases of septicemic candidosis.

V. Summary

Thirty-four cases (33 autopsies and one lobectomy) with visceral candidosis from Cincinnati were reviewed. The material was collected from 1960-1974, all infections occurring in debilitated patients.

Details of organ involvement, sometimes only with fungemia, were listed and rare localization of the disease described.

Special tissue reactions were: granulomatous reaction in numerous cases without satisfactory causal explanation; membranous proliferative glomerulitis in renal candidosis; an allergic reaction with fibrinoid vasculitis and DIC.

The frequent presence of atypical fungus cells in tissues raised the question of infection with other fungal species or mixed fungal infections. However, the atypical fungus cells were interpreted in the majority of cases as degenerated secondarily deformed *Candida* elements (spheroplasts).

A few particular features of candidosis complicating basic diseases were found.

A single portal of entry of infection on the basis of anatomic evidence was determined in less than 50% of the cases.

Further prospective studies, including extensive culture work, are recommended to clarify diagnostic problems in opportunistic fungal infections.

References

Aterman, K.: Pathology of Candida infection of the umbilical cord. Amer. J. clin. Path. **49**, 798-804 (1968).

Bader, G., Bader, N.G.: Morphologie der Gewebsformen von Erregern visceraler Mykosen. (5. Mitt.) Mykosen **17**, 265-285 (1974).

Bodey, G.P., Luna, M.: Skin lesions associated with disseminated Candidiasis. J.A.M.A. **229**, 1466-1468 (1974).

Bridges, R.A., Berendes, H., Good, R.A.: A fatal granulomatous disease of childhood. Amer. J. Dis. Child. **97**, 387-408 (1959).

Fetter, B.F., Klintworth, G.K., Wilson, S.H.: Mycoses of the central nervous system. Baltimore: Williams and Wilkins 1967.
Goldman, J., Schwarz, J.: Chronic muco-cutaneous candidiasis of children. Acta. derm.-vener. (Stockh.) **42**, 314-322 (1962).
Goldman, L., Schwarz, J.: Cytology of four yeast like organisms in tissue explants. Mycopathologia (Den Haag) **29**, 161-167 (1966).
Hauser, F.V., Rothman, S.: Monilial granuloma: Report of a case and review of the literature. Arch. Derm. Syph. (Chicago) **61**, 297-310 (1950).
Hermanek, P.: Zur Mykologie der Galle bzw. Gallenblase. II. Mitteilung Histologische Untersuchungen. Langenbecks Arch. klin. Chir. **310**, 137-159 (1965).
Kuehner, U., Stroeder, J.: Die Candida albicans–Meningitis im Säuglingsalter. Mykosen **18**, 81-85 (1975).
Law, E.J., Kim, O.J., Stieritz, D.D., McMillan, B.G.: Experience with systemic candidiasis in the burned patient. J. Trauma. **12**, 543-552 (1972).
Peña, C.C.: Aspergillosis. In: Human Infections with Fungi, Actinomycetes and Algae, Baker, R.D. (ed.). Berlin-Heidelberg-New York: Springer 1971, pp. 762-831.
Robboy, S.J., Colman, R.W., Minna, J.D.: Pathology of Disseminated Intravascular Coagulation (DIC). Hum. Path. **3**, 327-343 (1972).
Salfelder, K., Mendelovici, M., Schwarz, J.: Multiple deep fungus infections: Personal observations and a critical review of the world literature. In: Current Topics in Pathology. Berlin-Heidelberg-New York: Springer 1973, Vol. 57, pp. 123-177.
Sauerteig, E., Liscano, T.R. de, Salfelder, K.: Candidosis aus pathologisch-anatomischer Sicht. Mykosen **14**, 283-298 (1971).
Seeliger, H.P.R., Tomsikova, A., Toeroek, I.: Immunologische Reaktionen durch *Candida albicans*. Mykosen **18**, 51-59 (1975).
Symmers, W.St.C.: The tissue reactions in deep seated fungal infections. The role of histological examination in mycological diagnosis. Ann. Soc. belge Med. trop. **44**, 869-886 (1964).
Symmers, W.St.C.: Septicaemic Candidosis. In: Symposium on Candida Infections, Winner, H.I., Hurley, R. (eds.). Edinburgh: Livingstone 1966, pp. 196-212.
Tomsikova, A., Dura, J., Novackova, D.: Beitrag zur experimentellen Candidose des Respirationstraktes. Mykosen **17**, 95-105 (1974).
Whitaker, J., Landing, B.H., Esselborn, V.M., Williams, R.R.: The syndrome of familial juvenile hypoadrenocorticism, hypoparathyroidism and superficial moniliasis. J. clin. Endocr. **16**, 1374-1387 (1956).
Winner, H.I.: Candidosis. In: Handbuch der speziellen pathologischen Anatomie und Histologie. Berlin-Heidelberg-New York: Springer 1971, Bd. III/5, pp. 731-761.

Subject Index

The numbers set in *italics* refer to those pages on which the respective catch-word is discussed in detail

Index to Volumes 57–63 Current Topics in Pathology

Current Topics in Pathology, Volume 64

Errata

The captions for the figures listed below failed to show the original magnifications after the negative number. These should be added as follows:

p. 31, Fig. 19	(Neg. 10553; 25360 x, . . .)
p. 32, Fig. 20	(Neg. 6825; 12312 x, . . .)
p. 33, Fig. 21	(Neg. 12196; 14760 x, . . .)
p. 35, Fig. 22	(Neg. 13469; 5882 x, . . .)
p. 36, Fig. 23	(Neg. 6732; 31806 x, . . .)
p. 37, Fig. 24	(Neg. 13473; 13680 x, . . .)
p. 38, Fig. 25	(Neg. 13474; 79800 x, . . .)
p. 39, Fig. 26	(Neg. 8666; 13680 x, . . .)
p. 40, Fig. 27	(Neg. 4466; 17100 x, . . .)

Springer-Verlag Berlin Heidelberg New York

MIX
Papier aus verantwortungsvollen Quellen
Paper from responsible sources
FSC® C105338

If you have any concerns about our products,
you can contact us on
ProductSafety@springernature.com

In case Publisher is established outside the EU,
the EU authorized representative is:
Springer Nature Customer Service Center GmbH
Europaplatz 3, 69115 Heidelberg, Germany

Printed by Libri Plureos GmbH
in Hamburg, Germany